Janine Cossy (Ed.)

Grignard Reagents and Transition Metal Catalysts

Also of interest

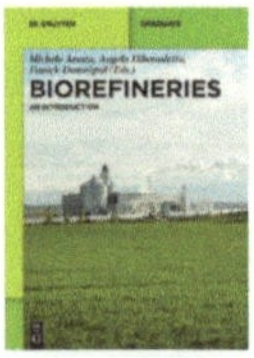

Biorefineries
Aresta, Dibenedetto, Dumeignil *(Eds.), 2015*
ISBN 978-3-11-033153-0, e-ISBN (PDF) 978-3-11-033158-5,
e-ISBN (EPUB) 978-3-11-038999-9

Flow Chemistry. Vol. 1: Fundamentals
Darvas, Hessel, Dorman *(Eds.), 2014*
ISBN 978-3-11-028915-2, e-ISBN (PDF) 978-3-11-028916-9,
e-ISBN (EPUB) 978-3-11-038875-6

Flow Chemistry. Vol. 2: Applications
Darvas, Hessel, Dorman *(Eds.), 2014*
ISBN 978-3-11-036707-2, e-ISBN (PDF) 978-3-11-036750-8,
e-ISBN (EPUB) 978-3-11-039260-9

Engineering Catalysis
Murzin, *2013*
ISBN 978-3-11-028336-5, e-ISBN (PDF) 978-3-11-028337-2

Chemical Photocatalysis
König *(Ed.), 2013*
ISBN 978-3-11-026916-1, e-ISBN (PDF) 978-3-11-026924-6

Heterocyclic Communications
Editor-in-Chief: Lucjan Strekowski
ISBN 978-3-11-028915-2, e-ISBN (PDF) 978-3-11-028916-9,
e-ISBN (EPUB) 978-3-11-038875-6

Janine Cossy

Grignard Reagents and Transition Metal Catalysts

Formation of C–C Bonds by Cross-Coupling

Edited by
Janine Cossy

DE GRUYTER

Author
Prof. Janine Cossy
Laboratoire de Chimie Organique
Institute of Chemistry, Biology and Innovation (CBI)
UMR 8231
ESPCI ParisTech/CNRS/PSL Research University
10 rue Vauquelin
75231 Paris Cedex 05
France
Janine.Cossy@espci.fr

ISBN 978-3-11-035266-5
e-ISBN (PDF) 978-3-11-035272-6
e-ISBN (EPUB) 978-3-11-038343-0
Set-ISBN 978-3-11-035273-3

Library of Congress Cataloging-in-Publication Data
A CIP catalog record for this book has been applied for at the Library of Congress.

Bibliographic information published by the Deutsche Nationalbibliothek
The Deutsche Nationalbibliothek lists this publication in the Deutsche Nationalbibliografie; detailed bibliographic data are available on the Internet at http://dnb.dnb.de.

Typesetting: Medienfabrik GmbH, Stuttgart
Printing and binding: CPI books GmbH, Leck
Cover image: von Fotograv. - Generalstabens Litografiska Anstalt Stockholm [Public domain], via Wikimedia Commons
∞ Printed on acid-free paper
Printed in Germany

www.degruyter.com

Contents

Contributing Authors

Gérard Cahiez
Institut de Recherche de Chimie Paris
CNRS – Chimie ParisTech, PSL Research University
11 rue Pierre et Marie Curie
75005 Paris
France

Catherine S. J. Cazin
EaStCHEM School of Chemistry
University of St Andrews
Purdie Building
North Haugh
St Andrews
Fife, KY16 9ST
UK

Janine Cossy
Laboratoire de Chimie Organique
Institute of Chemistry, Biology and Innovation (CBI)
UMR CNRS 8231
ESPCI ParisTech/CNRS/PSL Research University
10 rue Vauquelin
75231 Paris Cedex 05
France

Fabienne Fache
Université Lyon 1
ICBMS – UMR CNRS 5246
43, boulevard du 11 novembre 1918
69622 Villeurbanne cedex
France

Bruno Figadère
CNRS
BioCIS UMR 8076
Labex LERMIT
Université Paris Sud
and
CNRS
5 rue J. B. Clément
92296 Châtenay-Malabry
France

Corinne Gosmini
Laboratoire de Chimie Moléculaire
Ecole Polytechnique
UMR CNRS 9168
Route de Saclay
91128 Palaiseau Cedex
France

Julien Legros
Normandie Université
COBRA UMR 6014
Université Rouen
INSA Rouen
and
CNRS
1 rue Lucien Tesnière
76821 Mont-Saint-Aignan
France

Alban Moyeux
Institut de Recherche de Chimie Paris
CNRS – Chimie ParisTech, PSL Research University
11 rue Pierre et Marie Curie
75005 Paris
France
and
Université Paris 13
UFR SMBH
Sorbonne Paris Cité
74 rue Marcel Cachin
93017 Bobigny
France

David J. Nelson
WestCHEM Department of Pure & Applied Chemistry
University of Strathclyde
295 Cathedral Street
Glasgow
Lanarkshire, G1 1XL
UK

Prof. S. P. Nolan
Department of Inorganic and
Physical Chemistry
Ghent University
Krijgslaan 281 - S3, 9000 Gent, Belgium

Armelle Ouali
Laboratoire de Chimie de Coordination CNRS
UPR 8241
BP 44099
205 route de Narbonne
31077 Toulouse Cedex 04
France
and
Université de Toulouse
UPS, INPT, LCC
31077 Toulouse
France

Béatrice Pelotier
Université Lyon 1
ICBMS – UMR CNRS 5246
43, boulevard du 11 novembre 1918
69622 Villeurbanne cedex
France

Olivier Piva
Université Lyon 1
ICBMS – UMR CNRS 5246
43, boulevard du 11 novembre 1918
69622 Villeurbanne cedex
France

Alice Rérat
Laboratoire de Chimie Moléculaire
Ecole Polytechnique
UMR CNRS 9168
Route de Saclay
91128 Palaiseau Cedex
France

Marc Taillefer
CNRS
UMR CNRS 5253, ICG, AM2N, ENSCM
8, rue de l'Ecole Normale
34296 Montpellier Cedex 5
France

Introduction

Since the early 1900s, Victor Grignard had been of great impact in organic chemistry in developing organomagnesium reagents called "Grignard reagents" and, due to the importance of his work, he was awarded the Nobel Prize in 1912. Grignard reagents are often written as RMgX (R = organic part, X= halide), and these reagents can take various forms in both solution and solid states. In solution, they become a mixture of R_2Mg (called organomagnesium reagents) and MgX_2 coming from a redistribution of the ligand in RMgX.

In their education and training, almost all the chemists have had an experience of preparing and/or utilizing Grignard reagents. Grignard reagents are prepared from readily available organo halides and magnesium and more recently *i*PrMgX.LiCl, called the "Turbo Grignard", was used to prepare Grignard reagents from halides by a magnesium-halide exchange. The Grignard reagents are mainly used to synthesize alcohols from carbonyl derivatives but, one major drawback of the Grignard reagents is that they are not functional group tolerant. However, since the discovery of the reactivity of the Grignard reagents, their reactivity has been modified by the addition of transition metal complexes in the reaction media. It has been reported for the first time by Kharash *et al.*, around 1940, that C–C bonds can be formed by a cross-coupling reaction between a halide and a Grignard reagent catalyzed by metallic halides. Thirty years later in their early work and independly, Kochi, Kumada and Tamao as well as Corriu reported coupling reactions between organo halides and Grignard reagents catalyzed by palladium, nickel and iron salts. Since then a number of reports have been published to realize coupling reactions using, palladium, nickel, iron, cobalt, manganese, copper, silver and to a less extend titanium, chromium, rhodium, ruthenium, etc. showing that the reactivity of Grignard can be modified in the presence of metallic complexes, making them very chemoselective.

In this book, we mainly focus on the cross-coupling reactions between organo halides as well as pseudo-halides and Grignard reagents to form C–C bonds using palladium (Chapter 1), nickel (Chapter 2), iron (Chapter 3), cobalt (Chapter 4), manganese (Chapter 5), copper (Chapter 6) and silver (Chapter 7). We do not pretend to be exhaustive but to give a good overview of what has been achieved and what has to be realized.

I hope that this book will be a good source of inspiration for those who are planning future development in the field of C–C bond formation (or C–Heteroatom bond formation) using Grignard reagents, or how to solve synthetic problems, or simply having fun discovering new conditions to form C–C bonds (or C–Heteroatom bonds) in a very chemo-, regio-, stereo- and even enantioselective fashion.

Paris, 27th of July 2015
Janine Cossy

Acknowledgements: I would like to warmly thanks all the authors for their contribution to this book and the team at DeGruyter, especially Sabina Dabrowski and Silke Hutt for their helpful assistance during the preparation of this book, their patience and kindness.

David J. Nelson, Catherine S. J. Cazin, Steven P. Nolan

1 Grignard Reagents and Palladium

1.1 Introduction

One contributor to the success of palladium-catalyzed cross-coupling as a technique in academia and industry has been the ability to make use of a wide range of reagents that are commercially available and/or relatively straightforward to prepare. Grignard reagents [1] are some of the most common and widely known organometallic reagents in chemistry [2]. While these are relatively reactive species, they do not typically require the cryogenic conditions that their organolithium counterparts frequently demand. Yet, many coupling reactions of Grignard reactions occur at, or close to, room temperature.

The use of these valuable reagents in cross-coupling has been known for decades, since the studies of Kharasch *et al.* in the early 1940s [3, 4]. The cross-coupling of Grignard reagents with aryl halides is often referred to as the "Kumada-Tamao-Corriu reaction" (or simply, the "Kumada coupling"), but we have chosen not to ascribe a name to the reaction here.

In this chapter, we briefly examine the historical development of the cross-coupling of Grignard reagents, and discuss the modern applications of the palladium-catalyzed variants. While we focus here only on palladium catalysis, Shinokubo and Oshima have published a micro review that includes examples of catalysis with a wider range of metals [5], and Tamao has reviewed the development of the nickel-catalyzed variant [6]. We consider the reactions of different classes of Grignard reagent (aryl, vinyl, alkyl and alkenyl) with different classes of electrophile (aryl, vinyl, alkyl and alkenyl), with the aim of providing the chemist with an appreciation of what is possible using this method. We also briefly describe and discuss some examples from target syntheses, and from larger-scale industrial applications.

1.2 The Discovery and Development of Catalytic Cross-Coupling Reactions of Grignard Reagents

The earliest studies originated from the laboratory of Kharasch, who demonstrated that various metal salts catalyzed the coupling of some aryl halides (and one alkyl halide) with phenyl magnesium bromide (Fig. 1.1) [3]; $CoCl_2$ was shown to be the most efficacious (at loadings of 2.5–9 mol %), although $MnCl_2$, $FeCl_2$ and $NiCl_2$ also showed some (lesser) catalytic activity.

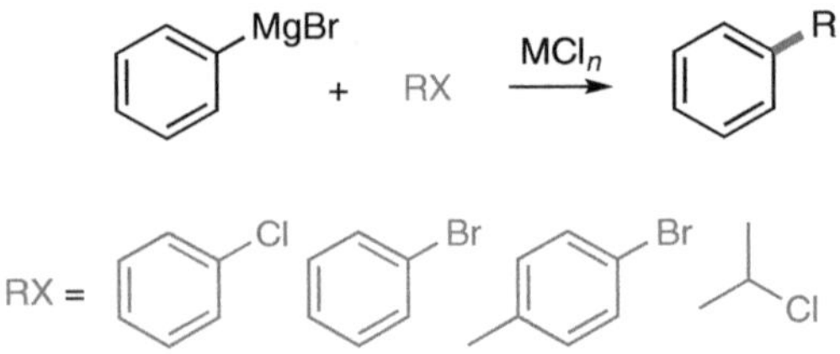

Fig. 1.1: Cross-coupling of phenyl magnesium bromide; MCl_n = $CoCl_2$, $MnCl_2$, $FeCl_2$, or $NiCl_2$ [3].

A subsequent study further probed the cobalt-catalyzed protocol, achieving the cross-coupling of phenyl-, cyclohexyl-, benzyl-, butyl- and naphthyl-magnesium bromide with several vinyl bromide and chloride compounds in the presence of typically 5 mol% $CoCl_2$ [4]. However, the corresponding styrenes were not the only product, and in many cases were not even the major product; polymers and protodehalogenated products were also obtained. Therefore, while these studies represented a significant step towards catalytic coupling of Grignard reagents, these protocols were not practically useful for synthetic processes. Alternative systems based on iron(II) and iron(III) chloride complexes were disclosed by Kochi, in which alkylmagnesium bromides were coupled with vinyl bromides [7].

A watershed moment in the development of Grignard cross-coupling was the use of catalytic quantities of nickel complexes, at loadings as low as 0.1 mol% in some cases, to selectively prepare the desired products. It was at this point that the technique began to emerge as a potentially general way to prepare a range of target molecules. In 1972, Corriu *et al.* disclosed the coupling of *trans*-β-bromostyrene with arylmagnesium bromide reagents, as well as examples of coupling between *trans*-1,2-dichloroethene or 1,4-dibromobenzene and arylmagnesium halides (Fig. 1.2) [8]. Reactions were catalyzed by [Ni(acac)$_2$] (acac = acetoacetonate).

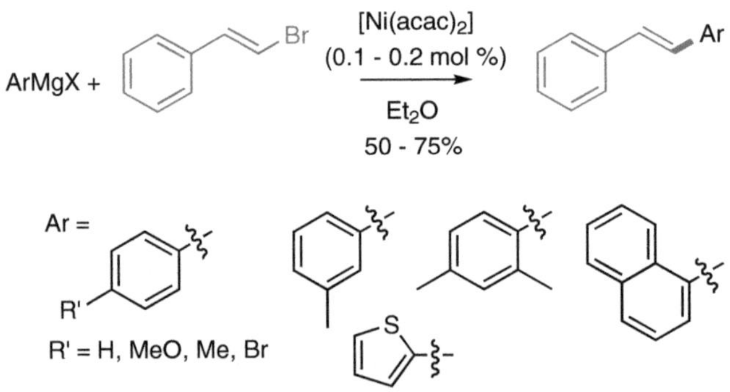

Fig. 1.2: Cross-coupling reactions catalyzed by [Ni(acac)2], disclosed by Corriu *et al.* [8].

In the same year, Kumada *et al.* published the use of [NiCl$_2$(dppe)] (dppe = 1,2-bis(diphenylphosphino)ethane) for the cross-coupling of alkyl and arylmagnesium bromide reagents with aryl and vinyl chloride species (Fig. 1.3) [9]. The catalyst in this case bore a bidentate phosphine ligand; dppe outperformed 1,3-bis(diphenylphosphino) propane (dppp) and 1,2-bis(dimethylphosphino)ethane (dmpe), which in turn outperformed complexes bearing two PPh_3, PEt_3 or PPh_2Me ligands.

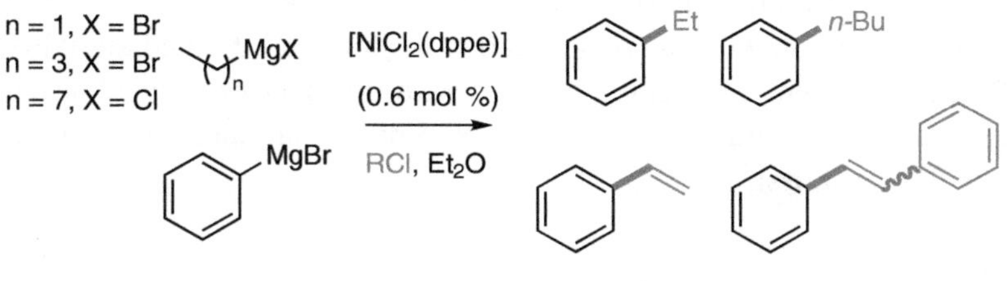

Fig. 1.3: Cross-coupling reactions catalyzed by [NiCl2(dppe)], disclosed by Kumada *et al.* [9].

In 1975, a palladium-catalyzed variant was developed by Murahashi *et al.* [10] for the coupling of vinyl halides with alkyl- and arylmagnesium iodides or bromides, catalyzed by $[Pd(PPh_3)_4]$ (Fig. 1.4). The authors specifically noted that catalytic cross-coupling of the corresponding lithium reagents was not possible, as "recycling of the palladium complex is slower than metallation of vinyl halide by the alkyllithium, resulting in formation of an acetylenic compound" under these conditions. It is worthwhile noting that Feringa *et al.* have since developed conditions for the cross-coupling reactions of a variety of aryllithium species with aryl bromides [11].

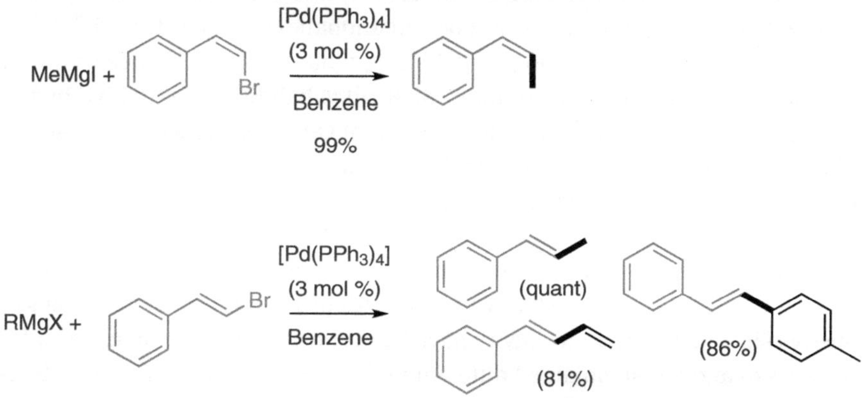

Fig. 1.4: Cross-coupling reactions catalyzed by [Pd(PPh3)4], disclosed by Murahashi *et al.* [10].

Subsequent efforts have allowed the cross-coupling of Grignard reagents to produce a wide range of compounds. While nickel is most often employed for this class of cross-coupling, a variety of palladium-catalyzed examples are also known. In this chapter, we examine the applications of palladium-catalyzed reactions where a Grignard reagent is one of the reactants.

1.3 The Mechanism of the Cross-Coupling of Grignard Reagents Catalyzed by Palladium Complexes

The mechanism of the cross-coupling of Grignard reagents with halides by palladium complexes follows a typical Pd(0)/Pd(II) couple (Fig. 1.5). The active Pd(0) complex undergoes oxidative addition with the organohalide to yield an intermediate Pd(II) species; the Grignard reagent subsequently reacts with this species to yield a new Pd(II) complex and a magnesium salt. Reductive elimination produces the product and re-forms the Pd(0) catalyst.

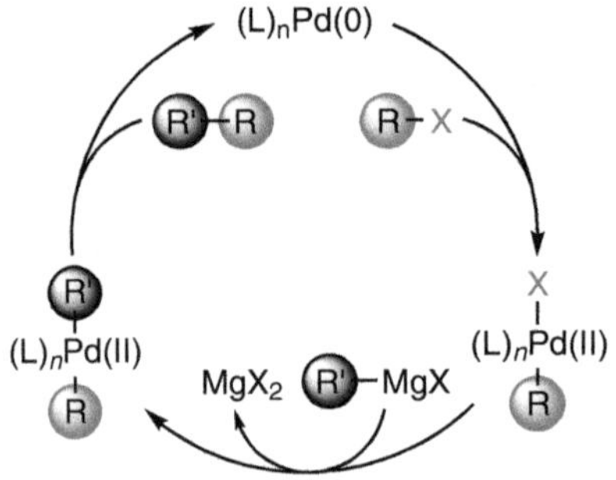

Fig. 1.5: General scheme for the catalytic cross-coupling of organohalides and Grignard reagents.

This reaction sequence can therefore be affected by the properties of the catalyst and substrate at various points. Electron-poor organohalides and electron-rich Pd(0) species will result in the fastest oxidative addition, while steric hindrance will slow down most steps of the reaction. In addition, when alkyl halide species (with β-hydrogens) are deployed, competing β-hydride elimination can interfere with the reaction and lead to degradation of both catalyst and substrate.

The palladium-catalyzed Grignard cross-coupling differs in some aspects from the nickel-catalyzed variant. For example, Sinou *et al.* have shown that, in the cross-coupling of aryloxy-functionalized carbohydrate derivatives, [$NiCl_2$(dppe)] leads to complete inversion of stereochemistry, while [$PdCl_2$(dppf)] completely retains the stereochemistry [12]. Differences in reactivity have been shown in some applications; for example, Janesko and Stefan obtained different molecular weight polymers from Grignard metathesis polymerization (GRIM), with nickel leading to a 'living' polymerization, and palladium to step growth polymerization [13], which was recently studied computationally by Bahri-Laleh *et al.* [14]. The available pool of electrophiles is somewhat different; nickel will interact with a wider range of phenol derivatives and organofluorine compounds than palladium. While this means that a smaller set of substrates might be amenable to palladium catalysis, it does overcome the challenge of selectivity in molecules with multiple potential sites for cross-coupling. Some examples of palladium-catalyzed cross-couplings with organofluorine molecules are known [15, 16].

Hoffmann *et al.* have investigated the mechanism of the palladium-, nickel-, iron- and cobalt-mediated variants of the cross-coupling of an enantioenriched secondary alkylmagnesium bromide compound (ee = 91 %) with vinyl bromide (Fig. 1.6) [17]. In each case, the pre-catalyst was reduced with EtMgCl at 0 °C, before the reaction was conducted at –78 °C. The products from the palladium- and nickel-catalyzed reactions were significantly enantioenriched (ee = 88–89 %), while much lower enantiomeric excesses were achieved with iron or cobalt (ee = 53–55 %). This result suggests that Single Electron Transfer (SET) processes are significant with iron and cobalt, but not with palladium or nickel.

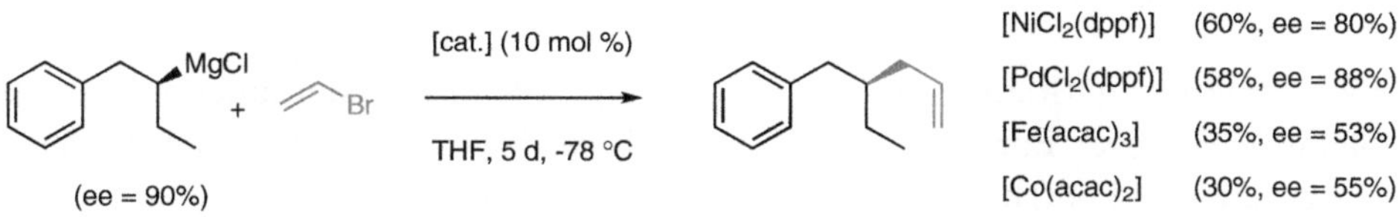

Fig. 1.6: SET in iron- and cobalt-catalyzed variants of Grignard cross-coupling reactions [17].

1.4 Applications of Palladium-Catalyzed Grignard Cross-Coupling Reactions

In this section, various categories of cross-coupling reactions are considered, with salient examples from the literature used to illustrate the scope and limitations of these reactions. Sections are organized first according to the nature of the Grignard reagent, and then by the nature of the electrophile (aryl, vinyl, alkyl or alkynyl). For the purposes of this chapter, halide, sulfonyl and other electrophiles are not considered separately.

While the coupling of aryl and vinyl units is often relatively straightforward, the coupling of alkyl fragments remains a task that is frequently quite challenging, and a key frontier in the area. Competing side-processes such as β-hydride elimination lead to competing side products and low yields of the desired cross-coupling processes [18, 19]. Lautens *et al.* have recently reviewed the deployment of secondary halide electrophiles in cross-couplings with a range of metals [20], while Kambe *et al.* have reviewed the use of alkyl halides in palladium-catalyzed cross-couplings [21].

1.4.1 Coupling of Arylmagnesium Compounds

1.4.1.1 With Aryl Electrophiles

Katayama *et al.* achieved selective cross-coupling of alkyl and aryl Grignards with dichlorobenzenes (albeit using the latter in excess) [22]. Phenylmagnesium chloride underwent reaction with *ortho*- or *meta*-dichlorobenzene (Fig. 1.7). The palladium/dppf (dppf = 1,1'-bis(diphenylphosphino)ferrocene) catalyst system is commonly used throughout the literature for the cross-coupling of Grignard reagents and organohalides.

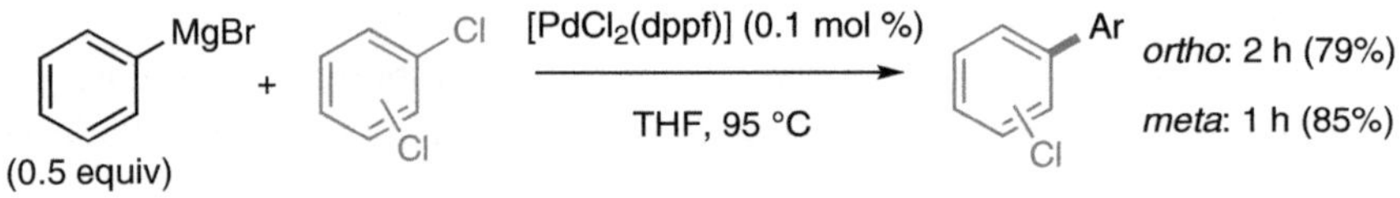

Fig. 1.7: Cross-coupling with dichlorobenzene substrates [22].

A key weakness of Grignard reagents is their typically rather poor functional group tolerance. However, Bumagin *et al.* successfully carried out cross-coupling reactions of aryl halides (Br or I) bearing amino, alcohol and carboxylic acid functional groups, simply by using extra equivalents of the Grignard reagent (2–3 equiv; one equiv per functional group, plus 1 equiv for cross-coupling) [23]. Some aryl-aryl examples are displayed in Fig. 1.8; the use of alkyl and vinyl Grignard reagents was also demonstrated.

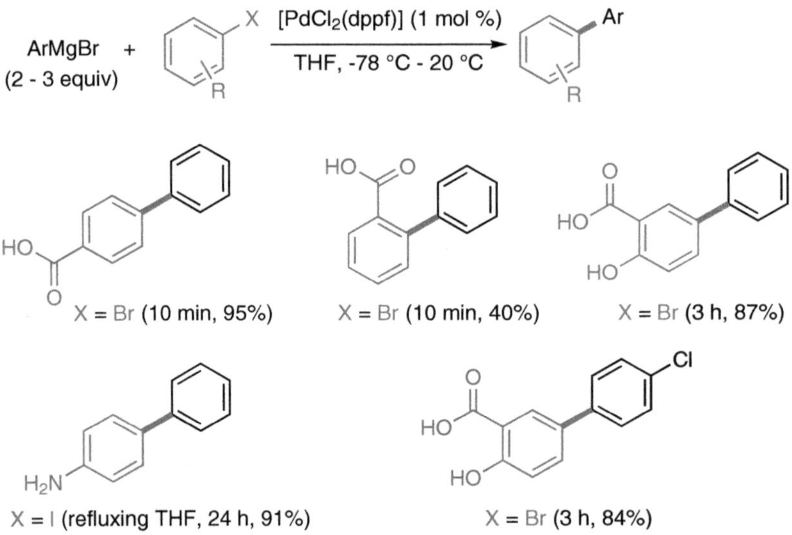

Fig. 1.8: Cross-coupling of anilines, halophenols, and halobenzoic acids [23].

In 1999, Nolan *et al.* reported the first Pd/NHC system for the catalytic cross-coupling of aryl Grignard reagents with aryl halides [24]. NHCs have since risen to become key tools in palladium-catalyzed cross-coupling chemistry [25, 26], amongst other fields [27, 28]. The catalyst system was prepared *in situ* from the reaction of [$Pd_2(dba)_3$]

(dba = dibenzylideneactone), which is a Pd(0) source, with the imidazolium salt. The use of a ligand was shown to be essential for catalytic turnover, while IPr (1,3-bis(2,6-di*iso*propylphenyl)imidazol-2-ylidene) was found to be a more efficient ligand than IMes (1,3-bis(2,4,6-trimethylphenyl)imidazol-2-ylidene). A range of substrates could be coupled (Fig. 1.9), including aryl chlorides rather than only bromides or iodides. However, the system could not couple two *ortho*-disubstituted substrates, which remained a challenge until relatively recently (*vide infra*).

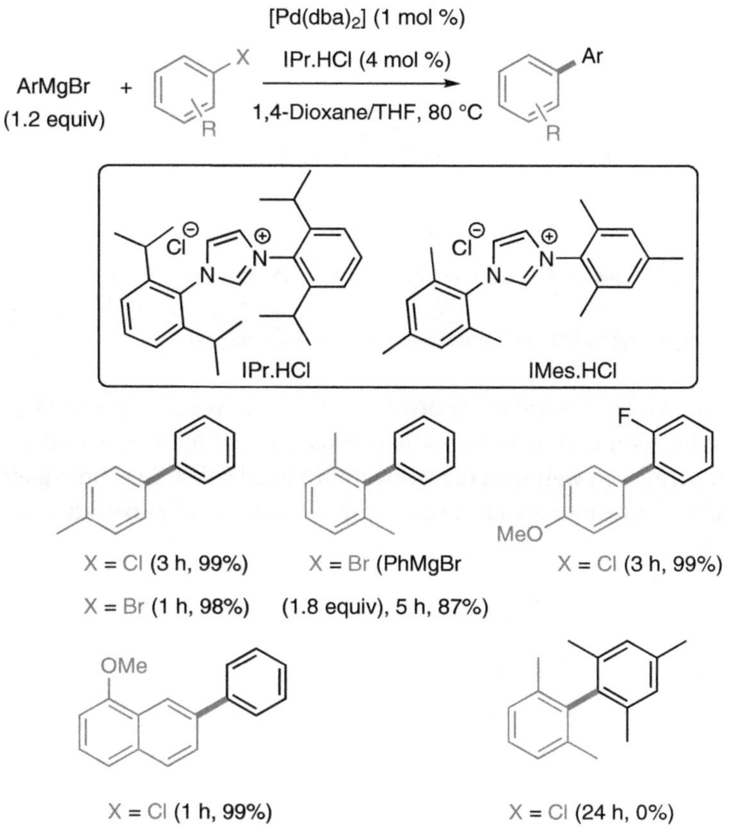

Fig. 1.9: The first Pd/NHC system for the cross-coupling of Grignard reagents [24].

Knochel *et al.* reported the cross-coupling of pyridyl halides with arylmagnesium halide reagents to generate biaryl compounds, at low temperatures (−40 to 0 °C) [29, 30]. A relatively simple palette of compounds was generated (some examples are shown in Fig. 1.10). Notably, the use of 3,5-dibromopyridine allowed for the isolation of the monoarylated product, which bears a site ripe for further functionalization. Heteroaromatic scaffolds of this type are of interest to researchers in the pharmaceutical and agrochemical industries.

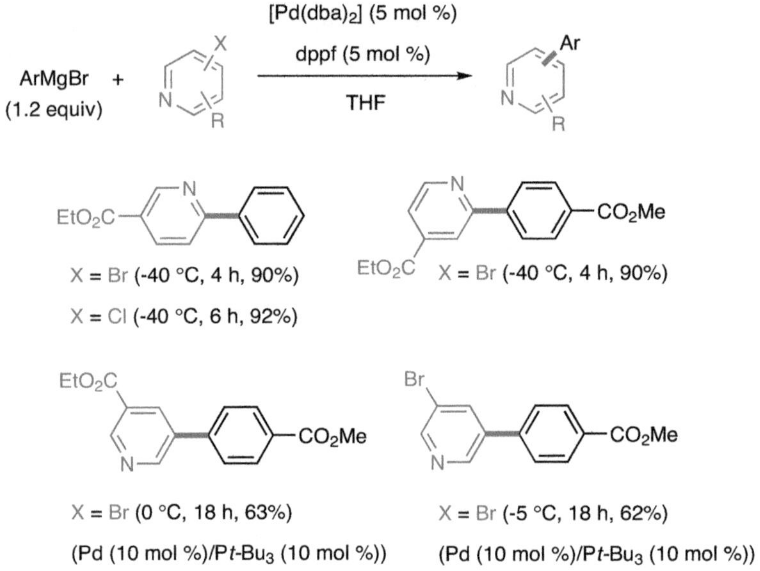

Fig. 1.10: Cross-coupling of pyridyl halides with organomagnesium reagents [29].

Li *et al.* have deployed palladium catalysts bearing phosphinous acid ligands (**1–3**) to couple aryl chlorides with aryl magnesium halide reagents at room temperature. This catalyst system performed well with relatively unhindered substrates, but yields decreased when bulkier substrates were used. Only a small set of substrates was considered (Fig. 1.11).

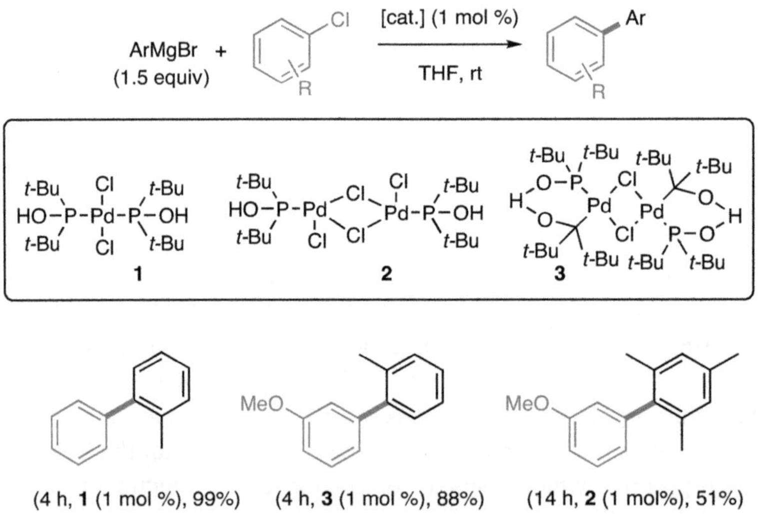

Fig. 1.11: Phosphinous acid ligands for the cross-coupling of Grignard reagents [31].

Naso *et al.* have used Grignard cross-coupling to prepare conjugated polymers [32]. The polymers were obtained in 40–73 % yield, with M_n (number average molecular masses) of *ca.* 4000 and M_w (weight average molecular masses) of *ca.* 6000–7000, indicating polymerization degrees of *ca.* 10 (Fig. 1.12). Very long reaction times were necessary to achieve these results (6 days at reflux).

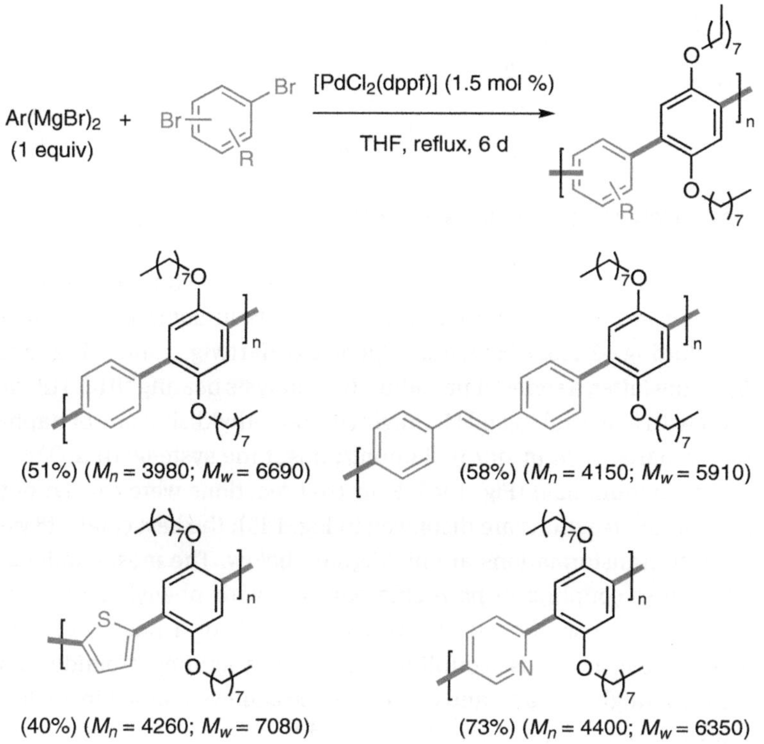

Fig. 1.12: Polymerization via Grignard cross-coupling [32].

Mongin *et al.* reported the cross-coupling of lithium tri(quinolinyl)magnesiates using a simple $Pd(dba)_2$/dppf system (Fig. 1.13) [33, 34]. These reagents are different from simple organomagnesium halide species. (Hetero)aryl bromides underwent cross-coupling with 0.35 equiv of the lithium magnesiate (generated *in situ*) in poor to moderate yields, albeit at room temperature. However, these species allow access to organomagnesium compounds that are otherwise inaccessible, as the 3-quinolinylmagnesium halide compounds cannot be formed due to the electrophilicity of quinolines.

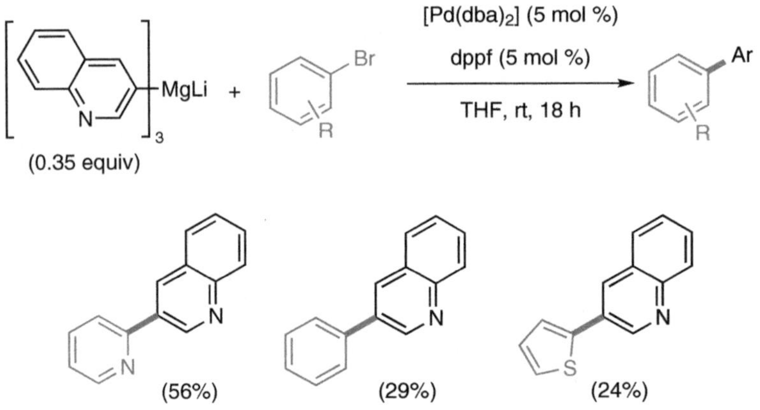

Fig. 1.13: Cross-coupling of lithium tri(quinolinyl)magnesiates [33, 34].

Both phosphine and NHC ligands are widely used for the cross-coupling of magnesium reagents. Beller *et al.* have compared these ligands in some prototypical cross-couplings, including Suzuki-Miyaura, Buchwald-Hartwig, and Grignard cross-coupling [35]. In the latter, a series of palladium(0) catalysts bearing NHCs (IPr or IMes) in combination with a neutral ligand (1,3-divinyltetramethyldisiloxane or naphthylquinone) were compared with *in situ* palladium/phosphine systems ([Pd(OAc)$_2$] plus PAdBu$_2$, P*t*-Bu$_3$ or CyJohnPhos) (Fig. 1.14). Four trial reactions were carried out (the three more challenging reactions are displayed in Fig. 1.15); the best catalyst system(s) for each of these transformations are highlighted below. The most challenging reaction was the cross-coupling of *para*-chloroanisole with phenylmagnesium bromide, which required the use of an IPr-bearing complex **7**, or a Buchwald-type ligand system (CyJohnPhos). The cross-coupling of the corresponding bromide was found to be much more straightforward, and could even be achieved (albeit in a relatively poor 35 % yield) without the use of an added ligand.

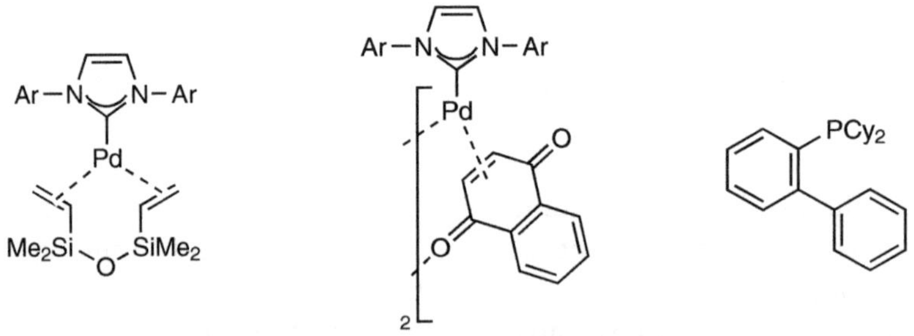

4, Ar = 2,4,6-trimethylphenyl
5, Ar = 2,6-di*iso*propylphenyl

6, Ar = 2,4,6-trimethylphenyl
7, Ar = 2,6-di*iso*propylphenyl

CyJohnPhos

Fig. 1.14: Catalysts used for NHC/phosphine comparisons [35].

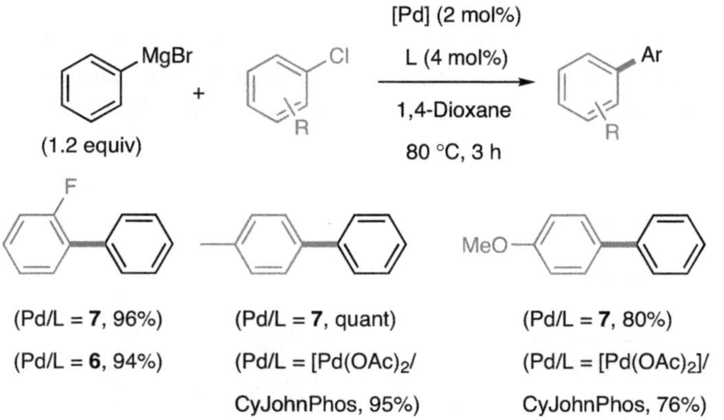

Fig. 1.15: Comparison of NHCs versus phosphine systems for the cross-coupling of aryl chlorides with phenylmagnesium bromide [35].

Sémeril and Matt have used diphosphinated calix[4]arenes **8** and **9** as ligands for palladium-catalyzed cross-couplings including the coupling of Grignard reagents [36]. The ligands consist of upper-rim functionalized calix[4]arenes, and were used in conjunction with $Pd(OAc)_2$ to arylate a small set of prototypical aryl halides (Fig. 1.16). The new ligands performed slightly better than a $[Pd(OAc)_2]/PPh_3$ system for aryl bromides, and were capable of arylating aryl chlorides, where the PPh_3-based system was not.

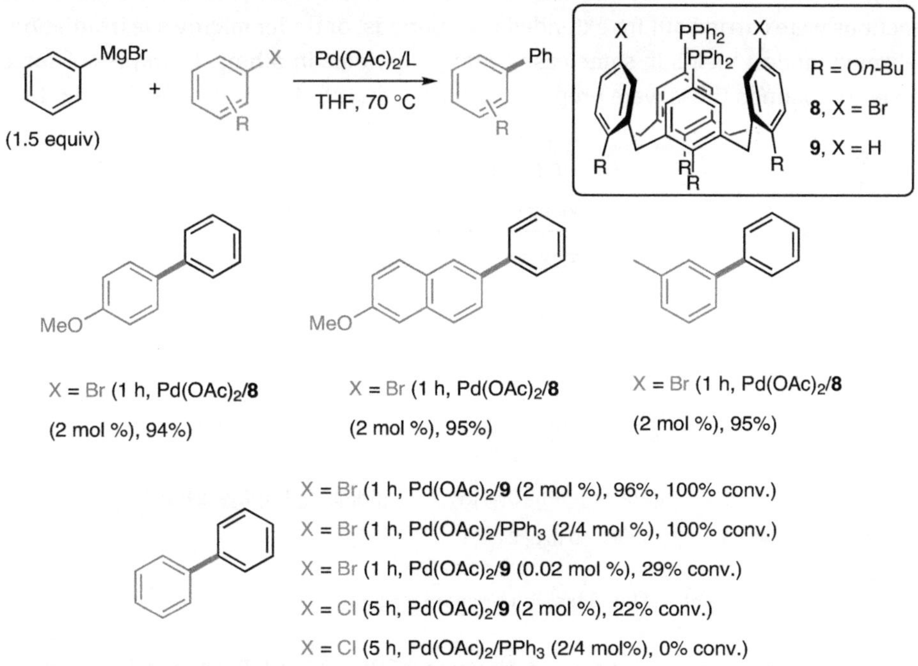

Fig. 1.16: Diphosphinated calix[4]arenes as ligands [36].

Hartwig and co-workers have deployed a bulky Pd(0) catalyst with ligand **10** to effect the cross-coupling of arylmagnesium reagents (and alkenyl- and alkylmagnesium reagents, *vide infra*) with aryl tosylates (and vinyl tosylates, *vide infra*) [37]. Reactions between aryl units typically proceeded in 2–4 h at rt (Fig. 1.17); more challenging reactions in which *ortho*-substituents were present required extended reaction times (up to 16 h) and/or increased temperatures.

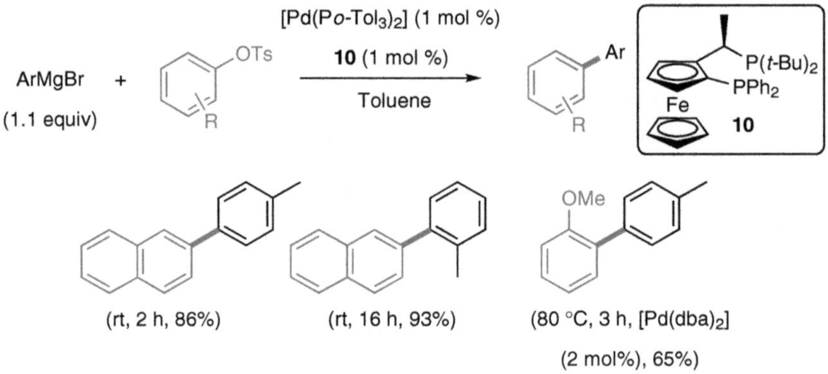

Fig. 1.17: Cross-coupling of aryl tosylates with arylmagnesium halides (isolated yields) [37].

Dankwardt *et al.* disclosed a series of palladium- and nickel-catalyzed cross-couplings of aryl fluoride substrates. Palladium does not normally react with C–F bonds. These reactions were carried out for extended time periods, or under microwave irradiation, delivered modest yields in some cases, and poor yields in others. Examples of some of the compounds those were prepared as part of this study are displayed in Fig. 1.18.

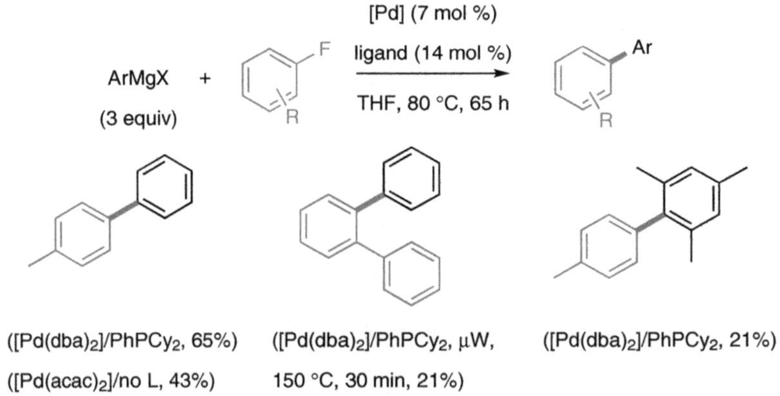

Fig. 1.18: Cross-coupling of aryl fluorides with Grignard reagents [37].

Ackermann *et al.* have developed the use of P = O and P–Cl bearing ligands for a number of palladium-catalyzed cross-couplings [38], particularly using

Heteroatom-Substituted Secondary Phosphine Oxide (HASPO) ligands. Reactions include the arylation of aryl tosylates using 2 equiv of ligand **11** (with respect to [Pd]) in combination with [Pd(dba)₂], reported in 2006 (selected examples in Fig. 1.19) [39]. A range of electron-rich, electron-poor and heteroaryl tosylates could be coupled with arylmagnesium halides, even those bearing *ortho*-methoxy groups.

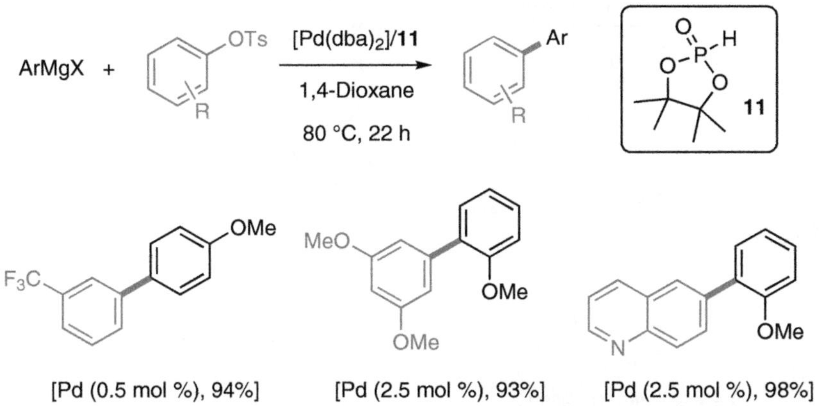

Fig. 1.19: Arylation of aryl tosylates [39].

This was subsequently extended to aryl and vinyl chlorides, using a TADDOL-derived ligand **12**; [40] a series of compounds were prepared in this manner (Fig. 1.20). Examples of the cross-coupling of silicon-, tin- and boron-based reagents were also demonstrated.

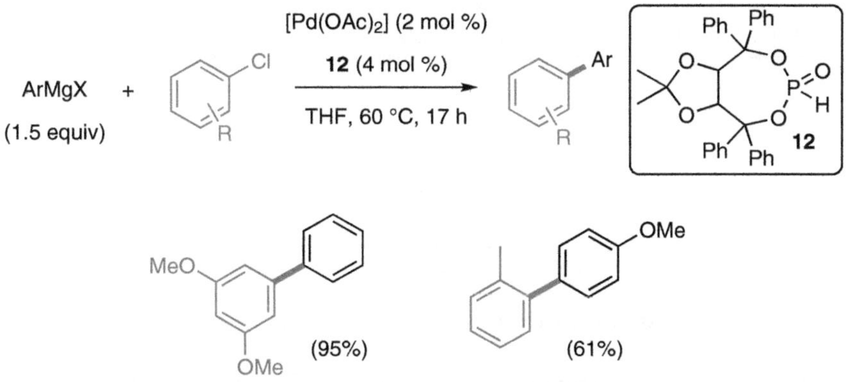

Fig. 1.20: Cross-coupling of aryl chlorides enabled by TADDOL-derived ligand 12 [40].

Hu *et al.* have developed a synthesis of fluorenes based on tandem Grignard cross-coupling/C–H activation, using palladium catalysts (Fig. 1.21) [41]. The scope of the reaction is quite limited, requiring a di*ortho*-methyl-substituted Grignard reagent; the

1,2-dihaloarenes tested possessed only methyl or methoxy functional groups, if any. However, it represents an interesting concept. Notably, when one of the halogens was a chloride, the reaction stopped at the corresponding 1-aryl-2-chloroarene intermediate, even if the reaction temperature was increased to 60 °C.

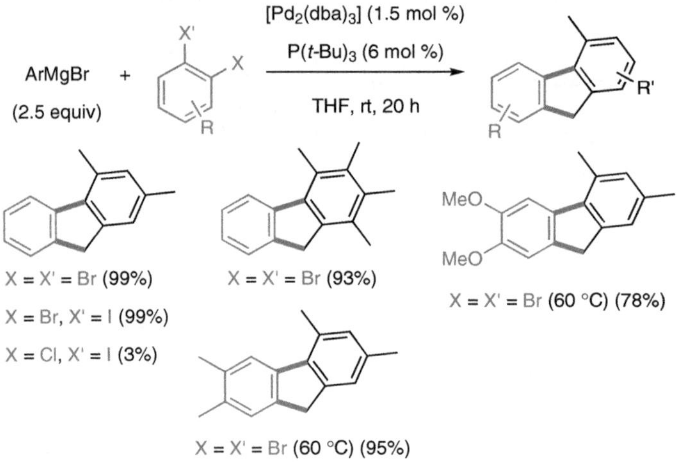

Fig. 1.21: Synthesis of fluorenes via Grignard cross-coupling/C-H activation [41].

In a subsequent study, the authors suggested that the reaction might proceed *via* an aryne intermediate coordinated to a palladium(II) center (Fig. 1.22) [42]. The use of a $[Pd(OAc)_2]$ catalyst at 60 °C allowed the method to be extended to aryl chlorides, while a subsequent study extended the scope further to fluorine-containing arenes, and arenesulfonates [43].

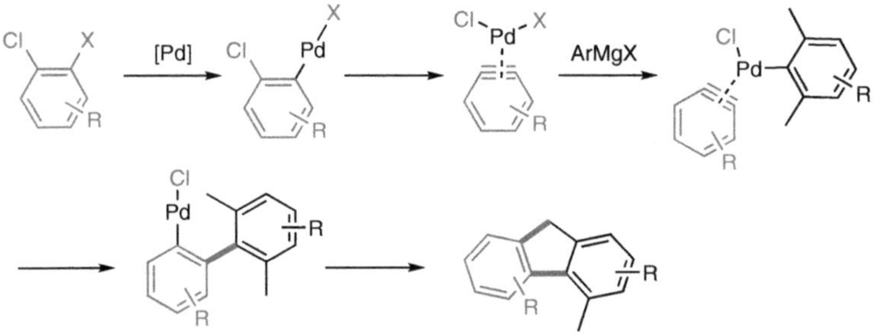

Fig. 1.22: Synthesis of fluorenes via Grignard cross-coupling/C-H activation [42].

Organ *et al.* have applied their 'PEPPSI' catalysts to the cross-coupling of aryl halides and arylmagnesium halide reagents (PEPPSI = pyridine-enhanced pre-catalyst preparation stabilization and initiation) [44]. The reactivity of $[PdCl_2(3\text{-}Cl\text{-}py)(IPr)]$ **13** was shown to be superior to that of systems based on triphenylphosphine, tricyclohexylphosphine and selected Buchwald-type biarylphosphines, in THF at room temperature with 2 mol % Pd. In addition, complexes such as **13** and **14**, which bear bulkier NHC ligands, were shown to be considerably more active than **15** and **16**, which bear smaller NHCs, through a benchmark reaction (Fig. 1.23). The theme of 'bulky-yet-flexible' NHCs has been a key research topic in catalytic cross-coupling in recent years [45–48].

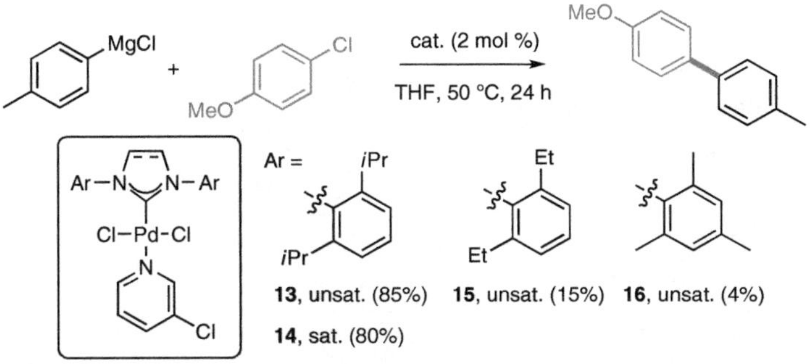

Fig. 1.23: Benchmarking of PEPPSI catalysts in the coupling of Grignard reagents; unsat. and sat. refer to the unsaturated or saturated nature of the NHC backbone [44].

A range of compounds could be prepared using a small set of protocols (THF, 2:1 THF/DMI, or 1:1 THF/DME; DMI = 1,3-dimethylimidazolidin-2-one, DME = 1,2-dimethoxyethane) at rt, using IPr-bearing **13** (Fig. 1.24). A sequential one-pot method was also developed by coupling a bromide with a chloride-functionalized Grignard reagent, followed by addition of a second Grignard reagent.

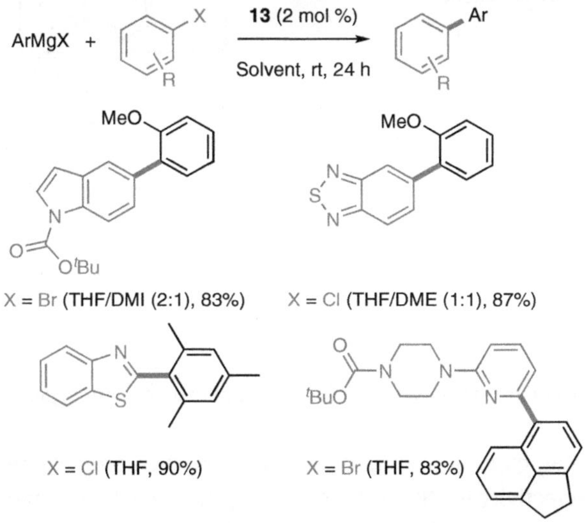

Fig. 1.24: Synthesis of biaryl compounds using a 'PEPPSI' catalyst [44].

Buchwald *et al.* have deployed Knochel's 'Turbo' Grignard reagents [49] in cross-couplings, using a Pd(0)/DavePhos **17** or SPhos **18** system (Fig. 1.25) [50]. A variety of aryl triflates and iodides could be cross-coupled with standard organomagnesium halide species using this catalyst system (at –20 to –65°C), while a range of (hetero)aryl 'Turbo' Grignard reagents could be cross-coupled at –20°C, including *ortho*-substituted reagents (Fig. 1.26).

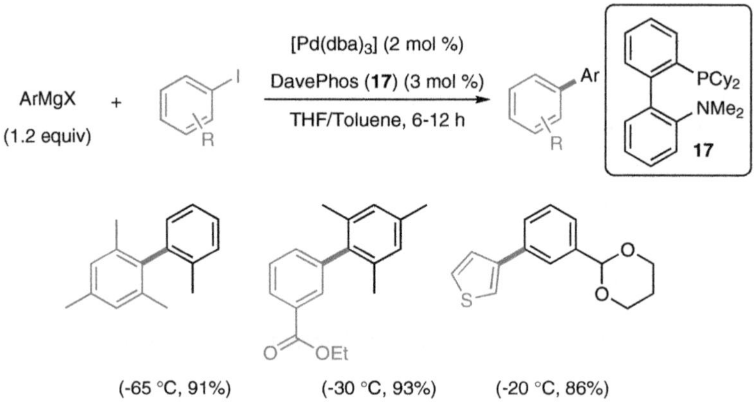

Fig. 1.25: Cross-coupling of Grignard reagents using a Pd/Buchwald phosphine system [50].

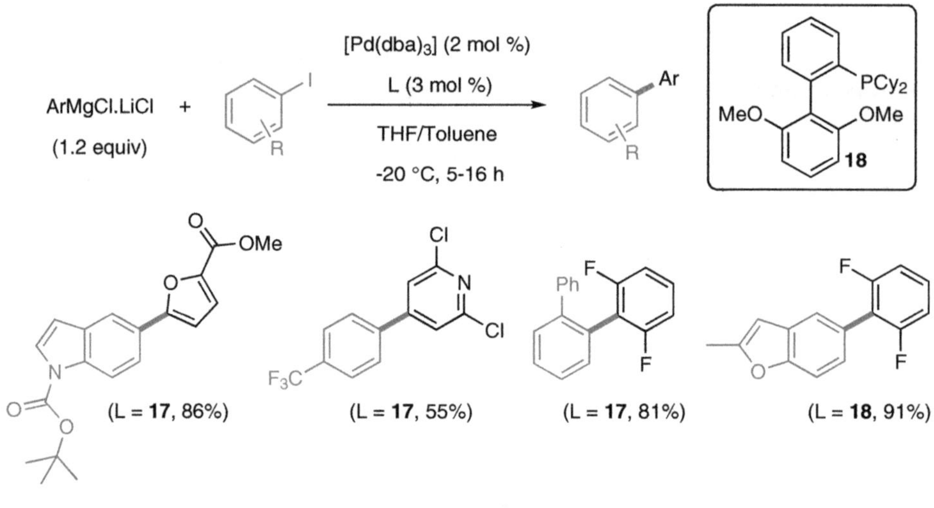

Fig. 1.26: Cross-coupling of 'Turbo' Grignard reagents using a Pd/Buchwald phosphine system [50].

Manabe *et al.* have explored the cross-coupling of *ortho*, *para*-dichlorobenzene species, reporting excellent site-selectivity for the *ortho*-position [51]. In each example, a hydroxyl, hydroxymethylene or amine/amide was present at the *ipso*-position, and was proposed to direct the arylation. A set of biaryl compounds was prepared (Fig. 1.27). Anisoles were unreactive in these reactions, suggesting that a protic functional group was required. Such selective reaction was not possible with the corresponding dibromo compounds using PCy_3, but was later achieved by use of a terphenyl ligand **19**, which was proposed to co-ordinate the hydroxyl/amine/amide moiety and direct oxidative addition to the *ortho* site [52–54].

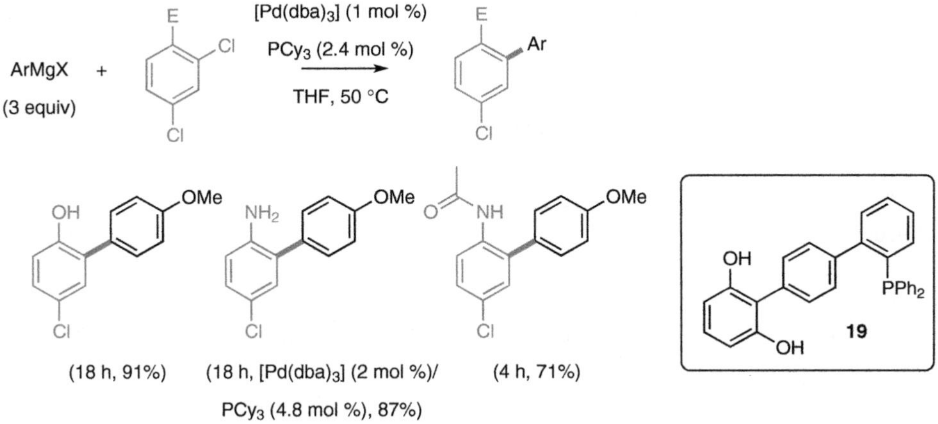

Fig. 1.27: Cross-coupling of *ortho, para*-dichlorobenzene compounds [51].

Wang *et al.* have compared the performance of palladium and nickel catalysts bearing *N,N,O*-chelate ligands **20** and **21** [55]; palladium complexes appeared to be more active, catalyzing reactions such as the cross-coupling of bromomesitylene and mesityl Grignard reagent in moderate to good yields (Fig. 1.28).

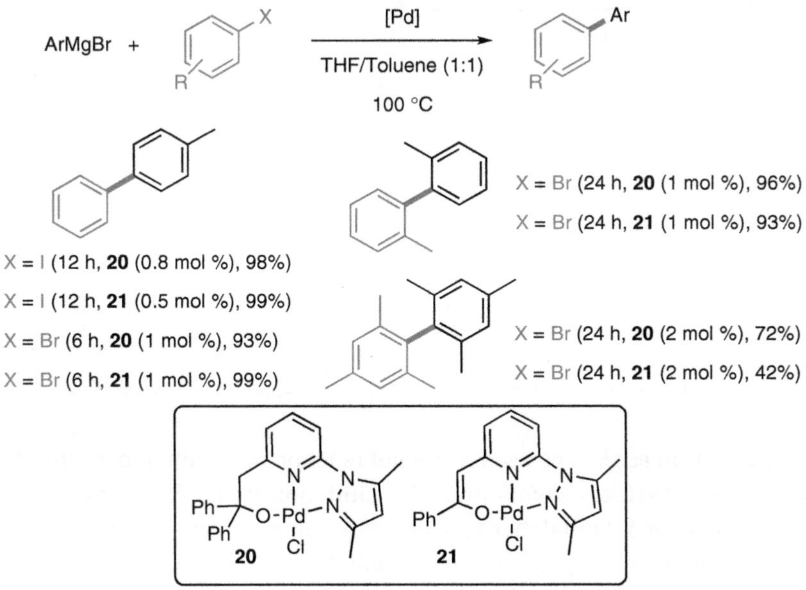

Fig. 1.28: Cross-coupling catalyzed by N,N,O-chelated palladium complexes [55].

Wolf *et al.* have documented the use of phosphinous-acid ligands in palladium catalysis, to form sterically crowded biaryls from aryl chlorides, bromides and iodides [56]. The corresponding nickel catalyst was also prepared and deployed, and showed similar activity at 5 mol % loadings to the palladium complex at 3 mol %. Examples of some of the molecules prepared can be found in Figs. 1.29 and 1.30.

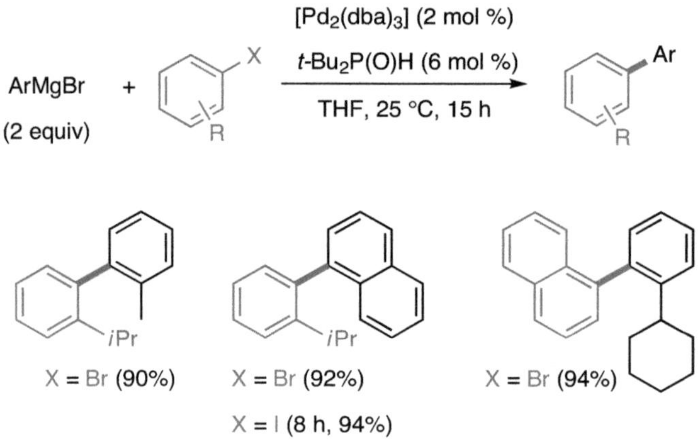

Fig. 1.29: Cross-coupling of Grignard reagents using a palladium/phosphinous acid system [56].

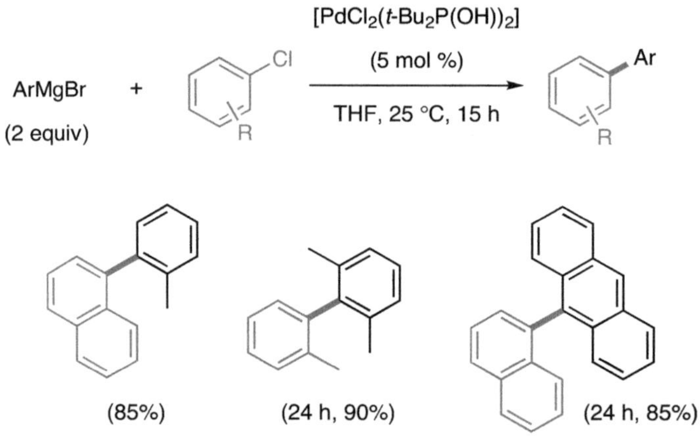

Fig. 1.30: Cross-coupling of Grignard reagents using a palladium/phosphinous acid system [56].

It is worth noting that Hu *et al.* have used a one-pot reaction of 1,2-dihalobenzenes to prepare sterically-hindered triaryl compounds [57]. Lithiation of a 1,3-dimethoxyarene, followed by addition of the 1,2-dihalobenzene and subsequent cross-coupling with an aryl Grignard, led to a set of triaryl compounds (Fig. 1.31).

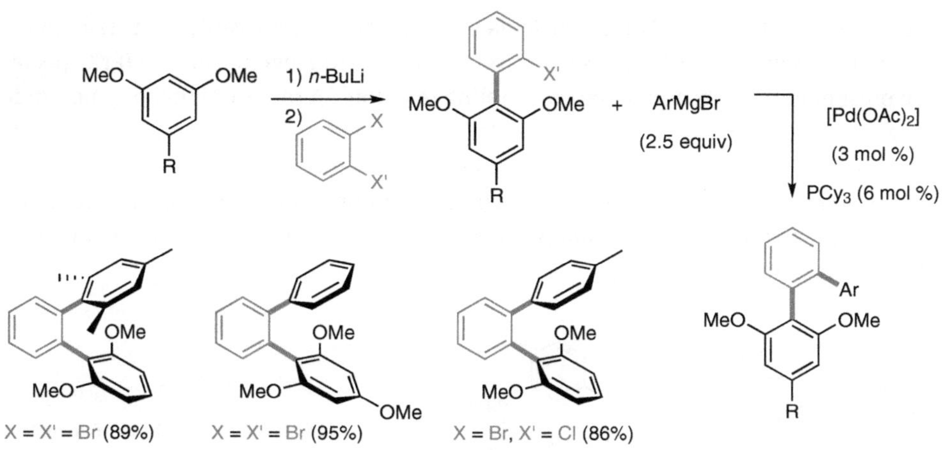

Fig. 1.31: Cross-coupling of Grignard reagents to prepare triaryl compounds [57].

Manabe *et al.* have achieved the cross-coupling of *ortho*-fluorophenols with aryl Grignard reagents (and one example of a vinyl Grignard reagent) [58]. Test reactions showed that cross-coupling occurred at the *ortho*-position of 4-chloro-2-fluorophenol, in preference to reaction with the *para*-chloro moiety. A small set of example compounds was prepared, examples of which can be found in Fig. 1.32.

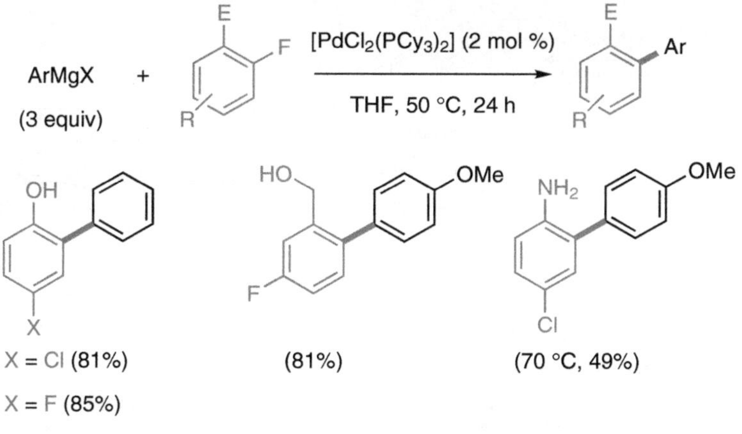

Fig. 1.32: Cross-coupling of *ortho, para*-dichlorobenzene compounds [58].

Knochel *et al.* have investigated the cross-coupling of Grignard reagents in some depth and suggested an alternative reaction mechanism based on radical intermediates in the presence of organohalide compounds [59]. When reactions were carried out in the presence of *iso*propyl iodide and bromide, the former led to a significant rate enhancement (Fig. 1.33). This is particularly important and relevant, as the reaction of aryl halides with *i*PrMgCl·LiCl is a key route to Grignard reagents, and so *i*PrX species are present in the subsequent cross-coupling reactions. A range of other organoiodide reagents gave similar rate enhancements. Reactions between organohalide reagents bearing a remote alkene led to some competing cyclization processes (Fig. 1.34) which were not observed with alkene-functionalized organomagnesium reagents, providing support for a radical mechanism. A range of aryl bromide reagents were coupled in the presence of *i*PrI (Fig. 1.35).

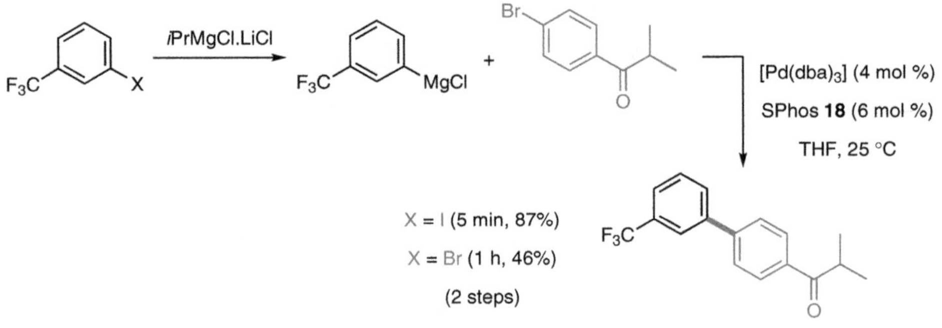

Fig. 1.33: Knochel *et al.*'s investigation of the cross-coupling of Grignard reagents [59].

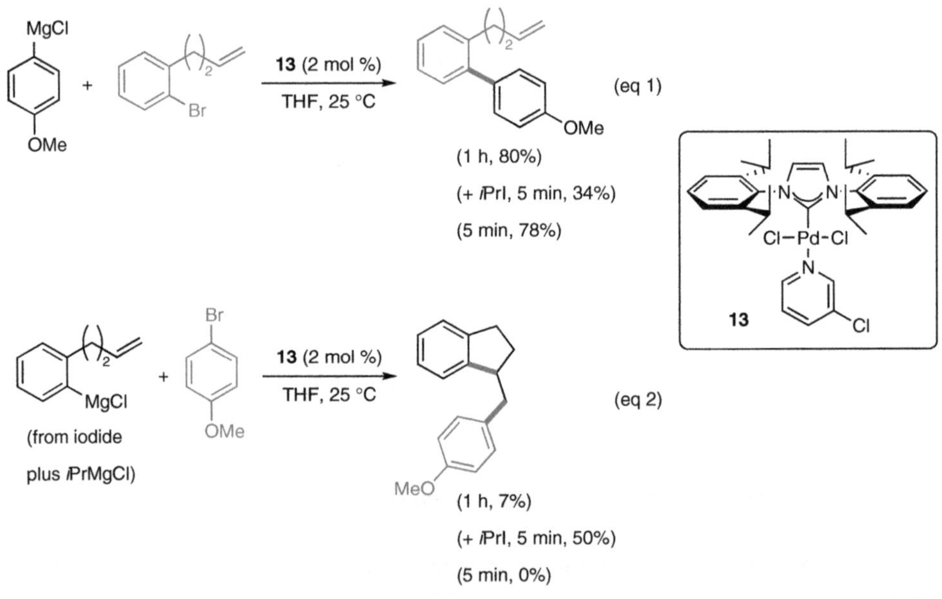

Fig. 1.34: Knochel *et al.*'s investigation of the cross-coupling of Grignard reagents [59].

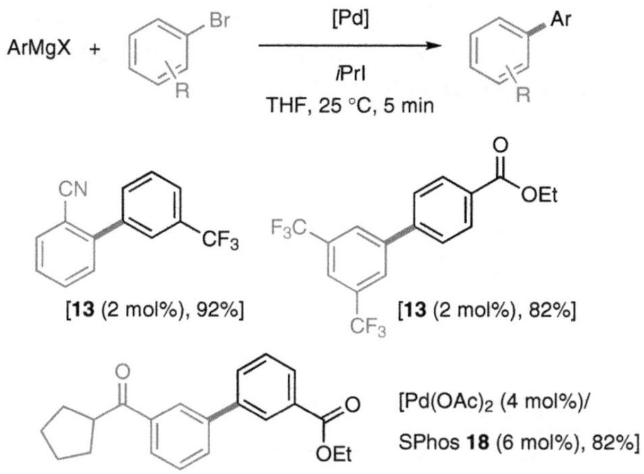

Fig. 1.35: Knochel *et al.*'s investigation of the cross-coupling of Grignard reagents [59].

Nolan and Cazin have utilized NHC-bearing palladium(II) chloride dimers of the form [PdCl(μ-Cl)(NHC)]$_2$ as catalysts for the cross-coupling of a range of arylmagnesium reagents with a series of aryl chlorides [60]. Of the four complexes tested (which bore IPr, SIPr, IMes or SIMes ligands), the SIPr derivative was found to be most active; this was then used to explore the scope and limitations of the method (Fig. 1.36).

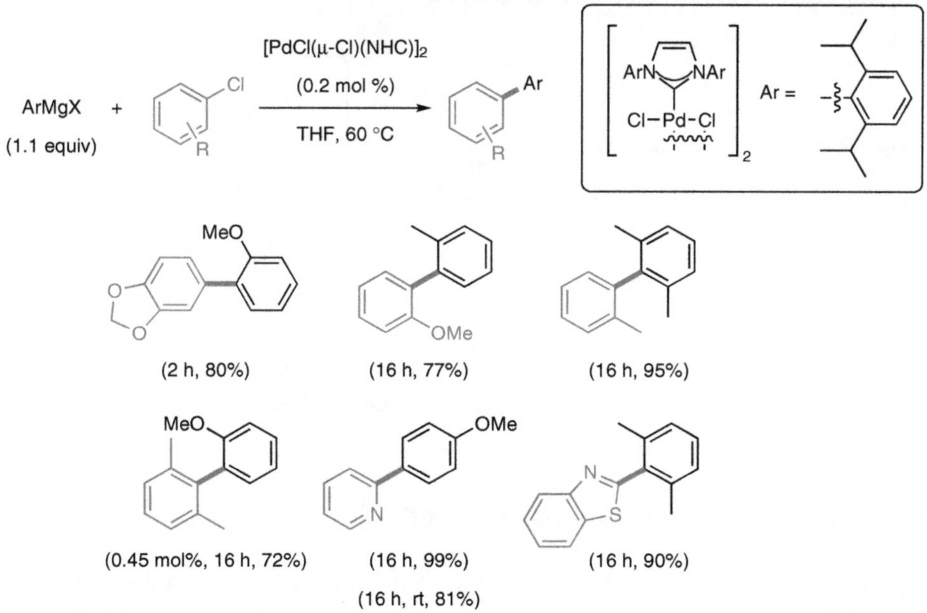

Fig. 1.36: Nolan and Cazin's use of Pd(II) dimers for the cross-coupling of arylmagnesium reagents with various aryl chlorides [60].

Van der Eycken and co-workers have utilized thioesters and thioesters (*vide infra*) as coupling partners in the cross-coupling of Grignard reagents catalyzed by palladium [61]. A set of heterocyclic products were prepared (Fig. 1.37).

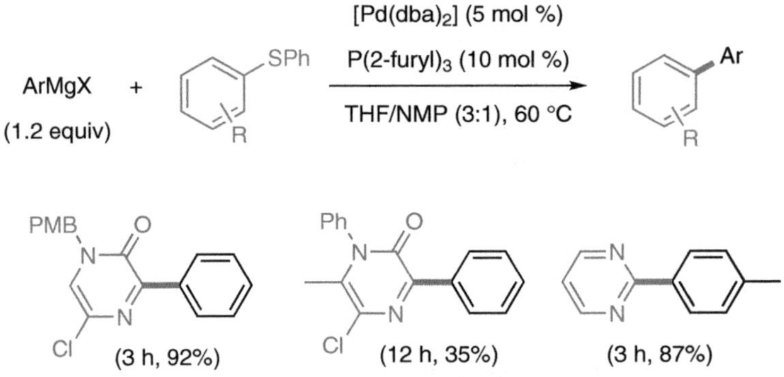

Fig. 1.37: Arylation of thioethers using palladium and Grignard reagents [61].

In 2010, the Ackermann group reported the arylation of 2-pyridyl Grignard reagents, initially using a combination of 2 mol% $[Pd(dba)_2]$ and 8 mol% (1-Ad)P(O)H (1-Ad = 1-adamantyl) in THF at 60 °C for 20 h [62]. However, heating $[Pd(OAc)_2]$ and (1-Ad)P(O)H together yielded a novel complex **22**, which was active in the cross-coupling reaction at lower loadings (Fig. 1.38).

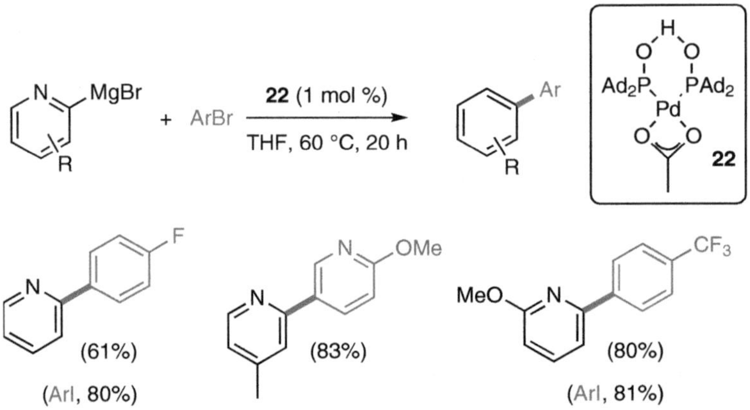

Fig. 1.38: Arylation using 2-pyridyl Grignard reagents [62].

Wu and co-workers showed that cyclopalladated ferrocenylimines could be used as catalysts for the cross-coupling of Grignard reagents with aryl halides [63]. A series of complexes were prepared, bearing various different ancillary ligands (Fig. 1.39).

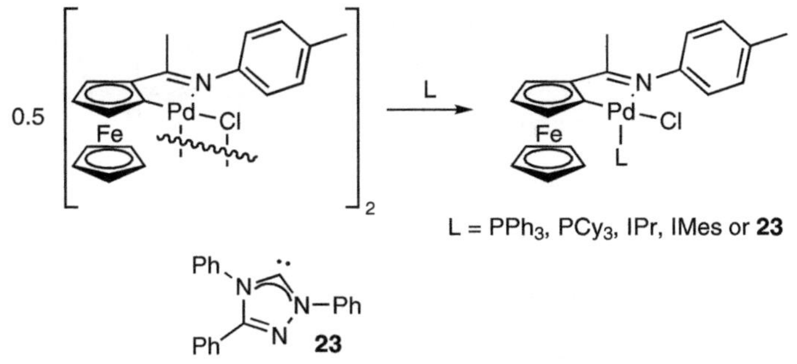

Fig. 1.39: Cyclopalladated ferrocenylimine complexes [63].

The triphenylphosphine- and triazole-bearing analogues were poorly active, while the other complexes were active in the cross-coupling of *para*-chlorotoluene and phenylmagnesium bromide. The IPr-bearing complex **24** was most active, so this was evaluated in the cross-coupling of a range of compounds including aryl chlorides and bromides (Fig. 1.40); lithium chloride was used as an additive. A number of *ortho*-trisubstituted biaryls could be prepared using only 0.5 mol % of **24**. The use of aryl chlorides required an increase in temperature to 60 °C, while the bromides could be coupled at room temperature.

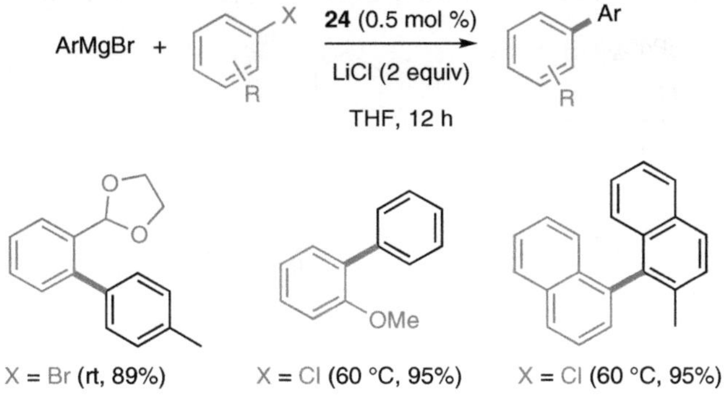

Fig. 1.40: Arylation catalyzed by cyclopalladated ferrocenylimine catalyst 24 [63].

Reeves and co-workers reported the palladium-catalyzed cross-coupling of aryltrimethylammonium triflate compounds with aryl Grignards in 2010 (Fig. 1.41). Reactions proceeded at room temperature with almost stoichiometric quantities of Grignard reagent, in contrast to a nickel-catalyzed variant that required an excess of Grignard reagent and elevated temperatures (3 equiv ArMgBr, benzene, 80 °C, 14 h). The requisite ammonium salts were prepared in one step from the corresponding dimethylanilines and methyl triflate. Competition experiments established that oxidative addition to the ammonium salts occurred faster than with aryl chlorides, bromides or triflates, but at a comparable rate to aryl iodides.

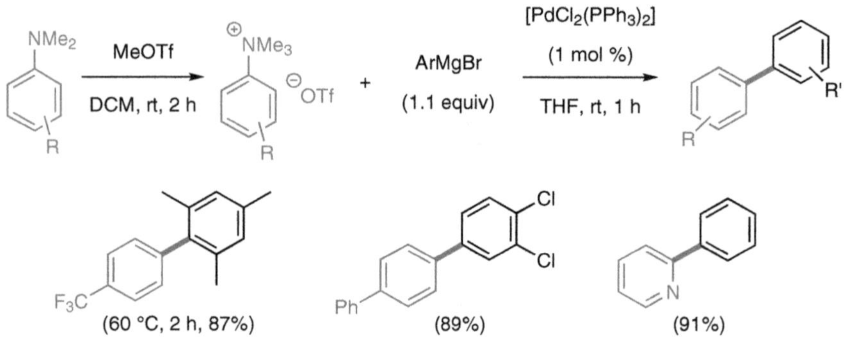

Fig. 1.41: Synthesis and cross-coupling of aryltrimethylammonium triflate salts with Grignard reagents [64].

Clarke and co-workers have conducted palladium-catalyzed Grignard cross-coupling reactions under very concentrated conditions, using a 5 mol L^{-1} solution of the Grignard in 2-methyltetrahydrofuran [65]; the high concentration and favorable solvent choice result in reductions in the cost and environmental impact of the solvent [66, 67]. In each case, palladium complexes bearing bidentate phosphine ligands were used (dippf = 1,1'-bis(di*iso*propylphosphino)ferrocene; dtbpf = 1,1'-bis(di*tert*butylphosphino) ferrocene). A handful of examples were evaluated (a sample of these is reported in Fig. 1.42).

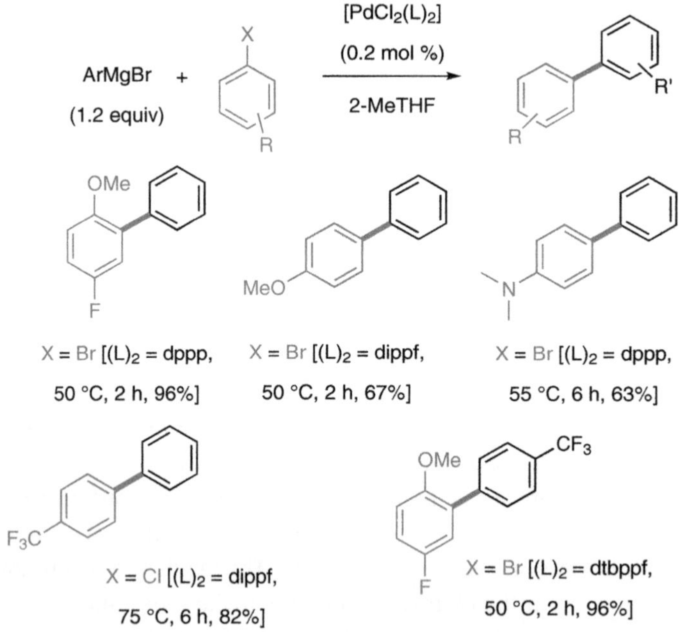

Fig. 1.42: Cross-coupling of Grignard reagents under high concentration conditions [65].

Sarkar *et al.* deployed pyrazole-functionalized phosphine ligands in a number of palladium-catalyzed processes, such as Stille and Hiyama cross-coupling, and the cross-coupling of Grignard reagents [68]. A set of molecules was prepared by this method (Fig. 1.43).

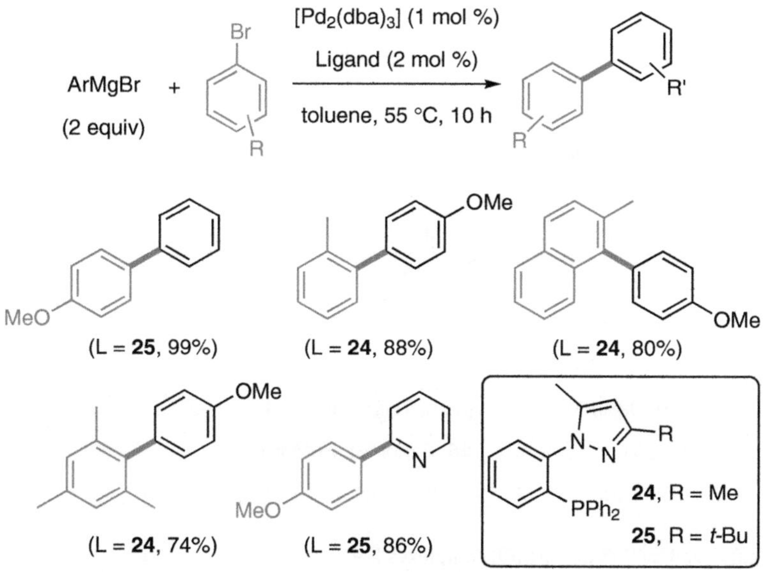

Fig. 1.43: Cross-coupling of Grignard reagents using pyrazole-functionalized phosphine ligands [68].

Jin and Fang have disclosed the use of [PdCl(Cp)(NHC)] complexes as pre-catalysts for the coupling of aryl Grignard reagents with aryl halides [69]. The catalysts were prepared in one pot from [$PdCl_2(NCPh)_2$], NHC. HX (X = Cl or BF_4) and NaCp (Fig. 1.44; SIMes = 1,3-bis(2,4,6-trimethylphenyl)-4,5-dihydroimidazol-2-ylidene; SIPr = 1,3-bis(2,6-di*iso*propylphenyl)-4,5-dihydroimidazol-2-ylidene); presumably the intermediate species is [PdCl(μ-Cl)(NHC)]$_2$ [70].

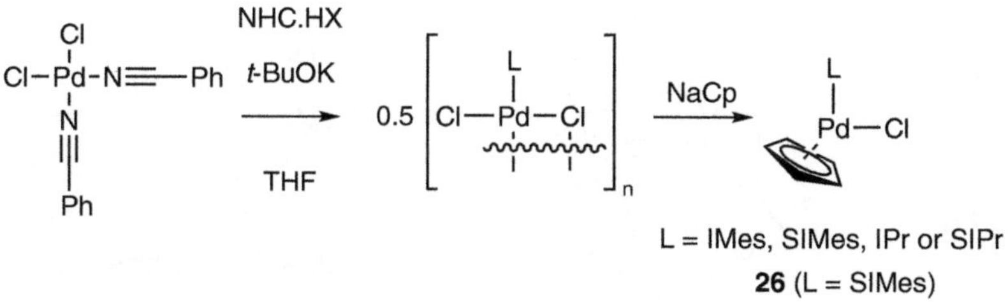

Fig. 1.44: Synthesis of [PdCl(Cp)(NHC)] complexes [69].

Complex **26**, bearing the SIMes ligand, was determined to be the most active, from a benchmark reaction (phenylmagnesium bromide and *para*-chloroanisole); the addition of 2 equiv LiCl was found to slightly improve isolated yields. A range of compounds were prepared from the corresponding aryl chlorides, using 1 mol % of **26** in THF (Fig. 1.45). More challenging substrates required an increase in reaction time.

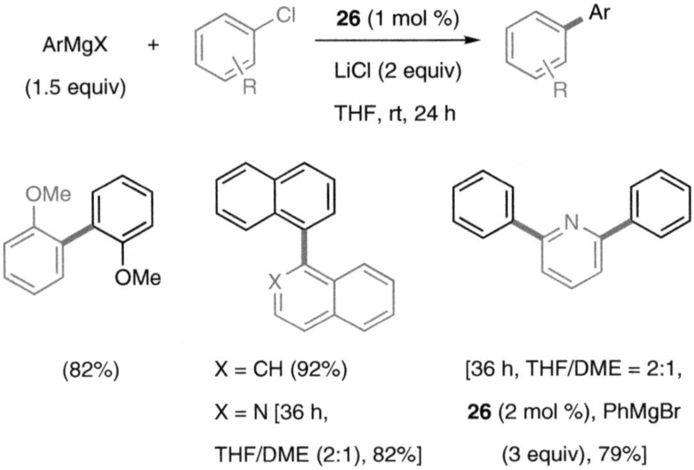

Fig. 1.45: Arylation catalyzed by [PdCl(Cp)(NHC)] complexes [69].

A series of well-defined mono- and dinuclear HASPO-Pd complexes were prepared (Fig. 1.46), and assessed in the coupling of Grignard reagents [71]. Complex **27** was found to be more active than the *in situ* generated species, while **28** and **29** were less reactive. A range of aryl and heteroaryl tosylates were arylated using Grignard reagents (0.2 mol % **27**, 1,4-dioxane, 80 °C, 24 h) in good to excellent yields (71–98 %). A series of 2-hydroxyaryl chlorides and tosylates could also be arylated in this manner, despite the presence of the relatively acidic O-H bond (Fig. 1.47).

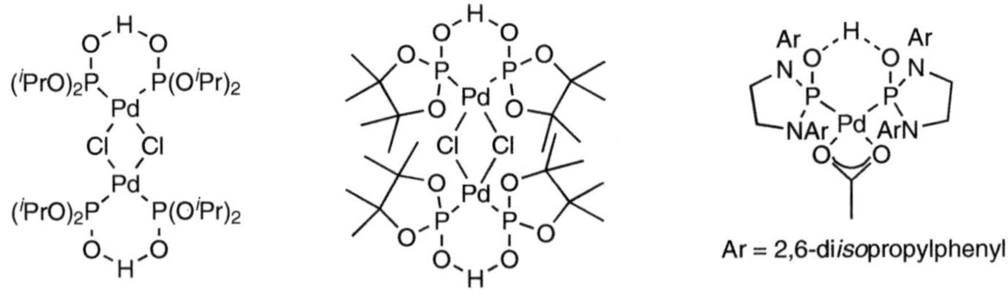

27 **28** **29**

Fig. 1.46: Well-defined HASPO-Pd complexes [71].

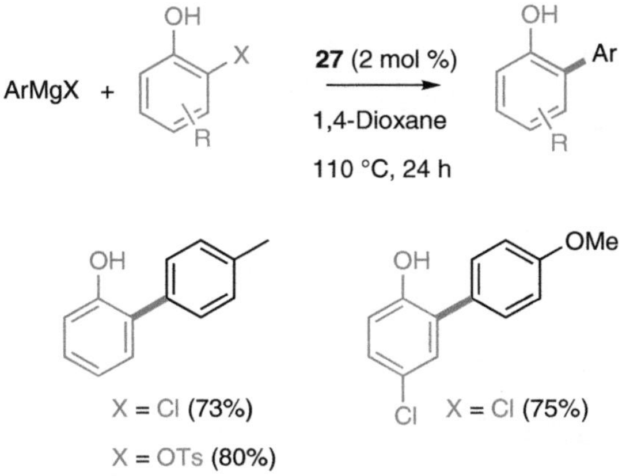

Fig. 1.47: Arylation of 2-chloro- and 2-tosylphenols [71].

The use of palladium-catalyzed cross-coupling of Grignard reagents is not limited to the synthesis of small molecules. McNeil *et al.* have used [$PdCl_2$(3-Cl-py)(IPr)] **13** to polymerize aryl units bearing both a magnesium chloride and a bromide unit (Fig. 1.48) [72]. The substrates were prepared from the reaction of the dibromo precursors with one equivalent of *iso*propylmagnesium chloride. The polymerization of the fluorene-based monomer was problematic; however, for the other two monomers molecular weights of 5–30 kDa could be achieved, with PDIs of less than 1.5.

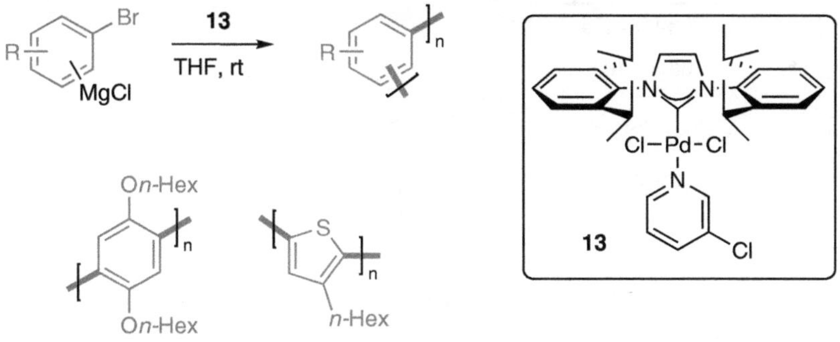

Fig. 1.48: Polymerization of Grignard reagents catalyzed by palladium [72].

Locklin *et al.* have also used palladium catalysis to functionalize surfaces; [Pd(P*t*-Bu_3)$_2$] was used to cross couple surface-bound aryl halides with Grignard reagents (Fig. 1.49) [73]. The use of 2-bromo-3-methylthiophenl-4-yl magnesium chloride allowed the construction of polymer chains at these sites.

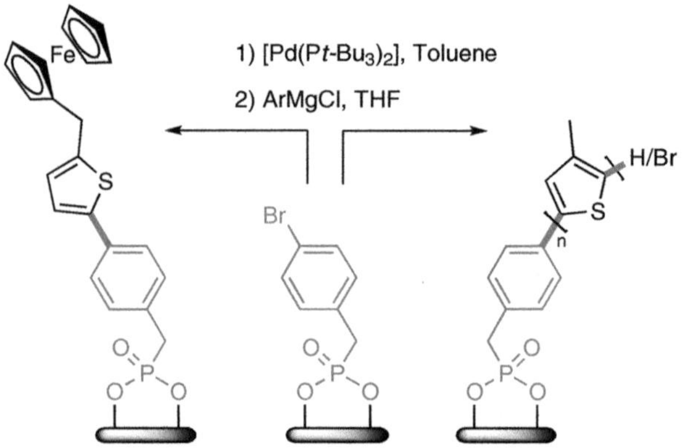

Fig. 1.49: Functionalization of surfaces using Grignard reagents, catalyzed by palladium [73].

Very recently, Cazin and co-workers developed a method for the synthesis of a range of tetrasubstituted biaryl compounds [74]. These targets are amongst the most challenging biaryl motifs to be prepared by cross-coupling. The most efficacious catalyst was found to be $[PdCl(\mu\text{-}Cl)(IPr^*)]_2$ **28** (IPr* = 1,3-bis(2,6-diphenylmethyl-4-methylphenyl) imidazol-2-ylidene) [46], which was capable of coupling a range of reagents at catalyst loadings of 0.1 mol %; a sample of the scope of the reaction is presented in Fig. 1.50. This catalyst loading is approximately an order of magnitude lower than those achieved in the majority of the previous studies.

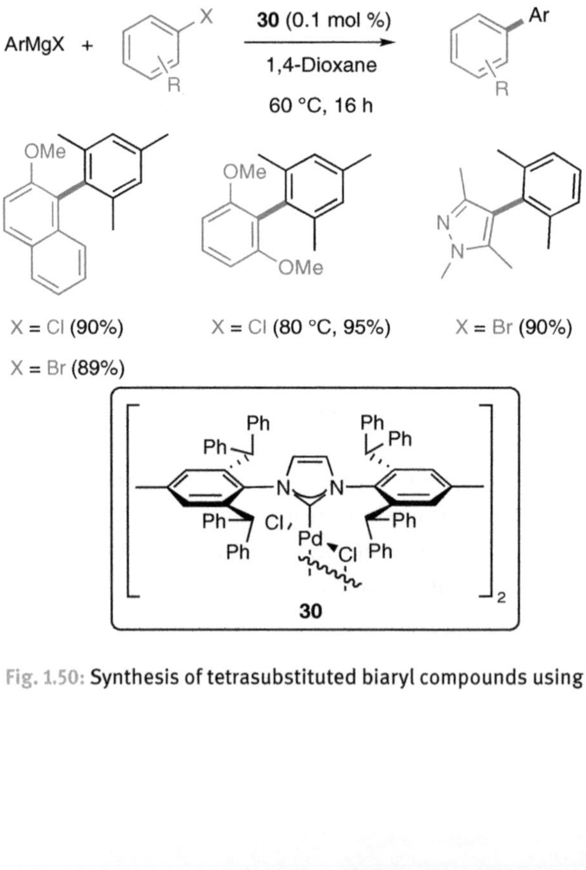

Fig. 1.50: Synthesis of tetrasubstituted biaryl compounds using palladium catalysis [73].

1.4.1.2 With Vinyl Electrophiles

Alami *et al.* reported the cross-coupling of chloro-dienes and chloro-enynes with a range of magnesium reagents, to furnish conjugated systems with typically excellent retention of stereochemistry [75]. A simple catalyst system was used to effect these reactions at room temperature in 20 min. Some examples of the products formed can be found in Fig. 1.51. Triethylamine (8 equiv) was found to be a useful additive here; in optimization experiments, DME was found to erode the yield (*versus* no additive), while sulfolane, $Et_2N(iPr)$, Et_3N and TMEDA all resulted in improved yields. Notably, both alkene isomers undergo successful cross-coupling.

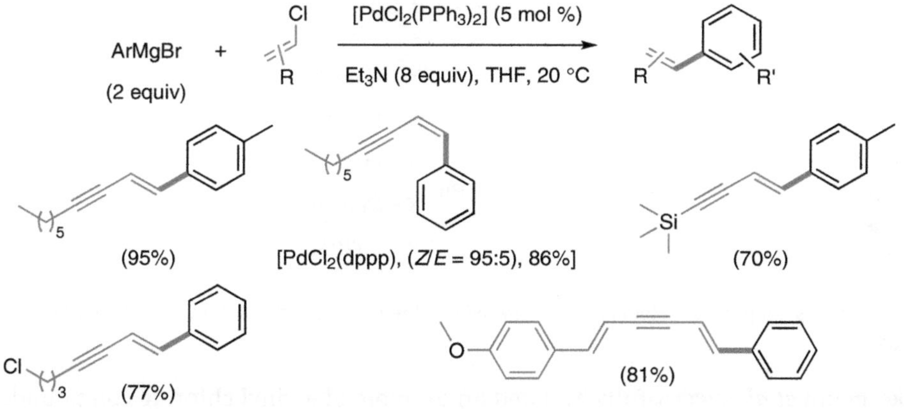

Fig. 1.51: Cross-coupling of chloro-dienes and chloro-enynes with arylmagnesium chlorides [75].

Hartwig *et al.* have shown that ferrocene-derived ligands are capable of facilitating the cross-coupling of vinyl tosylates with *para*-tolylmagnesium bromide (Fig. 1.52; see also Fig. 1.17 above). The use of a more electron-donating ligand bearing two dialkylphosphine sites was often necessary to achieve some of these reactions.

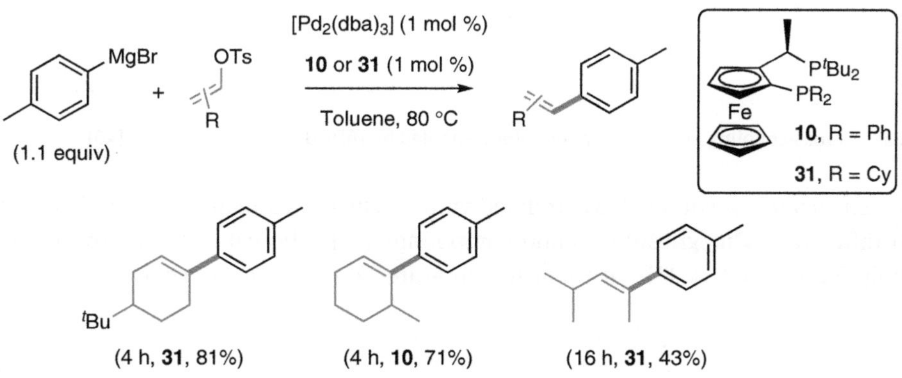

Fig. 1.52: Cross-coupling of vinyl tosylates with *para*-tolylmagnesium bromide [37].

Shi *et al.* have shown that selenium-containing compounds can be substrates for cross-coupling with aryl Grignard reagents; following cross-coupling of the vinyl halide substrate, oxidation of the selenide leads to elimination to yield a functionalized 1,3-butadiene motif (Fig. 1.53) [76]. The use of a simple palladium catalyst system prevented further coupling reactions of the *para*-chlorophenyl derivative. Other transformations on these starting materials, such as Negishi and Suzuki cross-couplings, were also demonstrated.

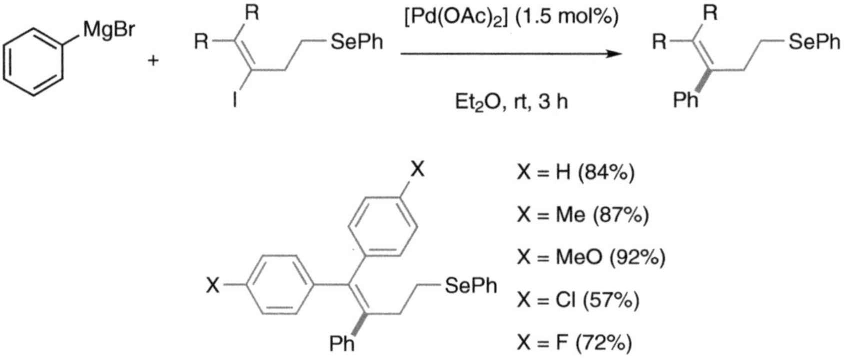

Fig. 1.53: Cross-coupling of selenium-bearing vinyl halides with aryl Grignard reagents (isolated yields) [76].

Ackermann *et al.* successfully arylated an example of a vinyl chloride compounds using a TADDOL-derived ligand **12** [40]; only one example was presented, alongside a series of other reactions catalyzed by the same palladium/ligand system (Fig. 1.54).

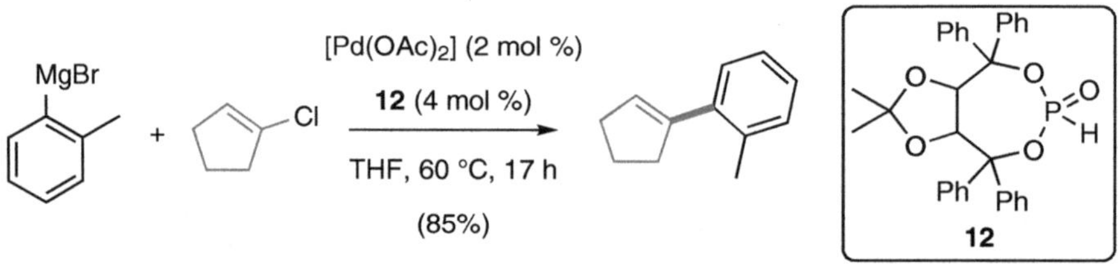

Fig. 1.54: Cross-coupling of 1-chlorocyclopentene enabled by TADDOL-derived ligand **12** [40].

Skrydstrup and co-workers have arylated various vinyl phosphate compounds using two different arylmagnesium halide compounds [77]. The catalyst employed was simply $PdCl_2$, used in THF at room temperature. A selection of cyclic and acyclic phosphates was deployed (Fig. 1.55). Most of these structures were simple hydrocarbons, without functional groups.

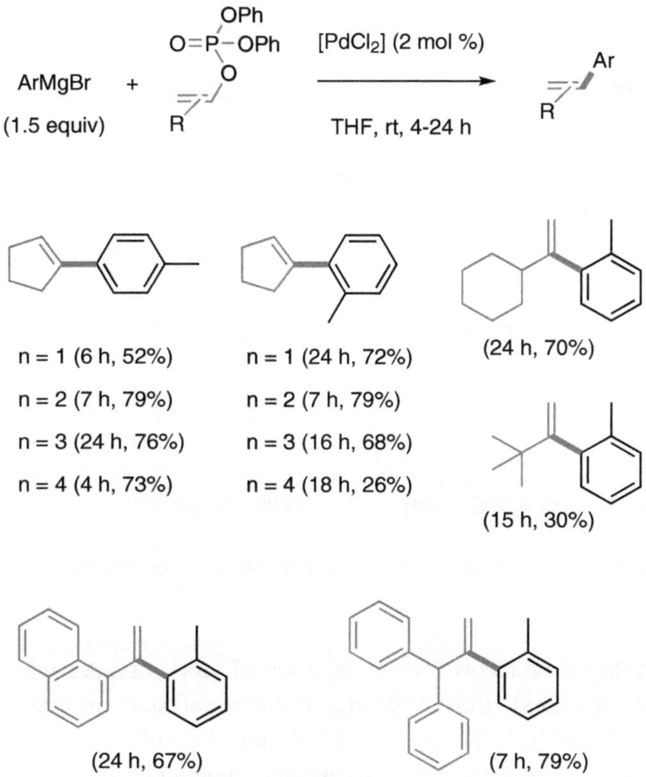

Fig. 1.55: Cross-coupling of vinyl phosphates [77].

In the same year, Brown *et al.* cross-coupled vinyl phosphates as part of a strategy to access tetrasubstituted alkene targets [78]. These electrophiles were prepared *via* a one-pot process in which the α,α-disubstituted aryl ester was treated with *n*-butyl lithium, to generate an equilibrium concentration of ketene, followed by subsequent nucleophilic addition of an organolithium reagent and trapping of the alkoxide with $ClP(O)(OEt)_2$ (Fig. 1.56, eq 1). The desired reagent was achieved with good *E*/*Z* ratios, and subsequent cross-coupled with Grignard reagents to produce tetrasubstituted alkenes (Fig. 1.56, eq 2). Cross-coupling with alkyl Grignard reagents was also undertaken (*vide infra*).

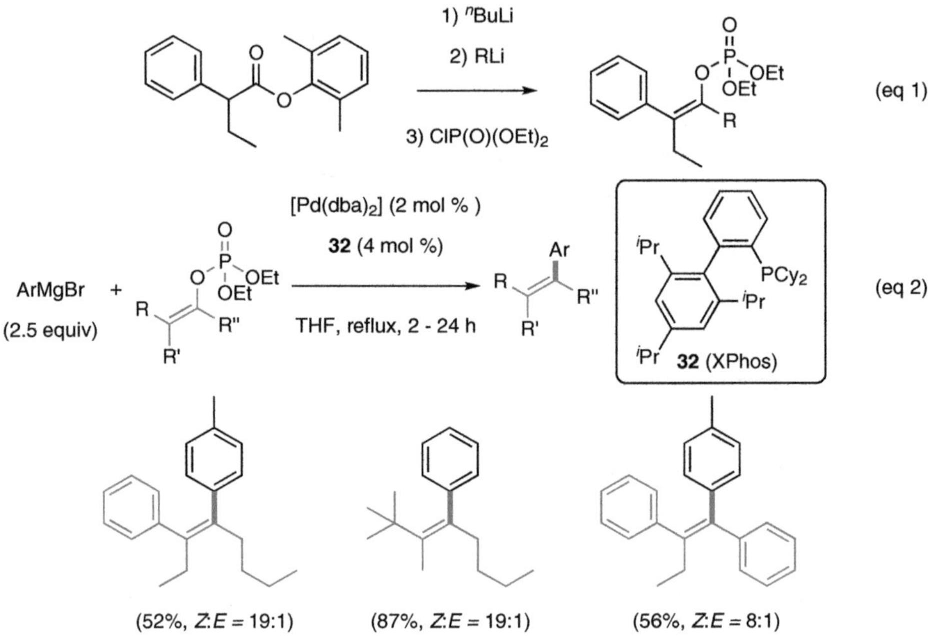

Fig. 1.56: (a) Synthesis of heavily substituted vinyl phosphates, and (b) their subsequent cross-coupling reactions [78].

Ackermann *et al.* have developed protocols for the arylation of vinyl tosylates using $[Pd_2(dba)_3]$ and the previously-discussed 'HASPO' ligands [71]. A range of cyclic and acyclic tosylates could be arylated using either 1.25 mol% $[Pd_2(dba)_3]$ plus 2.5 mol% **11** at room temperature or 0.2 mol% of well-defined complex **27** at elevated temperatures (Fig. 1.57).

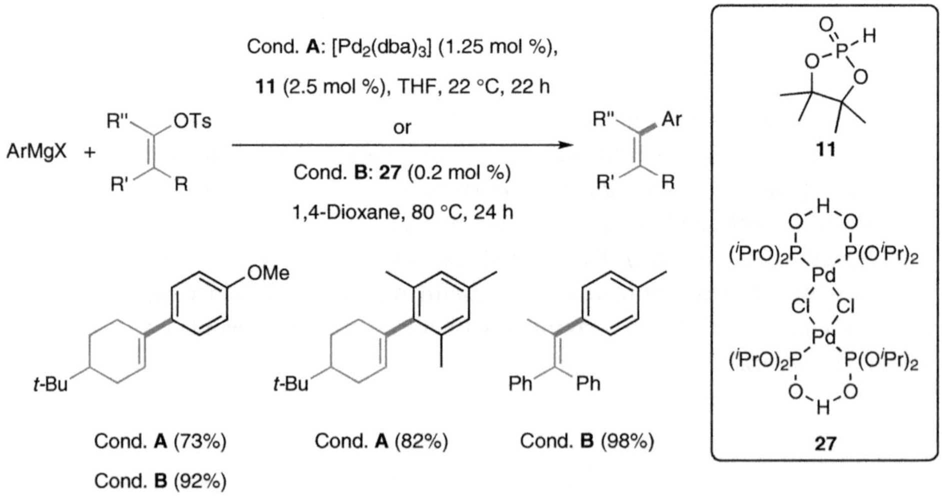

Fig. 1.57: Arylation of vinyl tosylates using HASPO-Pd complexes [71].

Zimmer and Reissig cross-coupled vinyl phosphate and vinyl nonaflate compounds as part of a strategy to access compounds such as chiral 1,4-amino alcohols [79]. A small number of examples of a few different reactions were demonstrated (including Sonagashira cross-coupling and Miyaura borylation). The cross-couplings of Grignard reagents proceeded in moderate yield (Fig. 1.58), and poorer yields were obtained with diphenylphosphate-based electrophiles.

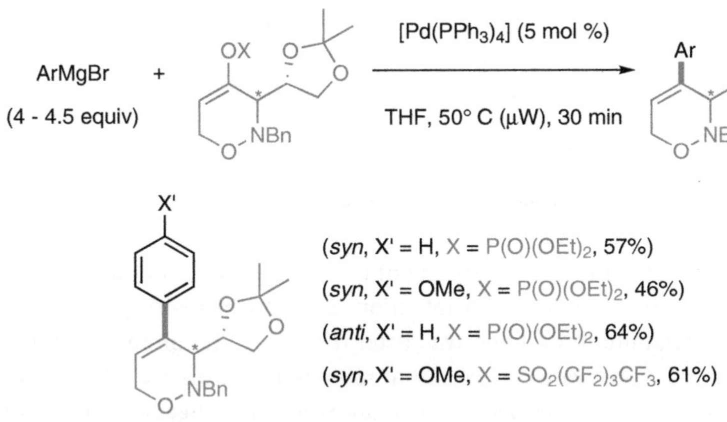

Fig. 1.58: Arylation of vinyl phosphates and vinyl tosylates [79].

Palladium is not typically capable of cross-coupling with organofluoride reagents; such reactions are typically catalyzed by nickel instead. However, Wu and Cao have recently reported both palladium- and nickel-catalyzed protocols for the cross-coupling of fluoro- and difluorostyrene reagents [16]. A number of arylmagnesium halide reagents were used (Figs. 1.59 and 1.60); cross-coupling was also achieved with alkylmagnesium chloride species (*vide infra*).

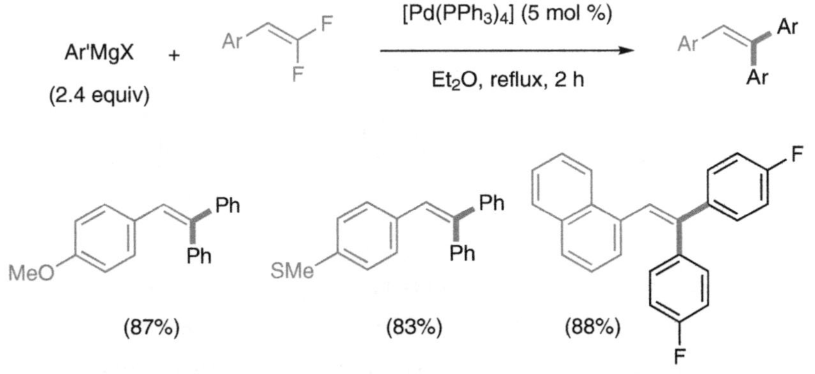

Fig. 1.59: Palladium-catalyzed cross-coupling of difluorostyrenes with arylmagnesium reagents [16].

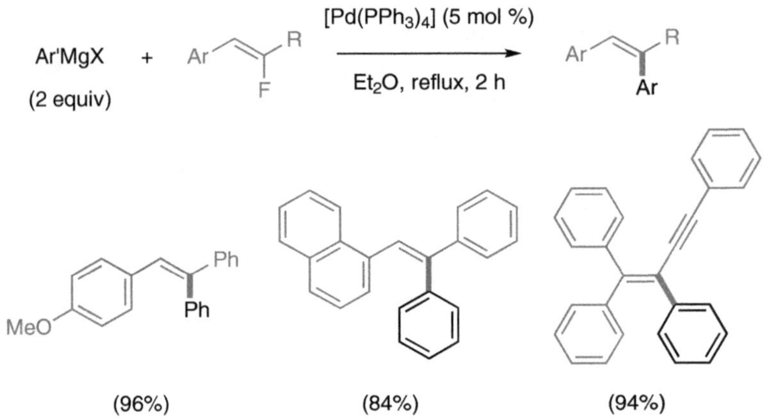

Fig. 1.60: Palladium-catalyzed cross-coupling of vinyl fluorides with arylmagnesium reagents [16].

Lipshutz *et al.* have reported the cross-coupling of (hetero)aryl Grignard and 'Turbo' Grignard reagents with vinyl halides, with retention of stereochemistry [80]. Where 1-chloro-3-(methylphenylamino)prop-1-ene was used ($E/Z = 1{:}1$), the product had the same E/Z ratio. Most reactions were carried out at room temperature for 8 h, and all used palladium dichloride complexes bearing bidentate phosphine ligands. A range of products were obtained (Fig. 1.61). Alkylmagnesium, alkenylmagnesium and alkynylmagnesium reagents could also be used (*vide infra*).

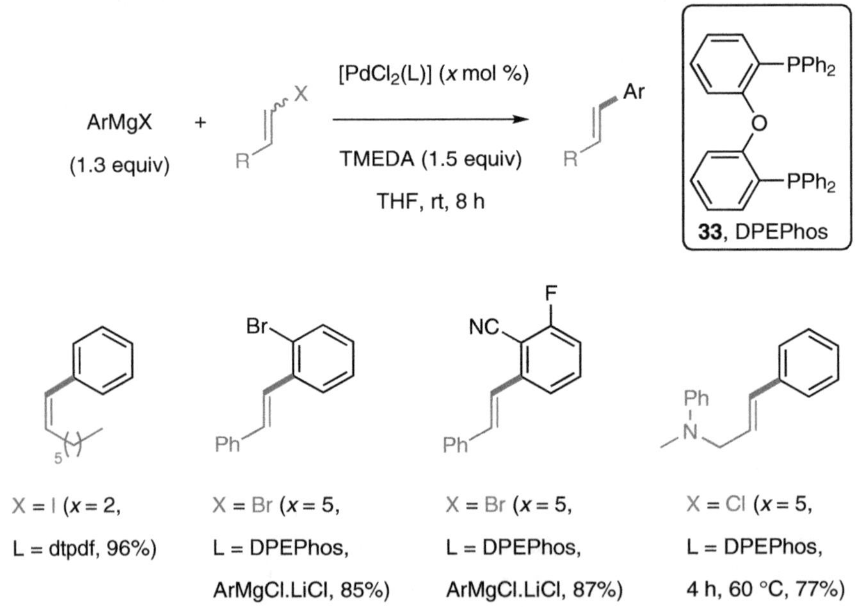

Fig. 1.61: Arylation of vinyl halides at room temperature [80].

1.4.1.3 With Alkyl Electrophiles

Sinou and co-workers cross-coupled *para-tert*-butylphenol glycopyranosides with a selection of aryl (and some alkyl; *vide infra*) Grignard reagents. Chelating bis(phosphine) ligands were found to be crucial for this transformation. Notably, this involved oxidative addition into a C–O bond, which is most often the domain of nickel catalysts. Examples of the products obtained are detailed in Fig. 1.62 [81].

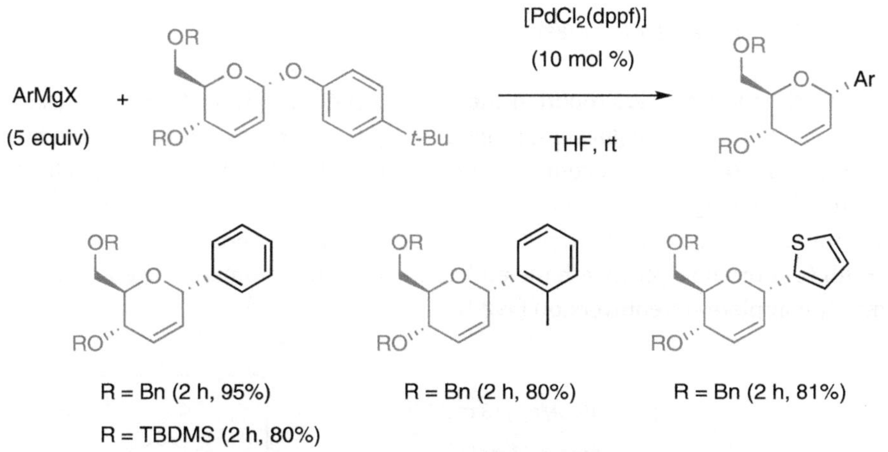

Fig. 1.62: Cross-coupling of *para-tert*-butylphenyl glycopyranosides [81].

Beller *et al.* disclosed one of the first coupling reactions of arylmagnesium halides with alkyl chlorides in 2002. A system composed of 4 mol% $Pd(OAc)_2$ and 4 mol% PCy_3 in NMP (1-methyl-2-pyrrolidinone) enabled the cross-coupling of a range of coupling partners (Fig. 1.63) [82].

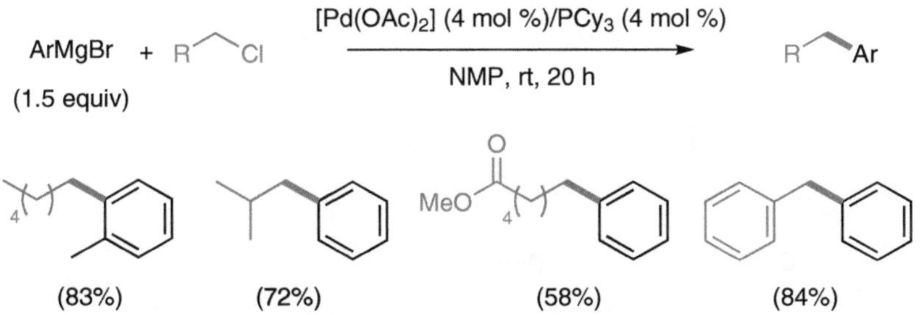

Fig. 1.63: Arylation of alkyl chlorides using a $Pd(OAc)_2/PCy_3$ catalyst system [82].

Kambe *et al.* reported the cross-coupling of alkyl bromides and alkyl tosylates with alkyl Grignard reagents, although one example was reported with an aryl Grignard (Fig. 1.64) [83]. 1,3-Butadiene was used as an additive.

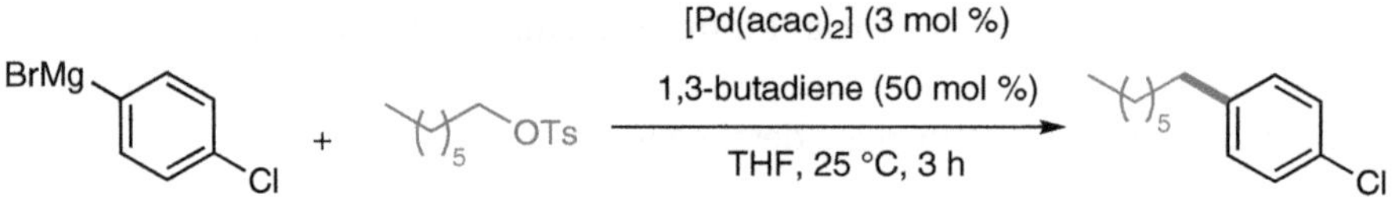

Fig. 1.64: Arylation of an alkyl tosylate [83].

In 2009, Adrio and Carretero reported the use of a $[PdCl_2(NCMe)_2]$/Xantphos system to couple vinyl- and arylmagnesium bromide reagents with secondary benzylic bromides (Fig. 1.65) [84]; the cross-coupling with cyclohexyl bromide, and attempts using 2-substituted arylmagnesium bromides were unsuccessful, but a variety of benzylic bromides and arylmagnesium bromides could be coupled. In addition, coupling reactions with (*S*)-1-bromo-1-phenylethane and arylmagnesium bromide reagents yielded essentially complete stereoinversion (>98 %).

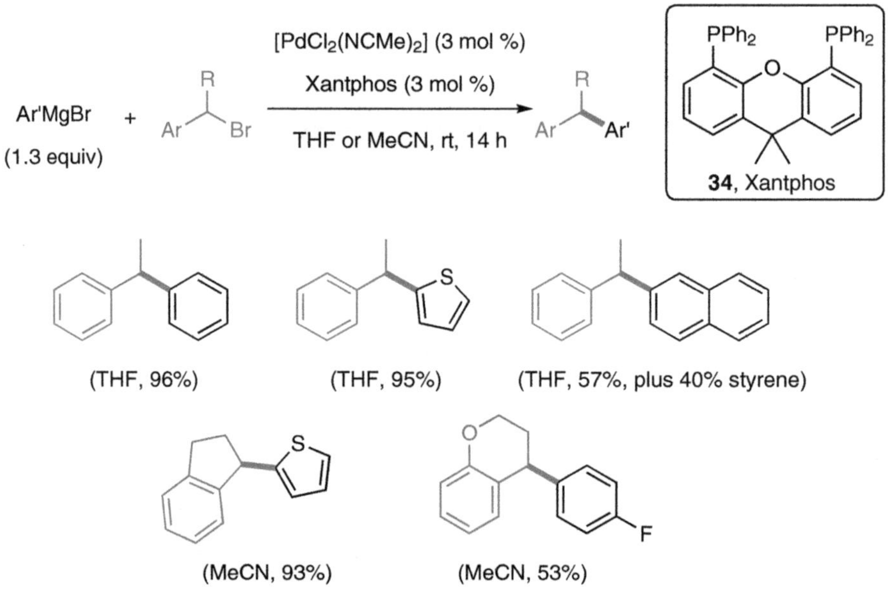

Fig. 1.65: Cross-coupling of benzylic bromides with arylmagnesium bromide reagents [84].

Van der Eycken *et al.* have achieved functionalization of thioesters and thioethers (*vide supra*) *via* palladium-catalyzed cross-coupling with Grignard reagents [61]. A limited set of products were prepared (Fig. 1.66).

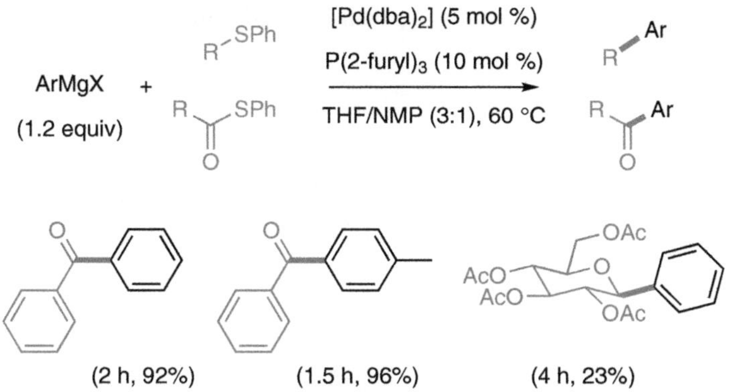

Fig. 1.66: Arylation of thioethers and thioesters using palladium and Grignard reagents [61].

Vogel *et al.* documented the use of allylsulfonyl chlorides and esters as electrophiles for cross-coupling with aryl Grignard reagents (Fig. 1.67). Selectivity for the two possible regioisomers could be influenced to some extent by changing the catalyst; [$PdCl_2(PPh_3)_2$] gave selectivity for the linear product, while [$Pd(PPh_3)_4$] led to a 1:1 mixture of linear and branched products. The reaction was proposed to proceed *via* Pd(η^3-allyl) intermediates, with an outer sphere mechanism for the palladium(II) catalyst, and an inner-sphere mechanism for the palladium(0) catalyst [85].

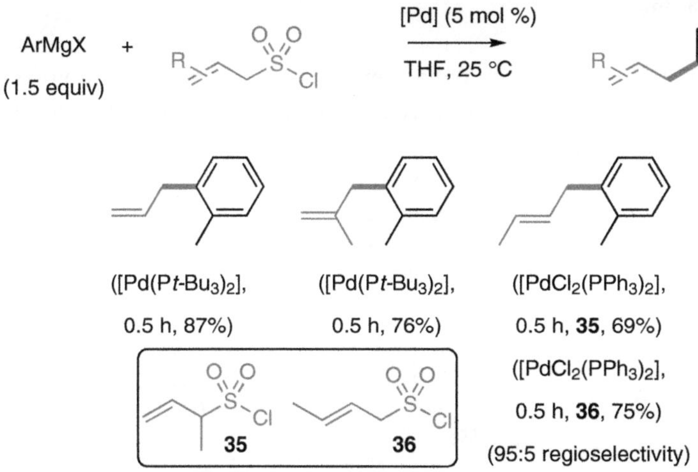

Fig. 1.67: Arylation of allyl sulfonyl compounds [85].

Ackermann and co-workers have employed secondary phosphine oxide and chloride ligands in the cross-coupling of aryl Grignard reagents and alkyl chlorides, typically at 25°C in NMP with 4 mol% $Pd(OAc)_2$ and 4 mol% of ligand (Fig. 1.68) [86]. Yields ranged from moderate to excellent. Alternatively, well-defined complexes could be used at lower catalyst loadings (1 mol %).

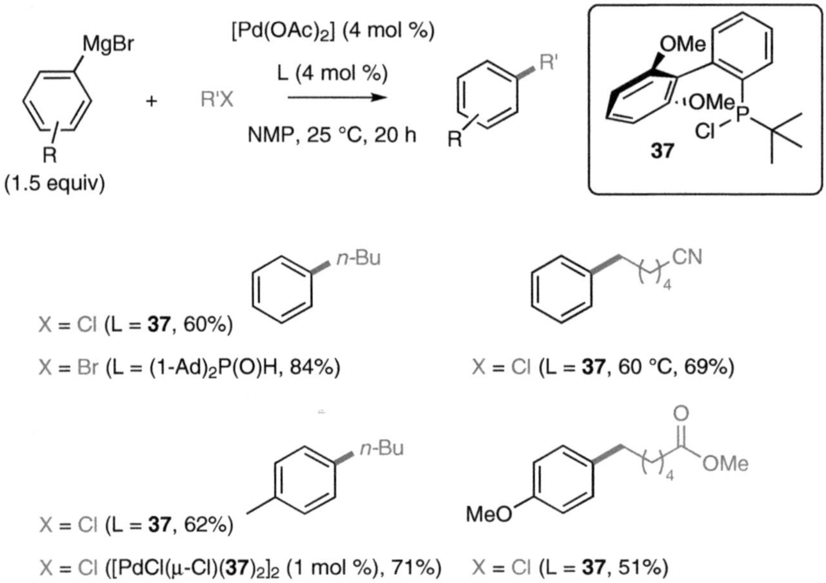

Fig. 1.68: Cross-coupling of arylmagnesium halides and alkyl halides, using secondary phosphine oxide and chloride ligands [86].

Wu and Cao have recently reported both palladium- and nickel-catalyzed protocols for the cross-coupling of fluoro- and difluorostyrene reagents with alkylmagnesium reagents (Figs. 1.69 and 1.70) [16].

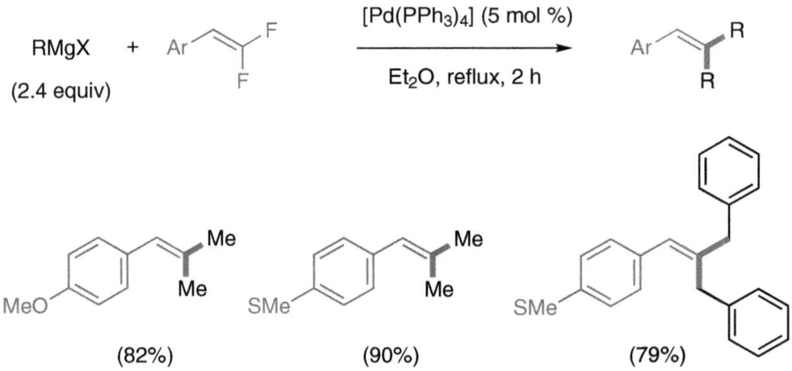

Fig. 1.69: Palladium-catalyzed cross-coupling of vinyl fluorides with alkylmagnesium reagents [16].

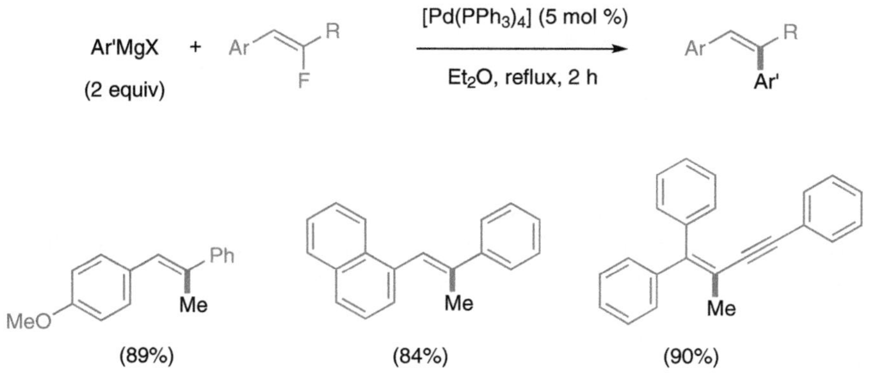

Fig. 1.70: Palladium-catalyzed cross-coupling of vinyl fluorides with alkylmagnesium reagents [16].

1.4.1.4 With Alkynyl Electrophiles

Song and Li have utilized a simple [Pd(PPh$_3$)$_4$]/TMEDA system for the cross-coupling of alkynyl iodide substrates with arylmagnesium species (and methylmagnesium bromide; *vide infra*) [87]. A series of (hetero)aryl acetylene and alkyl acetylene derivatives were deployed, giving moderate to very good yields, of which some examples can be found in Fig. 1.71. Some reactions were also trialed with organobromide and organochloride compounds, with slightly lower isolated yields.

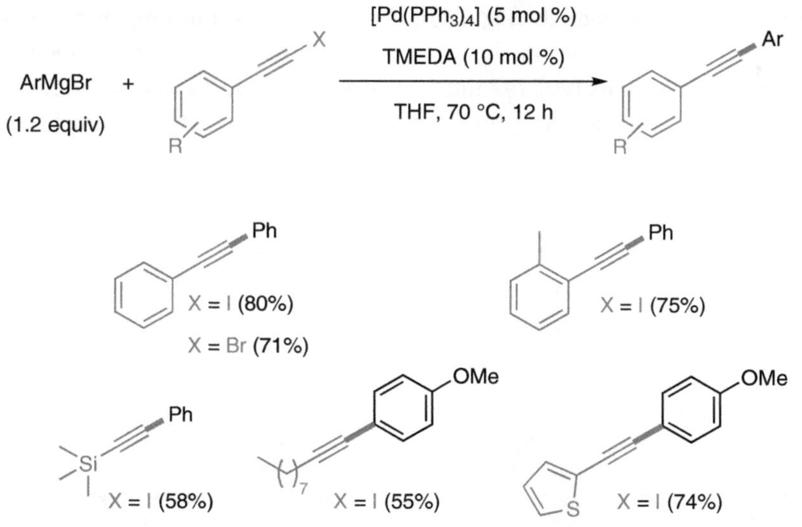

Fig. 1.71: Cross-coupling of alkynyl halide compounds with arylmagnesium bromides [87].

1.4.2 Coupling of Vinylmagnesium Compounds

1.4.2.1 With Aryl Electrophiles

Bumagin's method to cross-couple aryl halides (Br or I) bearing amino, alcohol and carboxylic acid functional groups, simply by using extra equivalents of the Grignard reagent (2–3 equiv; one equiv per functional group, plus 1 equiv for cross-coupling), has been demonstrated with vinyl Grignard reagent (Fig. 1.72) [23].

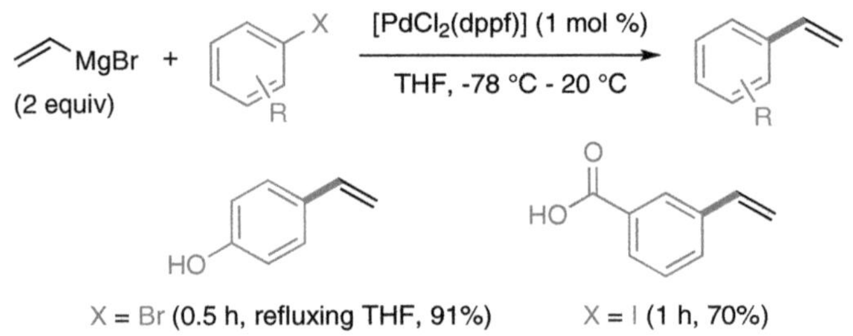

Fig. 1.72: Cross-coupling of halophenols and halobenzoic acids [23].

Itami and Yoshida developed an elegant one-pot method to synthesize some tetrasubstituted alkenes [88, 89]. The silylacetylene starting material first underwent a copper-catalyzed reaction with an organomagnesium reagent, to yield an intermediate vinylmagnesium species; this then participated in a palladium-catalyzed cross-coupling reaction to deliver the products (Fig. 1.73). The compounds thus prepared bear a pyridylsilane at one position of the alkene, which is proposed to function as a directing group to enforce regioselectivity [90]. *E*/*Z* ratios were typically >90/10. The silane could then be replaced with pinacolborane, providing a site for further cross-coupling [89].

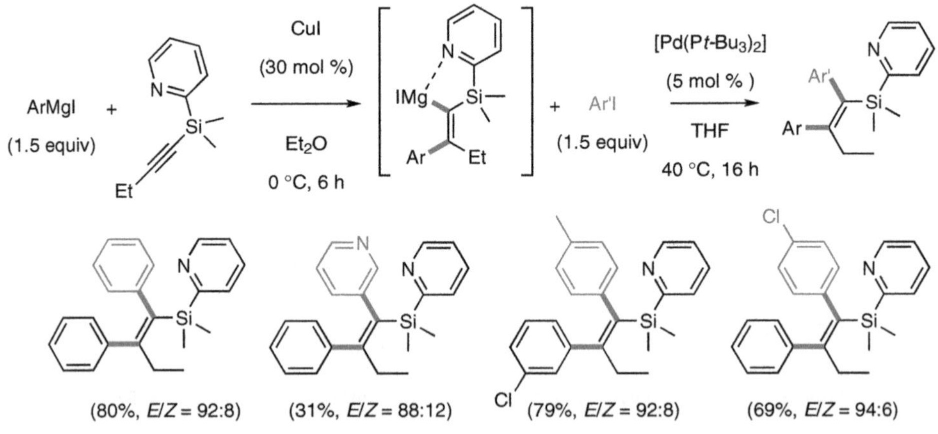

Fig. 1.73: Synthesis of tetrasubstituted alkenes *via* a one-pot copper/palladium-catalyzed sequence [88].

Hartwig's ferrocene-derived ligand has been applied to cross-couple an aryl tosylate with a vinylmagnesium bromide species (Fig. 1.74; see also Figs. 1.17 and 1.52 above). The use of a more electron-donating ligand bearing two dialkylphosphine sites was necessary to achieve this reaction.

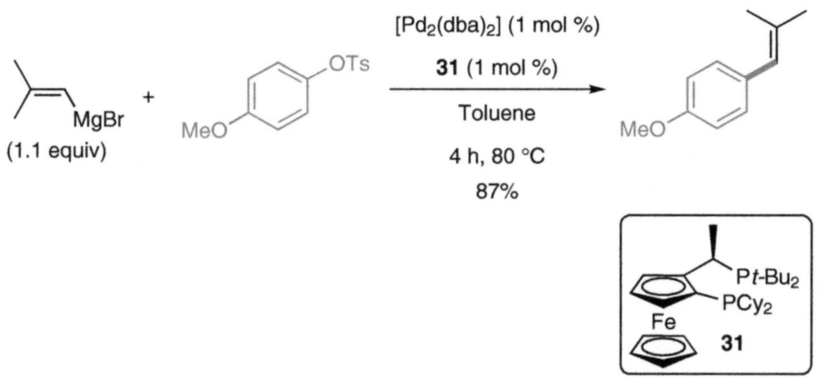

Fig. 1.74: Cross-coupling of aryl tosylates with vinylmagnesium halides [37].

Nolan *et al.* have applied a 'bulky-yet-flexible' complex **38** to reactions including the coupling of Grignard reagents, although only a very limited substrate scope was presented, consisting of the cross-coupling of one magnesium reagent with three different aryl bromides (Fig. 1.75) [91].

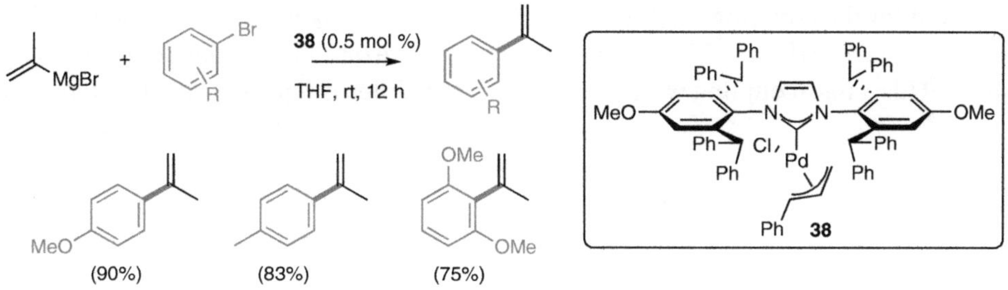

Fig. 1.75: Cross-coupling of a Grignard reagent catalyzed by [Pd(cin)Cl(IPr*^{OMe})] [91].

1.4.2.2 With Vinyl Electrophiles

Alami *et al.*'s system was shown to be practical for the cross-coupling of chloro-dienes and chloro-enynes with vinylmagnesium reagents [75]. Some examples can be found in Fig. 1.76.

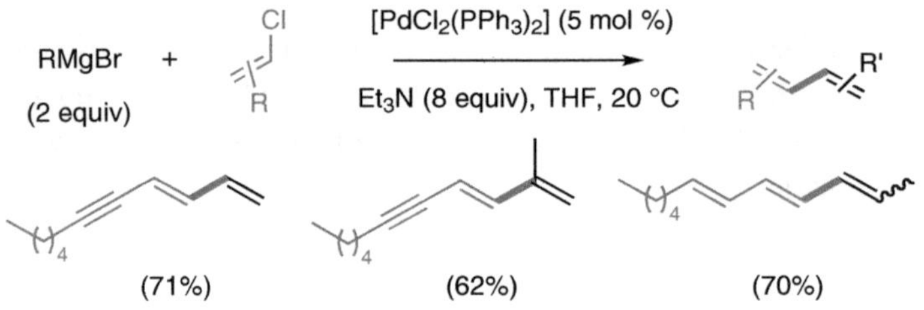

Fig. 1.76: Cross-coupling of chloro-dienes and chloro-enynes with arylmagnesium chlorides [75].

Lipshutz *et al.* have used palladium-catalyzed cross-coupling to prepare a conjugated diene (Fig. 1.77), using a method that has been shown to be applicable to a wide range of compounds [80].

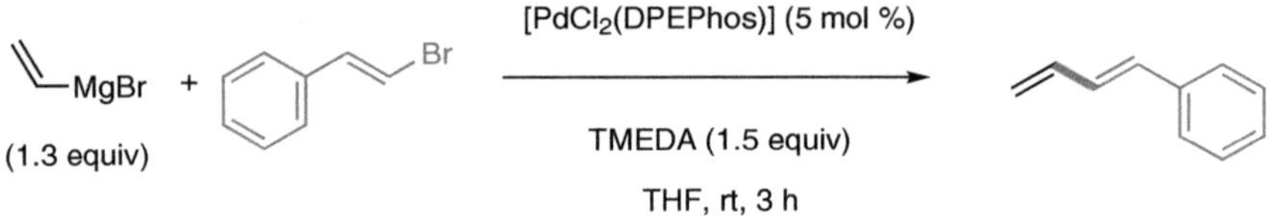

Fig. 1.77: Cross-coupling of an vinylmagnesium compound with a vinyl halide [80].

1.4.2.3 With Alkyl Electrophiles

Adrio and Carretero's system ([$PdCl_2(NCMe)_2$]/Xantphos), discussed previously, also allowed for the coupling of vinylmagnesium bromide reagents with secondary alkyl halides (Fig. 1.78) [84]. While only four examples were disclosed, in moderate to excellent yields, cross-coupling with secondary alkyl halides is typically very challenging.

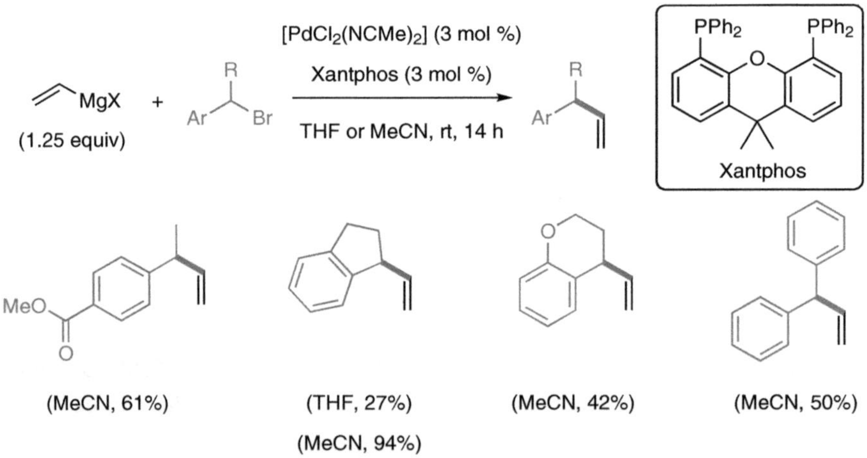

Fig. 1.78: Cross-coupling of vinylmagnesium halides and secondary alkyl halides [84].

Takanami *et al.* have used the palladium-catalyzed coupling of silyl-bearing alkylmagnesium reagents with bromide-functionalized porphyrins to enable access to a variety of motifs (Fig. 1.79) [92]. The silyl group could be further elaborated by e. g. oxidation or reaction with electrophiles (including DAST, to yield a fluoride group).

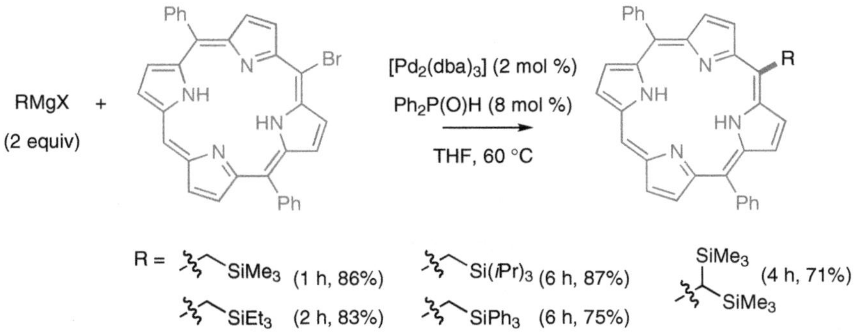

Fig. 1.79: Functionalization of porphyrin compounds [92].

1.4.3 Coupling of Alkylmagnesium Compounds

1.4.3.1 With Aryl Electrophiles

Hayashi *et al.* used [$PdCl_2$(dppf)] to couple *sec*-butylmagnesium chloride with aryl and vinyl electrophiles. Moderate to excellent yields were achieved with aryl halides (Fig. 1.80) [93].

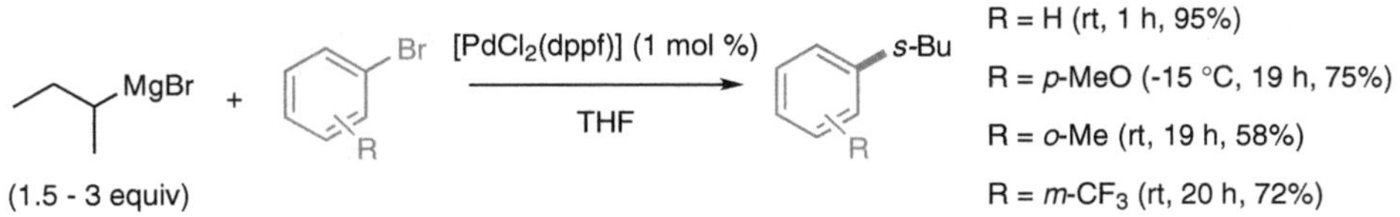

Fig. 1.80: Cross-coupling with *sec*-butylmagnesium chloride [93].

Katayama *et al.* achieved selective cross-coupling of some primary and secondary alkyl Grignards with dichlorobenzenes (albeit using the latter in excess) (Fig. 1.81). The second chloride substituent is a useful subsequent functional group to handle.

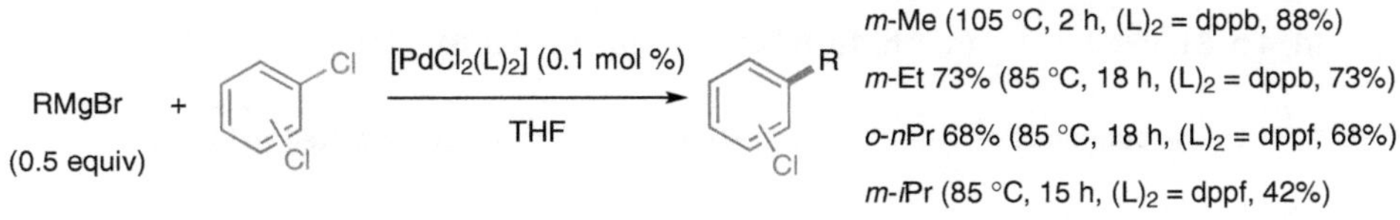

Fig. 1.81: Cross-coupling with dichlorobenzene substrates [22].

Bumagin *et al.* have cross-coupled aryl halides (Br or I) bearing amino, alcohol and carboxylic acid functional groups with (an excess of) alkyl Grignard reagents (Fig. 1.82) [23].

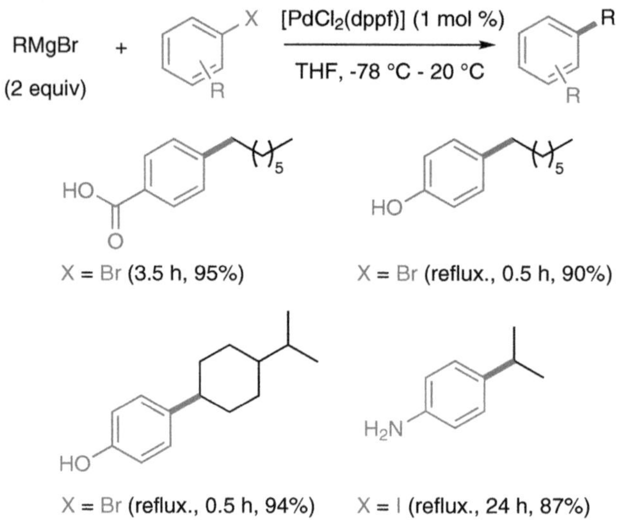

Fig. 1.82: Cross-coupling of halobenzoic acids, halophenols and halobenzoic acids [23].

Hartwig *et al.* have employed ferrocene-derived ligands to cross-couple aryl tosylates with alkylmagnesium bromide species (Fig. 1.83; see also Figs. 1.17, 1.52 and 1.74 above) [37]. The use of a more electron-donating ligand bearing two dialkylphosphine sites was often necessary to achieve this reaction.

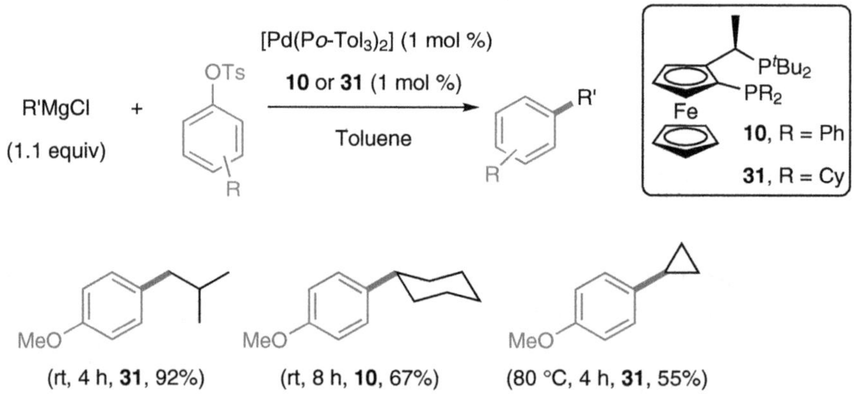

Fig. 1.83: Cross-coupling of aryl tosylates with alkylmagnesium halides (isolated yields) [37].

Using the [PdCl(μ-Cl)(**37**)]$_2$ catalyst system described previously, Ackermann *et al.* demonstrated the cross-coupling of alkylmagnesium chlorides and bromides with a selection of aryl halides (Fig. 1.84) [86]. Notably, this same catalyst system is active for the reaction between arylmagnesium halides and alkyl halides.

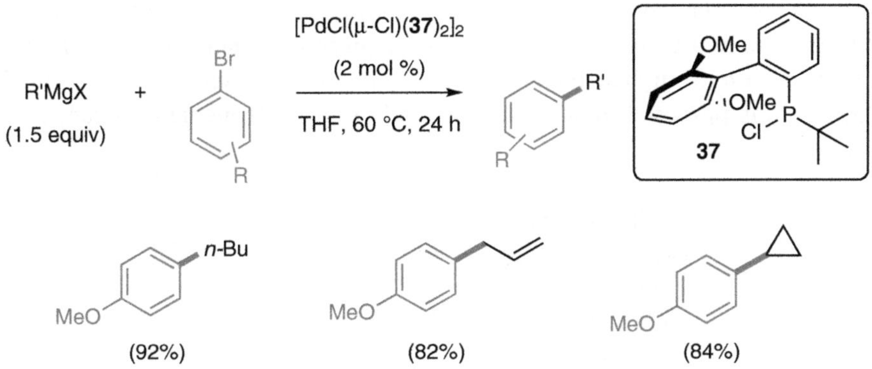

Fig. 1.84: Arylation of alkylmagnesium halide compounds [86].

Douglas *et al.* have applied 'PEPPSI'-type catalysts to the synthesis of sterically-demanding acene derivatives, although the required catalyst loadings were rather high (3 mol % [PdCl$_2$(3-Cl-py)(IPr)] **13** per cross-coupling site) (Fig. 1.85) [94]. Notably, all six positions of hexachlorobenzene underwent cross-coupling.

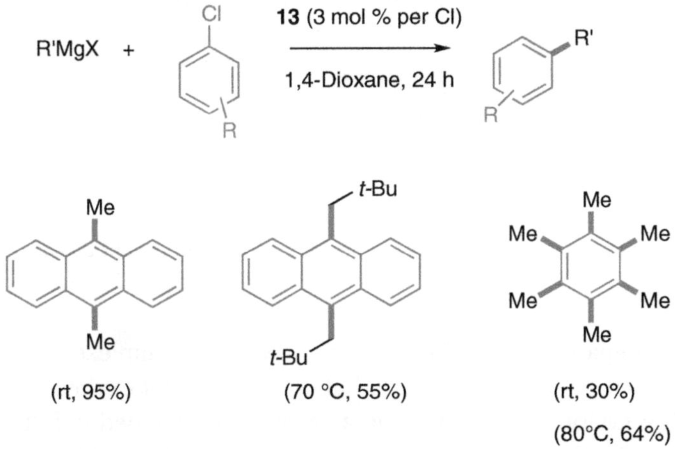

Fig. 1.85: Alkylation of acene derivatives [94].

While it is known that Grignard reagents will react with zinc halides to form Negishi reagents of the form RZnX, Wang and co-workers developed a protocol that uses catalytic quantities of zinc bromide (15–30 mol %), which was found to be sufficient to 'soften' cyclopropyl magnesium bromide enough to allow smooth cross-coupling [95]. A selection of other alkylmagnesium halide compounds could be coupled with aryl halides using the same procedure. A selection of the molecules prepared can be found in Fig. 1.86. Notably, attempts to use *iso*-propylmagnesium bromide led to the *n*-propyl product, and *tert*-butylmagnesium chloride produced the *sec*-butyl-substituted product; this was proposed to occur *via* a palladium hydride-mediated scrambling pathway.

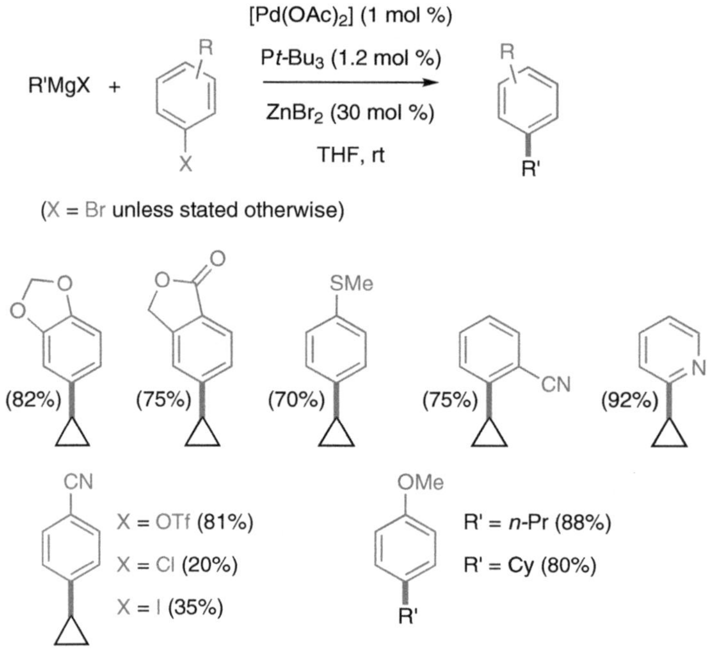

Fig. 1.86: Cross-coupling of alkylmagnesium halides in the presence of catalytic quantities of zinc bromide [95].

Liu and co-workers have prepared and studied a series of NHC-Pd complexes [96]; monomeric species **39** failed to cross-couple aryl halides with cyclohexylmagnesium bromide, producing cyclohexene and arene as a result of competing β-hydride elimination (Fig. 1.87). However, homobimetallic complex **40** yielded *ca.* 80 % conversion to the desired cross-coupling product, presumably by altering the geometry of the complex and preventing the opening up of the necessary vacant co-ordination site.

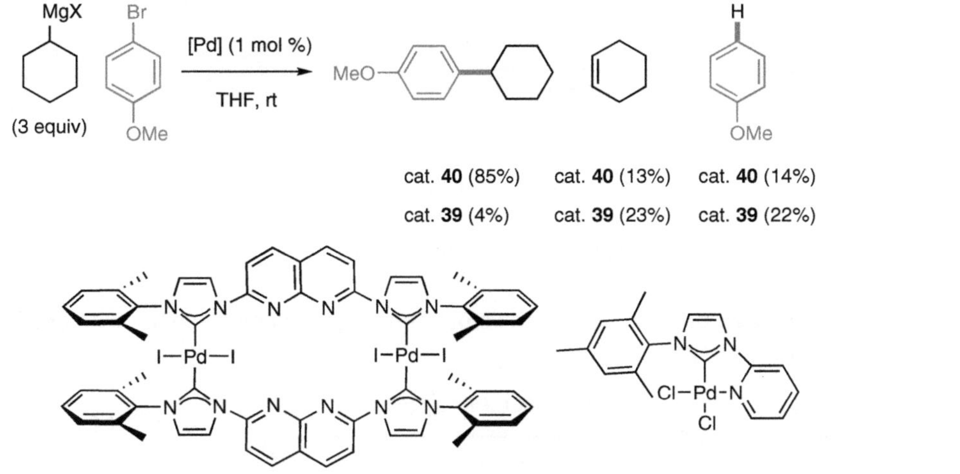

40 **39**

Fig. 1.87: Catalyst structure influences the rates of competing processes [96].

Bull *et al.* have developed a method for the cross-coupling of aziridinylmagnesium chloride compounds with aryl bromides (Fig. 1.88) [97]; the former were generated by sulfonyl-magnesium exchange. However, as the subsequent cross-coupling was carried out in the presence of 1.5 equiv $ZnCl_2$, it is most likely that the corresponding zinc species undergoes transmetalation, rather than the organomagnesium halide.

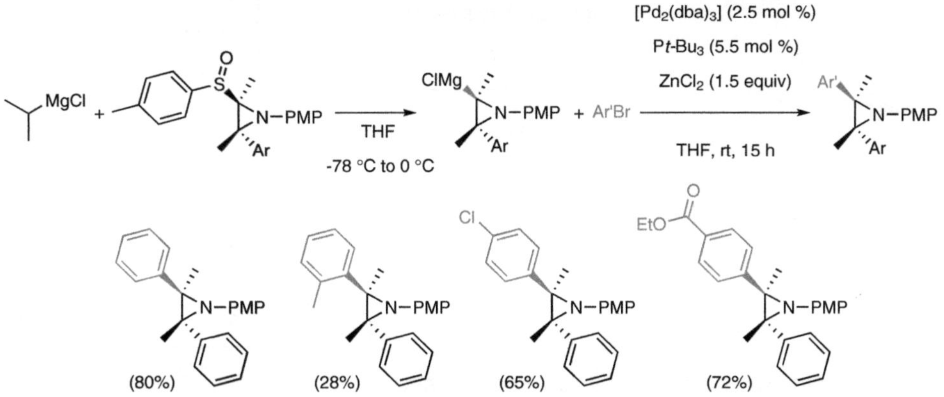

Fig. 1.88: Arylation of aziridines [97].

1.4.3.2 With Vinyl Electrophiles

Hayashi *et al.*'s [$PdCl_2$(dppf)]-catalyzed method was also used to couple *sec*-butylmagnesium chloride with two model vinyl electrophiles, with very good to excellent yield (Fig. 1.89) [93].

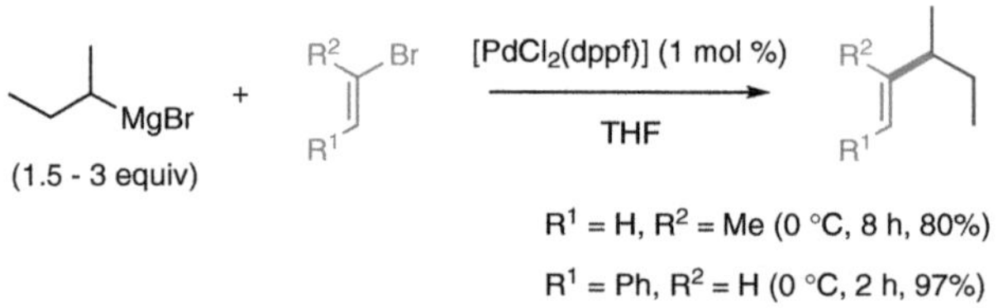

Fig. 1.89: Cross-coupling with *sec*-butylmagnesium chloride [93].

Alami *et al.* reported one example of the cross-coupling of a chloro-enyne with octylmagnesium chloride (Fig. 1.90) [75].

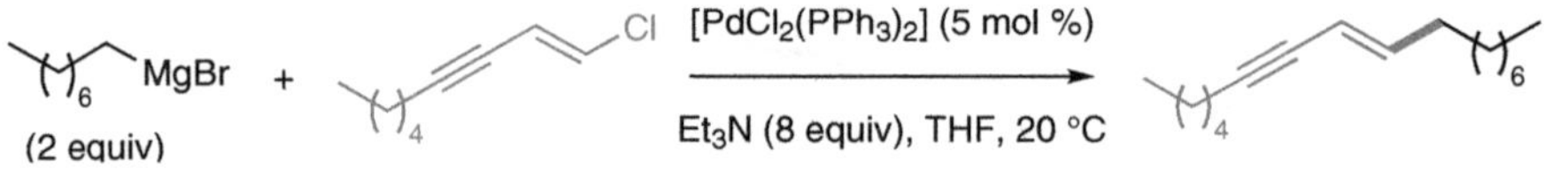

Fig. 1.90: Cross-coupling of chloro-dienes and chloro-enynes with arylmagnesium chlorides [75].

Aoyama *et al.* have used an axially chiral *P,N*-ligand **41** in palladium-catalyzed cross-couplings of 1-phenylethylmagnesium chloride with β-bromostyrenes [98, 99]. The configuration at the alkene was retained in each case, while the ee of the products varied from 71–80 %. The *S* configuration was obtained from each reaction. Some examples of the products prepared can be found in Fig. 1.91.

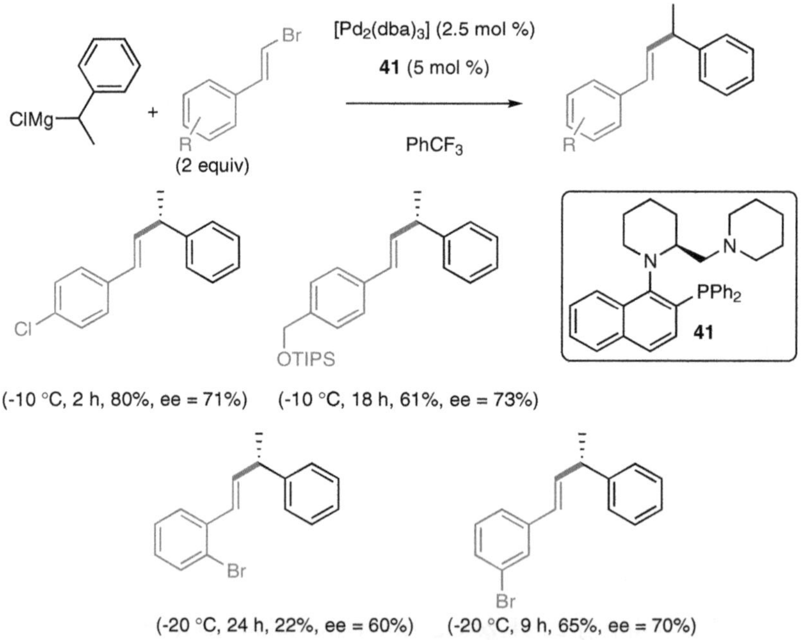

Fig. 1.91: Enantioselective Grignard cross-coupling [98].

Hartwig *et al.*'s ferrocene-derived ligands have been used to enable the cross-coupling of vinyl tosylates with a prototypical alkylmagnesium chloride reagent (Fig. 1.92; see also Figs. 1.17, 1.52, 1.74 and 1.83 above). While only a limited number of compounds were explored, the study suggests that a wide range of cross-couplings ought to be possible.

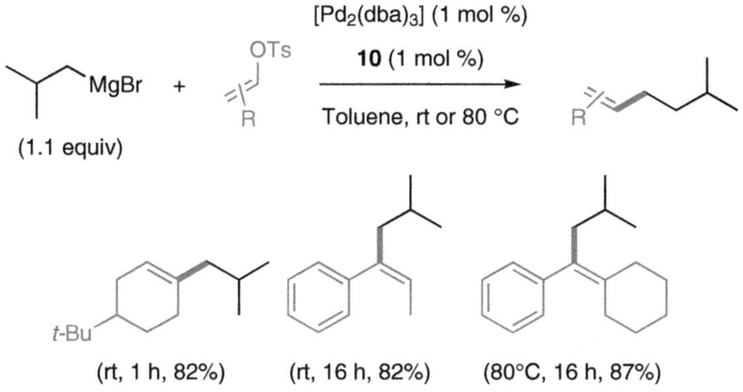

Fig. 1.92: Cross-coupling of aryl tosylates with alkylmagnesium halides [37].

Brown *et al.*'s method to cross-couple vinyl phosphates to access tetrasubstituted alkenes has also been realized with alkyl Grignard reagents [78]. The desired reagent products were obtained with very good overall *E*/*Z* ratios (Fig. 1.93).

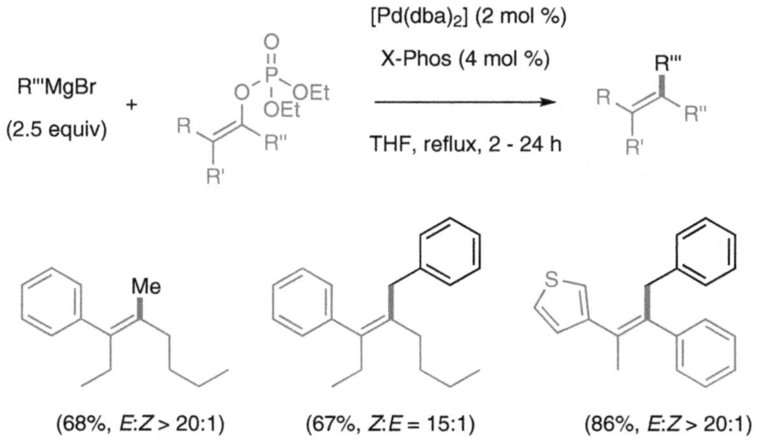

Fig. 1.93: (a) Synthesis of heavily substituted vinyl phosphates, and (b) their subsequent crosscoupling reactions [78].

Lipshutz *et al.* have applied a [$PdCl_2$(dtbpf)] complex to cross-couple a range of vinyl iodides and vinyl bromides with alkylmagnesium halide reagents (Fig. 1.94) [80].

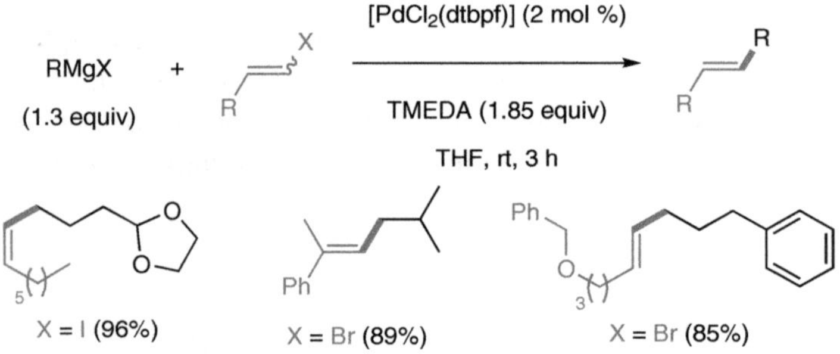

Fig. 1.94: Arylation of vinyl halides at room temperature [80].

1.4.3.3 With Alkyl Electrophiles

Sinou and co-workers have cross-coupled *para-tert*-butylphenol glycopyranosides with two examples of alkyl Grignard reagents. The products obtained are displayed in Fig. 1.95 [81].

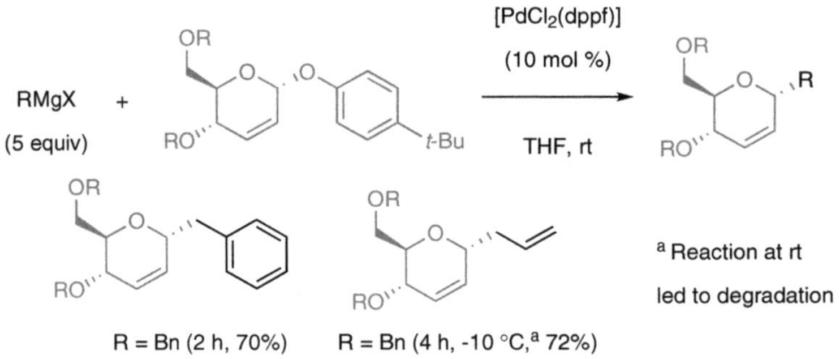

Fig. 1.95: Cross-coupling of *para-tert*-butylphenyl glycopyranosides [81].

Kambe *et al.* reported the cross-coupling of alkyl bromides and alkyl tosylates with alkyl Grignard reagents (Fig. 1.96) [83]. 1,3-Butadiene was used as an additive.

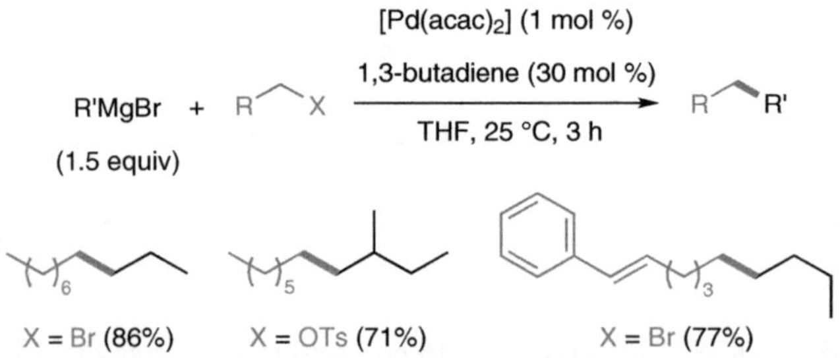

Fig. 1.96: Alkyl-alkyl cross-coupling [83].

1.4.3.4 With Alkynyl Electrophiles

Song and Li's [Pd(PPh_3)$_4$]/TMEDA has been shown to catalyze the cross-coupling of alkynyl iodide substrates with methylmagnesium bromide [87]; three examples were reported (Fig. 1.97).

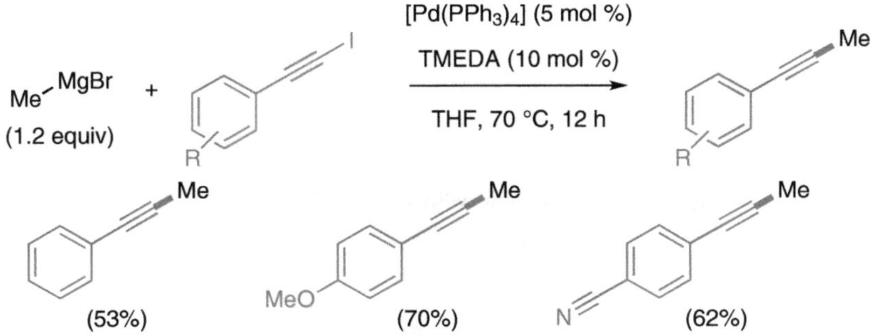

Fig. 1.97: Cross-coupling of alkynyl halide compounds with MeMgBr [87].

1.4.4 Coupling of Alkynylmagnesium Compounds

1.4.4.1 With Vinyl Electrophiles

Lipshutz *et al.* have used palladium-catalyzed cross-coupling to prepare one example of an enyne compound (Fig. 1.98), which appears to be a rare example of this type of cross-coupling reaction.

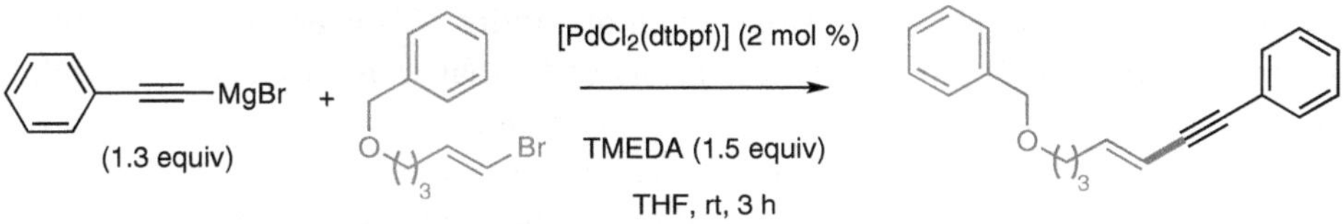

Fig. 1.98: Cross-coupling of an alkynylmagnesium compound with a vinyl halide [80].

1.4.4.2 With Alkyl Electrophiles

Luh *et al.* have used a simple Pd(0)/PPh_3 system to couple alkynyl Grignard reagents with alkyl halides [100]. Initial optimization of the reaction focused on reducing the quantities of side products derived from homocoupling; in the absence of phosphine, only low conversion to the homocoupling products was obtained (10 %). The optimized conditions could be used to prepare a small set of alkynes (Fig. 1.99).

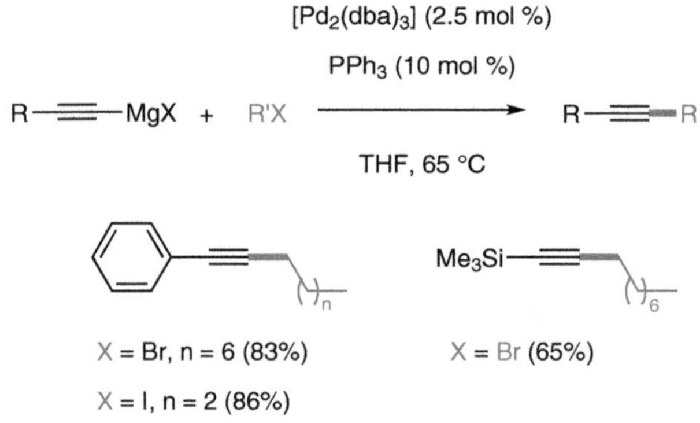

Fig. 1.99: Cross-coupling of alkynyl Grignard reagents with alkyl halides [100].

1.5 Industrial Deployment of Palladium-Catalyzed Grignard Cross-Couplings

Few examples of industrial applications of palladium-catalyzed Grignard cross-couplings have been reported, while several are known that utilize nickel or copper [101]. Key challenges during the scale-up of these processes typically involve control of heat transfer, limitation of impurity concentrations, minimization of the overall process cost (including heating, palladium catalyst, etc.), and simple and economical purification. In particular, the development of pharmaceuticals requires tight control of the level of palladium in the final material, typically to <10 ppm levels.

Zacuto and co-workers reported the allylation of 4-bromoisoindoline using allylmagnesium chloride (Fig. 1.100) [102]. Small quantities of isomer **42** were produced during initial screening reactions; the levels of **42** that were produced were minimized by carrying out a detailed factorial experimental design study in which the THF/toluene ratio of the solvent and the temperature were varied. Isolation of the product as the HCl salt allowed the palladium content to be reduced to ≤25 ppm, and the proportion of the styrene by-product to ≤1 %. The reaction was successfully conducted on a *ca.* 2.5 litre scale, by addition of 800 mL of a 1.7 $mol \cdot L^{-1}$ solution of the Grignard reagent in THF to a 1.8 L solution containing 100 g of the starting bromide. The work-up involved washing with aqueous citric acid, extraction into toluene and azeotropic distillation, before salt formation with HCl in *iso*propyl alcohol, filtration, washing and drying.

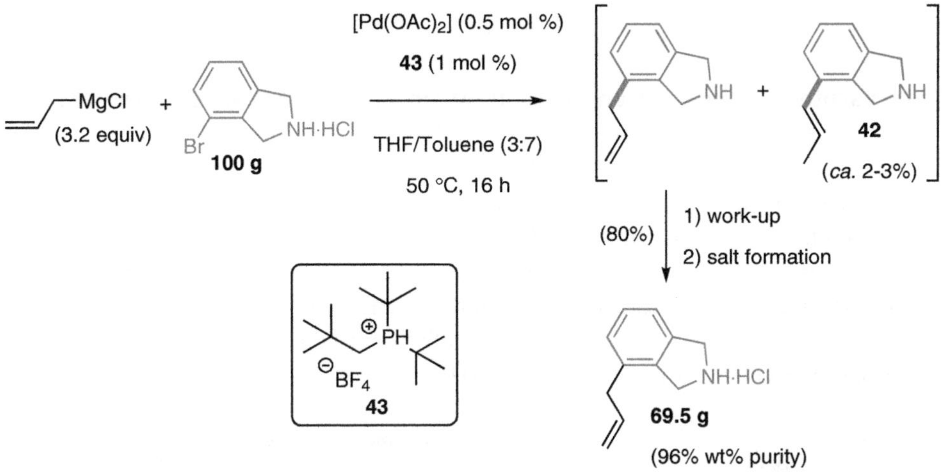

Fig. 1.100: Synthesis of 4-allylisoindoline by cross-coupling with allylmagnesium chloride [102].

Marzoni and Varney have utilized a Grignard cross-coupling to effect the methylation of 5-iodobenz[*cd*]indol-2(1*H*)-one (Fig. 1.101) [103]. A relatively high catalyst loading was necessary (10 mol%). The reaction started at room temperature, and the Grignard reagent was added over the course of 90 min (3.2 L of a 1.4 mol·L^{-1} solution in 3:1 toluene/THF added to a 5.6 L THF solution of the starting material). During the addition, the reaction temperature reached temperatures close to reflux. The pure product was obtained in moderate yield after recrystallization from chlorobenzene.

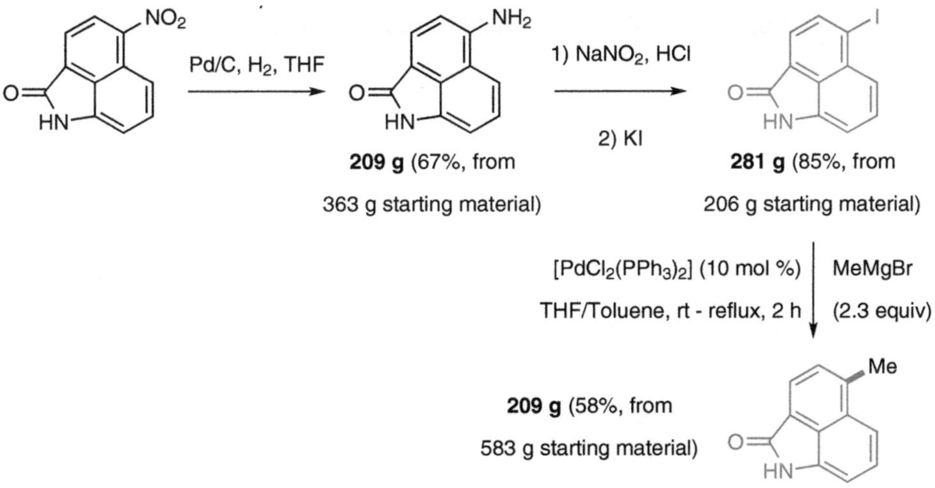

Fig. 1.101: Synthesis of 4-allylisoindoline by cross-coupling with allylmagnesium chloride [103].

1.6 Applications in Target Synthesis

Target synthesis represents an excellent way in which to develop and test new synthetic methods, often under the most challenging conditions. Selected examples of palladium-catalyzed cross-coupling of Grignard reagents are discussed here. Nicolaou *et al.* have previously reviewed the applications of palladium-catalyzed cross-coupling in total synthesis [104].

Hartmann and co-workers utilized Grignard cross-couplings to synthesise P450 17 inhibitors [105]. Aryl halides bearing imidazolemethylenic substituents underwent cross-coupling with a range of organometallic coupling reagents, including three examples of aryl Grignard species (Fig. 1.102).

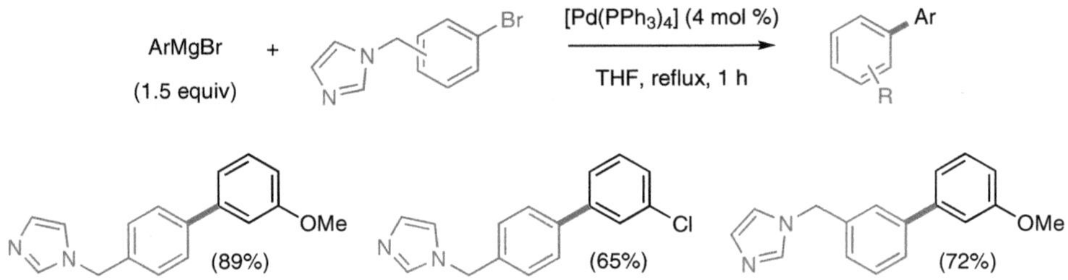

Fig. 1.102: Cross-coupling to form imidazole-bearing biaryls [105].

Müllen, Spiess, and co-workers have used a palladium-catalyzed Grignard coupling *en route* to hexa-*peri*-hexabenzacoronenes. The reaction was used to assemble symmetrical alkynes, which then underwent cobalt-catalyzed trimerization followed by iron-mediated oxidation to yield the desired product (Fig. 1.103).

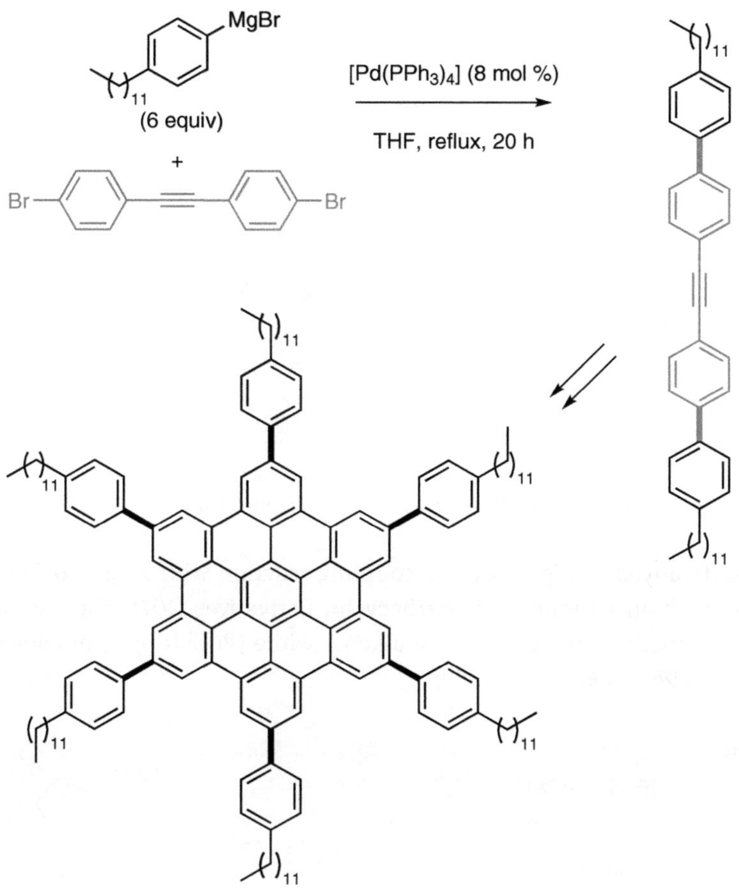

Fig. 1.103: Cross-coupling *en route* to *hexa-peri*-hexabenzacoronenes.

Danishefsky *et al.* used a Grignard cross-coupling to construct the left fragment of [18]dehydrodesoxyepothilone B (Fig. 1.104) [106]. The alkene stereochemistry was retained in this transformation; subsequent deprotection, coupling and ring-closing metathesis furnished the desired product. For the 17-membered ring analogue, a palladium-catalyzed Stille reaction was used to install the requisite allyl group.

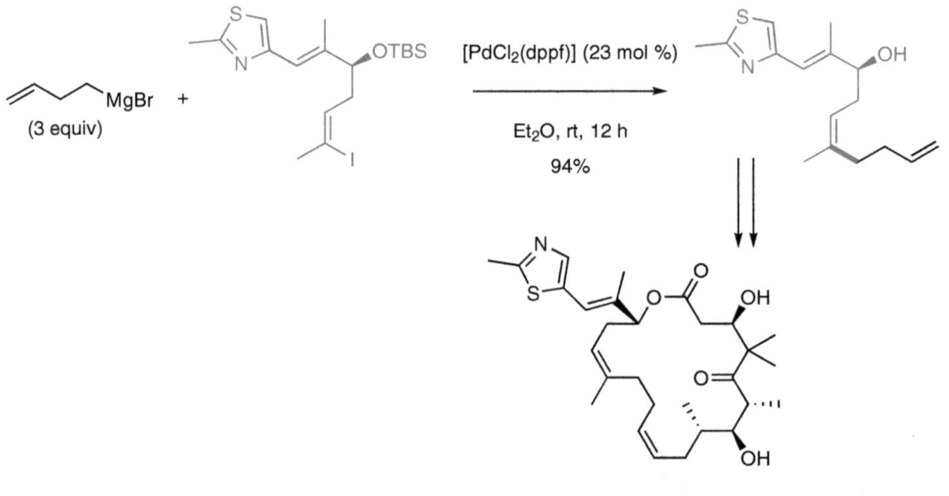

Fig. 1.104: Synthesis of the left fragment of [18]dehydrodesoxyepothilone B [106].

Mulzer *et al.* have deployed a Grignard cross-coupling reaction with a vinyl triflate to install a key side chain *en route* to isocarbacyclin derivatives [107]. The use of $[Pd(PPh_3)_4]$ led to competing formation of the alkene, while $[PdCl_2(dppf)]$ provided the desired product in 98 % yield (Fig. 1.105).

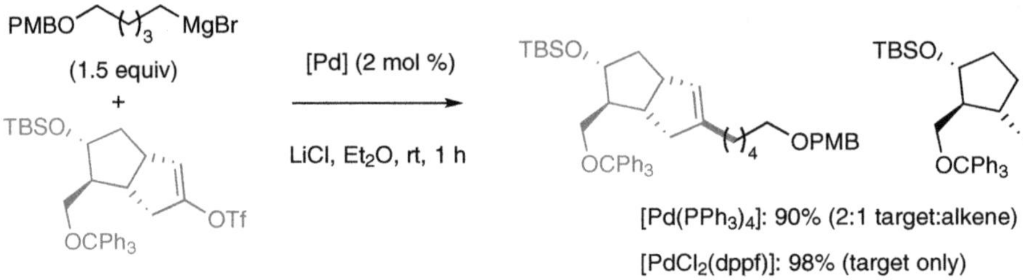

Fig. 1.105: Synthesis of isocarbacyclin derivatives [107].

Houpis *et al.* have utilized a palladium-catalyzed coupling of aryl bromides and Grignard reagents towards the synthesis of isochromanes (Fig. 1.106) [108]. An alkoxide additive was found to be beneficial in helping to avoid competing formation of the styrene byproduct.

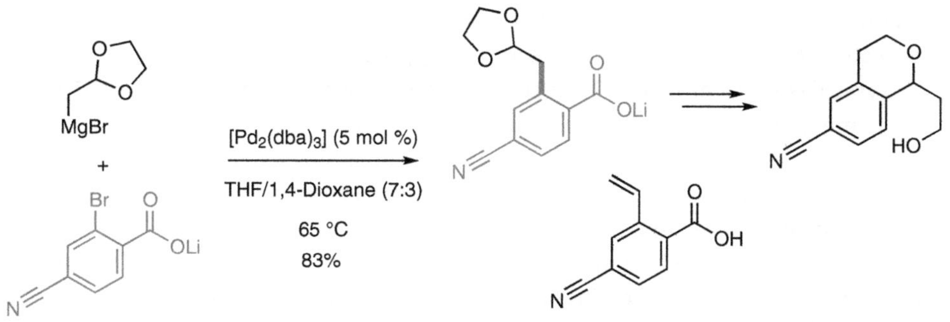

Fig. 1.106: Grignard cross-coupling towards isochromanes [108].

Floreancig *et al.* used a Grignard cross-coupling during the early stages of the synthesis of lactodehydrothyrsiferol, in order to install a vinyl group (Fig. 1.107) [109]. The cross-coupling with the vinyl iodide proceeded in high yield, with retention of the alkene stereochemistry. The alkene terminus thus installed was further elaborated by hydroboration and then by Suzuki-Miyaura cross-coupling.

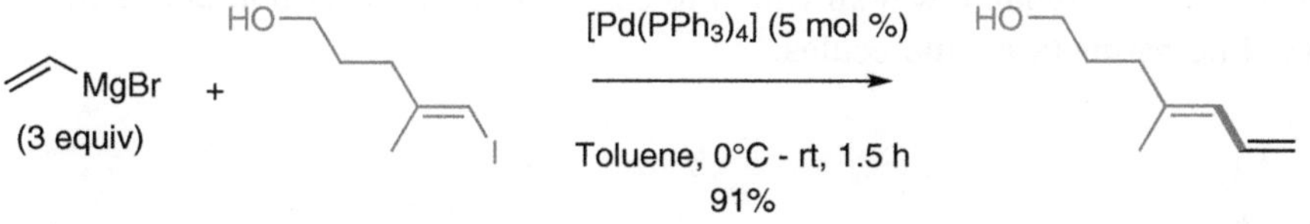

Fig. 1.107: Initial stages in the synthesis of lactodehydrothyrsiferol [109].

Honda *et al.* utilized a palladium-catalyzed Grignard coupling as the penultimate step in the synthesis of sequosempervirin A, to attach a benzylic fragment to the spirocyclic core (Fig. 1.108) [110].

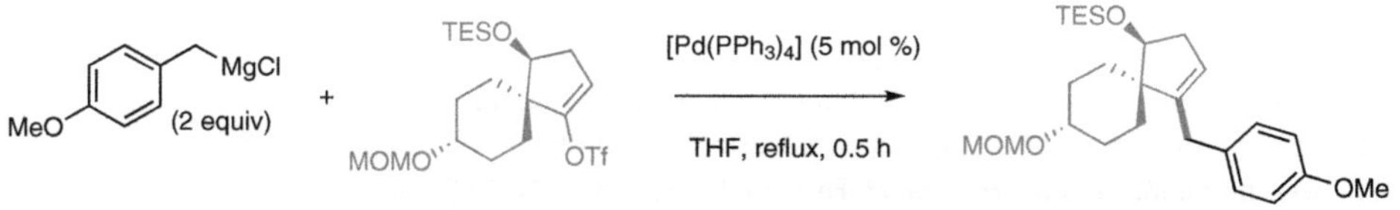

Fig. 1.108: Synthesis of sequosempervirin A [110].

1.7 Summary, Conclusions and Outlook

This Chapter details the capabilities and limitations of Grignard cross-coupling reactions catalyzed by palladium. Most of these transformations are mediated by palladium species or catalyst/ligand systems that are available within typical modern synthetic chemistry laboratories, rendering this technology highly accessible.

Typical conditions involve the use of relatively low temperatures (rt to *ca.* 70 °C in THF being amongst the most common), with straightforward catalyst systems based on triaryl or trialkylphosphines or chelating bisphosphines. NHC ligands have made some impact in this area. Additives are typically unnecessary, making these reactions straightforward to execute.

The majority of the literature focusses on the cross-coupling of aryl fragments, due to their ubiquity in medicinal chemistry, agrochemistry, and materials science. However, there are a significant number of examples of non-aryl sp^2–sp^2, sp^3–sp^3 and sp^2–sp^3 cross-coupling reactions. Couplings of alkynes are relatively rare, undoubtedly due at least in part to the ability of the Sonagashira reaction to couple alkynes directly, or with a copper co-catalyst. The scope of electrophiles is impressive, ranging from common iodides and bromides, to more challenging chlorides, *via* sulfonates, phosphates and even organofluorides.

The area of Grignard cross-coupling has benefited somewhat from the development of modern "Turbo" Grignard reagents and associated protocols for the preparation of organomagnesium reagents. Even without these developments, this reaction remains a relatively reliable way in which to construct carbon–carbon bonds in the pursuit of interesting target molecules.

References

[1] Grignard V. Compt Rend 1900; 130: 1322–4.
[2] Seyferth D. Organometallics 2009; 28: 1598–605.
[3] Kharasch MS, Fields EK. J Am Chem Soc 1941; 63: 2316–20.
[4] Kharasch MS, Fuchs CF. J Am Chem Soc 1943; 65: 504–7.
[5] Shinokubo H, Oshima K. Eur J Org Chem 2004; 2004: 2081–91.
[6] Tamao K. J Organomet Chem 2002; 653: 23–6.
[7] Tamura M, Kochi JK. J Am Chem Soc 1971; 93: 1487–9.
[8] Corriu RJP, Masse JP. J Chem Soc, Chem Commun 1972: 144a.
[9] Tamao K, Sumitani K, Kumada M. J Am Chem Soc 1972; 94: 4374–6.
[10] Yamamura M, Moritani I, Murahashi S-I. J Organomet Chem 1975; 91: C39–C42.
[11] Giannerini M, Fañanás-Mastral M, Feringa BL. Nat Chem 2013; 5: 667–72.
[12] Moineau C, Bolitt V, Sinou D. J Organomet Chem 1998; 567: 157–62.
[13] Bhatt MP, Magurudeniya HD, Sista P, *et al.* Journal of Materials Chemistry A 2013; 1: 12841–9.
[14] Bahri-Laleh N, Poater A, Cavallo L, Mirmohammadi SA. Dalton Trans 2014; 43: 15143–50.
[15] Saeki T, Takashima Y, Tamao K. Synlett 2005; 2005: 1771–4.
[16] Dai W, Xiao J, Jin G, Wu J, Cao S. J Org Chem 2014; 79: 10537–46.
[17] Holzer B, Hoffmann RW. Chem Commun 2003: 732–3.
[18] Castle PL, Widdowson DA. Tetrahedron Lett 1986; 27: 6013–6.
[19] Yuan K, Scott WJ. Tetrahedron Lett 1989; 30: 4779–82.
[20] Rudolph A, Lautens M. Angew Chem Int Ed 2009; 48: 2656–70.

[21] Kambe N, Iwasaki T, Terao J. Chem Soc Rev 2011; 40: 4937–47.
[22] Katayama T, Umeno M. Chem Lett 1991; 20: 2073–6.
[23] Bumagin NA, Luzikova EV. J Organomet Chem 1997; 532: 271–3.
[24] Huang J, Nolan SP. J Am Chem Soc 1999; 121: 9889–90.
[25] Fortman GC, Nolan SP. Chem Soc Rev 2011; 40: 5151–69.
[26] Valente C, Çalimsiz S, Hoi KH, Mallik D, Sayah M, Organ MG. Angew Chem Int Ed 2012; 51: 3314–32.
[27] Hopkinson MN, Richter C, Schedler M, Glorius F. Nature 2014; 510: 485–96.
[28] Díez-González S, Marion N, Nolan SP. Chem Rev 2009; 109: 3612–76.
[29] Bonnet V, Mongin F, Trécourt F, Quéguiner G, Knochel P. Tetrahedron Lett 2001; 42: 5717–9.
[30] Bonnet V, Mongin F, Trécourt F, Quéguiner G, Knochel P. Tetrahedron 2002; 58: 4429–38.
[31] Li GY. J Organomet Chem 2002; 653: 63–8.
[32] Babudri F, Colangiuli D, Farinola Gianluca M, Naso F. Eur J Org Chem 2002; 2002: 2785–91.
[33] Dumouchel S, Mongin F, Trécourt F, Quéguiner G. Tetrahedron 2003; 59: 8629–40.
[34] Dumouchel S, Mongin F, Trécourt F, Quéguiner G. Tetrahedron Lett 2003; 44: 3877–80.
[35] Frisch AC, Zapf A, Briel O, Kayser B, Shaikh N, Beller M. J Mol Cat A, Chem 2004; 214: 231–9.
[36] Sémeril D, Lejeune M, Jeunesse C, Matt D. J Mol Cat A, Chem 2005; 239: 257–62.
[37] Limmert ME, Roy AH, Hartwig JF. J Org Chem 2005; 70: 9364–70.
[38] Ackermann L. Synthesis 2006; 2006: 1557–71.
[39] Ackermann L, Althammer A. Org Lett 2006; 8: 3457–60.
[40] Ackermann L, Gschrei CJ, Althammer A, Riederer M. Chem Commun 2006: 1419–21.
[41] Dong C-G, Hu Q-S. Angew Chem Int Ed 2006; 45: 2289–92.
[42] Dong C-G, Hu Q-S. Org Lett 2006; 8: 5057–60.
[43] Dong C-G, Hu Q-S. Tetrahedron 2008; 64: 2537–52.
[44] Organ MG, Abdel-Hadi M, Avola S, *et al.* Chem Eur J 2007; 13: 150–7.
[45] Altenhoff G, Goddard R, Lehmann CW, Glorius F. J Am Chem Soc 2004; 126: 15195–201.
[46] Izquierdo F, Manzini S, Nolan SP. Chem Commun 2014; 50: 14926–37.
[47] Meiries S, Le Duc G, Chartoire A, *et al.* Chem Eur J 2013; 19: 17358–68.
[48] Luan X, Mariz R, Gatti M, *et al.* J Am Chem Soc 2008; 130: 6848–58.
[49] Ila H, Baron O, Wagner AJ, Knochel P. Chem Commun 2006: 583–93.
[50] Martin R, Buchwald SL. J Am Chem Soc 2007; 129: 3844–5.
[51] Ishikawa S, Manabe K. Org Lett 2007; 9: 5593–5.
[52] Ishikawa S, Manabe K. Tetrahedron 2011; 67: 10156–63.
[53] Ishikawa S, Manabe K. Chem Lett 2007; 36: 1302–3.
[54] Ishikawa S, Manabe K. Chem Lett 2007; 36: 1304–5.
[55] Wang Z-X, Chai Z-Y. Eur J Inorg Chem 2007; 2007: 4492–9.
[56] Wolf C, Xu H. J Org Chem 2008; 73: 162–7.
[57] He P, Dong C-G, Hu Q-S. Tetrahedron Lett 2008; 49: 1906–9.
[58] Manabe K, Ishikawa S. Synthesis 2008; 2008: 2645–9.
[59] Manolikakes G, Knochel P. Angew Chem Int Ed 2009; 48: 205–9.
[60] Hartmann CE, Nolan SP, Cazin CSJ. Organometallics 2009; 28: 2915–9.
[61] Mehta VP, Modha SG, Van der Eycken E. J Org Chem 2009; 74: 6870–3.
[62] Ackermann L, Potukuchi HK, Kapdi AR, Schulzke C. Chem Eur J 2010; 16: 3300–3.
[63] Ren G, Cui X, Wu Y. Eur J Org Chem 2010; 2010: 2372–8.
[64] Reeves JT, Fandrick DR, Tan Z, *et al.* Org Lett 2010; 12: 4388–91.
[65] Milton EJ, Clarke ML. Green Chem 2010; 12: 381–3.
[66] Henderson RK, Jimenez-Gonzalez C, Constable DJC, *et al.* Green Chem 2011; 13: 854–62.
[67] Prat D, Pardigon O, Flemming H-W, *et al.* Org Proc Res Dev 2013; 17: 1517–25.
[68] Pal A, Ghosh R, Adarsh NN, Sarkar A. Tetrahedron 2010; 66: 5451–8.
[69] Jin Z, Gu X-P, Qiu L-L, Wu G-P, Song H-B, Fang J-X. J Organomet Chem 2011; 696: 859–63.

[70] Viciu MS, Kissling RM, Stevens ED, Nolan SP. Org Lett 2002; 4: 2229–31.
[71] Ackermann L, Kapdi AR, Fenner S, Kornhaaß C, Schulzke C. Chem Eur J 2011; 17: 2965–71.
[72] Bryan ZJ, Smith ML, McNeil AJ. Macromolecular Rapid Communications 2012; 33: 842–7.
[73] Huddleston NE, Sontag SK, Bilbrey JA, Sheppard GR, Locklin J. Macromolecular Rapid Communications 2012; 33: 2115–20.
[74] Lesieur M, Slawin AMZ, Cazin CSJ. Org Biomol Chem 2014; 12: 5586–9.
[75] Ramiandrasoa P, Bréhon B, Thivet A, Alami M, Cahiez G. Tetrahedron Lett 1997; 38: 2447–50.
[76] Shi M, Liu L-P, Tang J. J Org Chem 2005; 70: 10420–5.
[77] Gauthier D, Beckendorf S, Gøgsig TM, Lindhardt AT, Skrydstrup T. J Org Chem 2009; 74: 3536–9.
[78] You W, Li Y, Brown MK. Org Lett 2013; 15: 1610–3.
[79] Moinizadeh N, Klemme R, Kansy M, Zimmer R, Reissig H-U. Synthesis 2013; 45: 2752–62.
[80] Krasovskiy AL, Haley S, Voigtritter K, Lipshutz BH. Org Lett 2014; 16: 4066–9.
[81] Moineau C, Bolitt V, Sinou D. J Org Chem 1998; 63: 582–91.
[82] Frisch AC, Shaikh N, Zapf A, Beller M. Angew Chem Int Ed 2002; 41: 4056–9.
[83] Terao J, Naitoh Y, Kuniyasu H, Kambe N. Chem Lett 2003; 32: 890–1.
[84] López-Pérez A, Adrio J, Carretero JC. Org Lett 2009; 11: 5514–7.
[85] Rao Volla CM, Dubbaka SR, Vogel P. Tetrahedron 2009; 65: 504–11.
[86] Ackermann L, Kapdi AR, Schulzke C. Org Lett 2010; 12: 2298–301.
[87] Zhang M-M, Gong J, Song R-J, Li J-H. Eur J Org Chem 2014; 2014: 6769–73.
[88] Kamei T, Itami K, Yoshida J-i. Adv Synth Catal 2004; 346: 1824–35.
[89] Itami K, Kamei T, Yoshida J-i. J Am Chem Soc 2003; 125: 14670–1.
[90] Itami K, Mitsudo K, Nokami T, Kamei T, Koike T, Yoshida J-i. J Organomet Chem 2002; 653: 105–13.
[91] Bastug G, Nolan SP. Organometallics 2014; 33: 1253–8.
[92] Sugita N, Hayashi S, Hino F, Takanami T. J Org Chem 2012; 77: 10488–97.
[93] Hayashi T, Konishi M, Kobori Y, Kumada M, Higuchi T, Hirotsu K. J Am Chem Soc 1984; 106: 158–63.
[94] Yagodkin E, Douglas CJ. Tetrahedron Lett 2010; 51: 3037–40.
[95] Shu C, Sidhu K, Zhang L, Wang X-j, Krishnamurthy D, Senanayake CH. J Org Chem 2010; 75: 6677–80.
[96] Chang Y-H, Liu Z-Y, Liu Y-H, Peng S-M, Chen J-T, Liu S-T. Dalton Trans 2011; 40: 489–94.
[97] Hughes M, Boultwood T, Zeppetelli G, Bull JA. J Org Chem 2013; 78: 844–54.
[98] Horibe H, Fukuda Y, Kondo K, Okuno H, Murakami Y, Aoyama T. Tetrahedron 2004; 60: 10701–9.
[99] Horibe H, Kazuta K, Kotoku M, *et al.* Synlett 2003; 2003: 2047–51.
[100] Yang L-M, Huang L-F, Luh T-Y. Org Lett 2004; 6: 1461–3.
[101] Magano J, Dunetz JR. Chem Rev 2011; 111: 2177–250.
[102] Zacuto MJ, Shultz CS, Journet M. Org Proc Res Dev 2010; 15: 158–61.
[103] Marzoni G, Varney MD. Org Proc Res Dev 1997; 1: 81–4.
[104] Nicolaou KC, Bulger PG, Sarlah D. Angew Chem Int Ed 2005; 44: 4442–89.
[105] Wachall BG, Hector M, Zhuang Y, Hartmann RW. Bioorg Med Chem 1999; 7: 1913–24.
[106] Rivkin A, Njardarson JT, Biswas K, Chou T-C, Danishefsky SJ. J Org Chem 2002; 67: 7737–40.
[107] Sheddan NA, Mulzer J. Org Lett 2005; 7: 5115–8.
[108] Houpis IN, Van Hoeck J-P, Tilstam U. Synlett 2007; 2007: 2179–84.
[109] Clausen DJ, Wan S, Floreancig PE. Angew Chem Int Ed 2011; 50: 5178–81.
[110] Ito Y, Takahashi K, Nagase H, Honda T. Org Lett 2011; 13: 4640–3.

Fabienne Fache, Béatrice Pelotier, Olivier Piva

2 Grignard Reagents and Nickel

2.1 Introduction

Grignard reagents, discovered more than one century ago, are still widely used to promote carbon–carbon bond formation. Over the time, their reactivity has been modulated by the presence of various additives, especially transition metals salts. In this area, nickel had played a major role in the development of new and selective cross-coupling reactions between organomagnesium derivatives and C(sp_2)-halides. It is generally admitted that the respective contributions of Kumada and Tamao in Japan [1] and Corriu in France [2] constitute a major breakthrough in the field of organometallic chemistry. Forty years after these two seminal contributions, the KumadaTamao-Corriu reaction and related processes are still investigated but also applied in industrial processes and used in the synthesis of complex molecules. Thanks to the design of more and more sophisticated ligands, including pincer and NHC ligands, this coupling reaction initially performed with halogen derivatives has been successfully applied to a large number of starting materials. Nowadays, the cleavage of carbon–oxygen and carbon–sulfur bonds has considerably enhanced the potential of this reaction in term of regioselectivity. Furthermore, the challenging sp_3–sp_3 cross-coupling has been addressed. Finally, the control of the stereochemistry has been considered, which also allows new perspectives for the future.

The mechanism generally admitted is depicted in Fig. 2.1. A nickel(0) species is generated *in situ* by reduction of a nickel(II) complex mainly by action of the Grignard species present in excess. Then, the oxidative addition results in the formation of a nickel(II) complex **A** which can undergo a transmetallation with the Grignard reagent leading to species **B**. An isomerization step *via* intermediate **C**, followed by a subsequent reductive elimination, release the coupled product **D** and regenerate the effective catalyst.

In the context of enantioselective processes, the coupling between an optically active Grignard reagent **1** and vinyl bromide was investigated and led to adduct **2** with almost complete retention of the configuration when conducted in the presence of

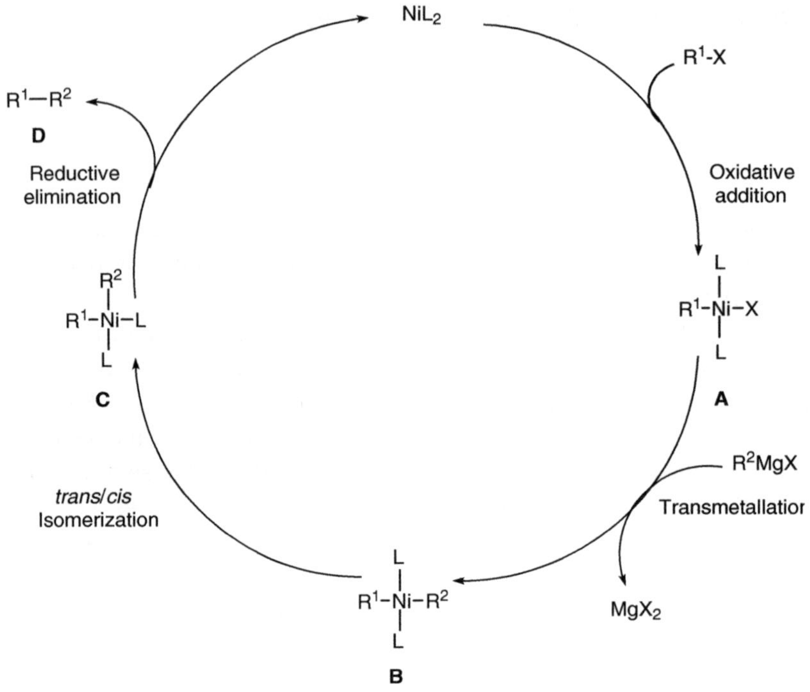

Fig. 2.1: Generally admitted mechanism for the Kumada-Corriu reaction.

$NiCl_2$dppf at low temperature (Fig. 2.2). This demonstrated that the key transmetallation step occurred surely according to a concerted polar mechanism [3]. In contrast, the same reaction promoted by iron or cobalt resulted in partial racemization of the final product which supports a radical based mechanism.

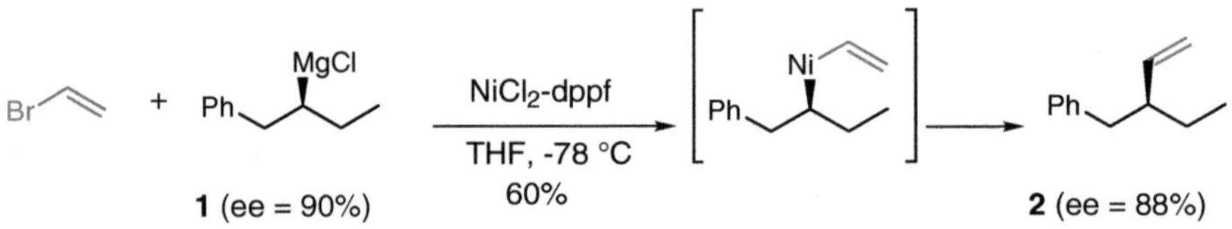

Fig. 2.2: Enantioselective transformation as a proof for a concerted polar mechanism.

The success of the different Ni-catalyzed cross-couplings [4] resulted from a large number of items such as:

- The ligands, especially mono- and diphosphines, which stabilize the organometallic intermediate to avoid a homodimerization process (Kharasch reaction).

- The elimination which usually occurred with alkyl Grignard reagents can be strongly reduced.
- In Csp_2–Csp_2 cross-couplings, the initial geometry of the vinyl derivatives is maintained.

2.2 Coupling reactions of aryl Grignard reagents

2.2.1 With aryl derivatives

2.2.1.1 With aryl bromides and chlorides

The formation of diaryl compounds has been extensively studied. The success of the couplings depends both on the structure of the substrates (nature of the leaving group, steric hindrance, etc.) but also on the nature of the nickel catalyst and principally the type of ligands directly linked to the metal.

Use of phosphine ligands

At the very beginning of the development of the Kumada-Corriu reaction [5–6], phosphine ligands were used to stabilize the nickel catalyst, the catalytic activity being significantly dependent on the nature of the phosphine ($NiCl_2$(dppp) > $NiCl_2$(dppe) > $NiCl_2(PR_3)_2$ = $NiCl_2$(dppb)). The Kumada-Tamao-Corriu reaction has attracted chemical companies to access fine intermediates such as substituted styrenes and biphenyls but also, at a later stage, for the final functionalization of the target molecules [7–8]. The reactions usually are highly yielding and selective. They are also highly attractive because of the large availability of chloro-derivatives as precursors of both Grignard reagents and cheap coupling partners. Of interest, industrial applications required low charge of catalyst of common and cheap diphosphine ligands. Moreover, these reactions are usually performed under suitable conditions of temperature. However, some drawbacks should be pointed out, for example Grignard reagents require anhydrous conditions, and they are not compatible with a large variety of (reductive) functionalities. Finally, the work-up of the reaction generates a large amount of magnesium salts and thus, requires a process for waste treatment.

PDE472 **4**, a selective inhibitor of the phosphodiesterase PDE4D isoenzyme has been prepared on pilot-plant scale taking advantage of the Kumada-Corriu coupling (Fig. 2.3). The coupling reaction of 4-chloropyridine in toluene, with the *p*-methoxyphenyl Grignard was performed to produce **3** in 74 % yield. The success of the reaction depends on the dryness of the chloropyridine [9].

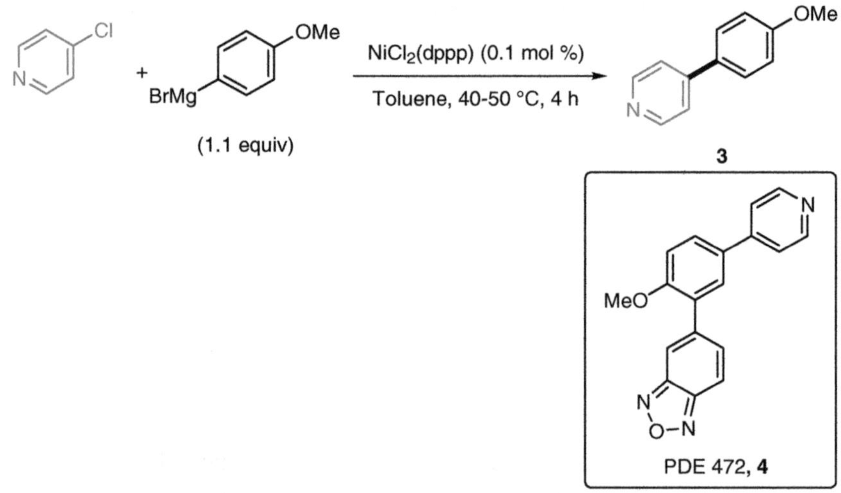

Fig. 2.3: Access to 4-arylpyridine **3**, a key intermediate for the synthesis of PDE 472.

To produce the HIV protease inhibitor atazanavir, chemists at Novartis have reported a multi-kg synthesis of substituted *p*-aryl benzaldehydes based on a Ni-catalyzed coupling between 2-bromopyridine and aryl Grignard derivatives (Fig. 2.4). The Ni(0) which is required for the oxidative addition was generated *in situ* by reduction of the catalyst by DiBAL-H. After removal of the protective group, aldehyde **5** was isolated in 90 % yield. When performed in the absence of DiBAL-H, the yield dropped dramatically to 68 % [10–11].

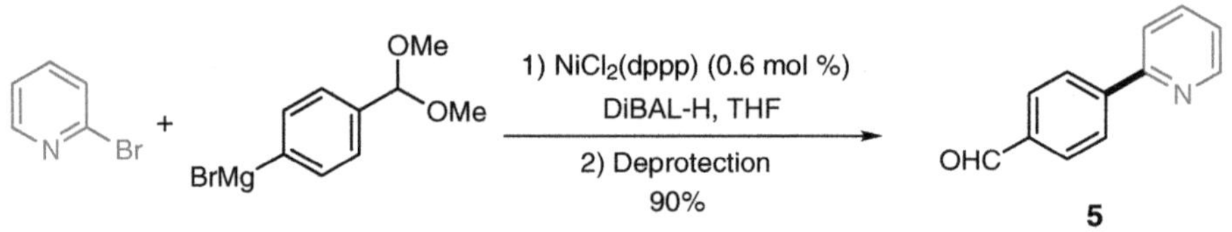

Fig. 2.4: Access to aldehyde **5**, precursor of atazanavir.

Catalyst $NiCl_2$(dppp), in combination with lithium triarylmagnesiate, instead of the classical organomagnesium reagents, allowed the Kumada-Corriu cross-coupling of a variety of aryl bromides and chlorides in good yields, and less than 10 % of the homocoupling products were formed (Fig. 2.5) [12].

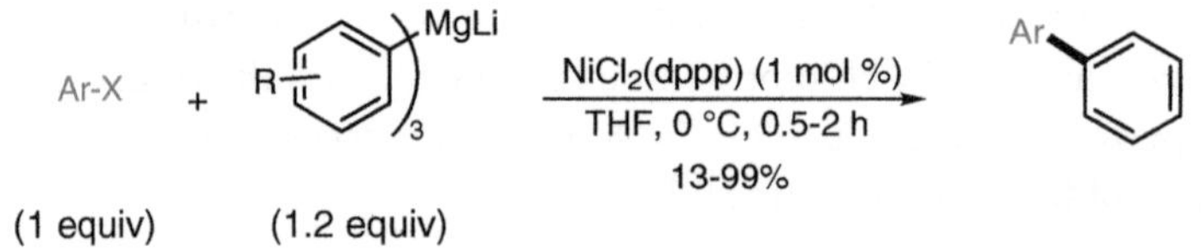

Fig. 2.5: Coupling of aryl halides with lithium triarylmagnesiates catalyzed by $NiCl_2$ and diphosphine.

Indole-derived air stable diphosphine ligands such as **L1** were also used successfully for the cross-coupling of a variety of aryl and heteroaryl chlorides with aryl Grignard reagents (Fig. 2.6) [13].

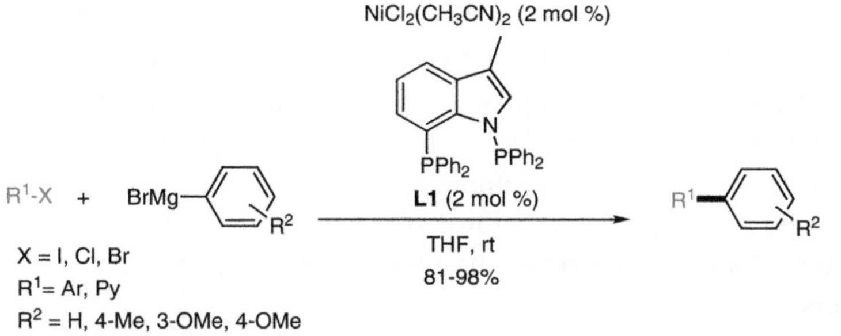

Fig. 2.6: Indole-derived air-stable diphosphine ligand **L1** in Kumada-Corriu reaction.

Calix[4]-diphosphine ligands were synthesized and proved to be efficient ligands, providing that the reactions were performed at 100 °C in dioxane with a PhMgBr/ArX ratio of 2/1. Compared to results obtained with dppp, which is considered in the field of diphosphines as one of the most efficient ligand [1], nickel complexes prepared from **L2** gave better results. The coupling was presumably facilitated by a temporary increase of the P-Ni-P angle and thus an increase of the steric pressure of the P-substituents on the two organic frameworks (Fig. 2.7) [14].

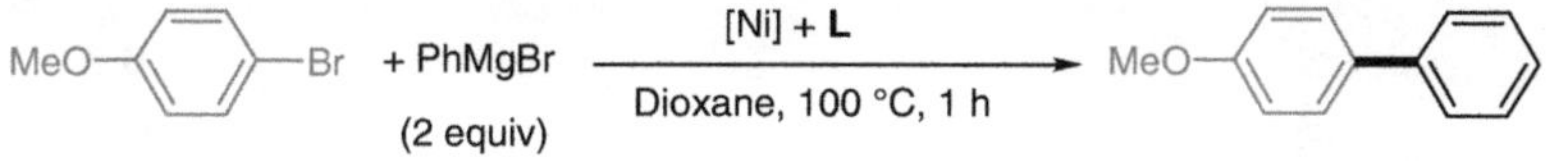

[Ni]	L	ratio ArBr/[Ni]	6 conversion	TOF
$Ni(cod)_2$ (10^{-3} mol %)	**L2** (10^{-3} mol %)	100 000	26.2 %	26 200
$Ni(cod)_2$ (10^{-3} mol)	PPh_3 (2. 10^{-3} mol %)	100 000	8.3 %	8 300
$NiCl_2$(d) (1 0^{-3} mol)		000	4 .1 %	2 3 0
$Ni(cod)_2$ (10^{-3} mol %)	**L3** (2. 10^{-3} mol %)	10 000	86.4 %	86 400
$Ni(cod)_2$ (10^{-3} mol %)	**L3** (2. 10^{-3} mol %)	100 000	31.3 %	31 300
$Ni(cod)_2$ (10^{-3} mol %)	**L4** (2. 10^{-3} mol %)	100 000	46.1 %	46 100

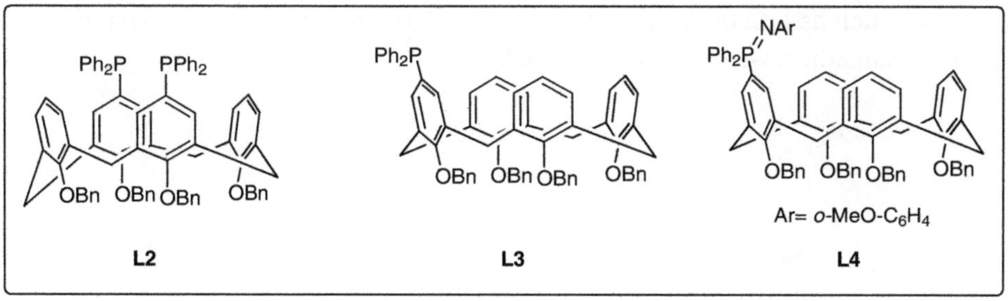

Fig. 2.7: Formation of bis-aryls using calixarene phosphine ligands.

Nevertheless, ligand **L3**, the monophosphine analog of **L2**, turned out to be more suitable and efficient at room temperature with aryl chlorides and four times more active than triphenylphosphine [15]. The orientation of the P–Ni bond towards the calixarene axis (and not outwards) seems to increase the ligand bulk and thus favored the formation of a mono-ligand Ni complex [16]. The mono-iminophosphorane analog **L4** (R = *o*-anisyl) allowed cross-coupling of aryl bromides in dioxane at 100 °C with a very low catalyst loading of 0.001 mol %. Thus, in 1 h, bromoanisole reacted with PhMgBr to give 46 % of the expected coupling product. In comparison with the cavityfree iminophosphorane $Ph_3P{=}N$(*o*-anisyl), the activity was ten fold higher, due to a more highly crowded metal environment in favor of a mono-ligated intermediate, more reactive than bis-ligated complexes and *endo*-located metal centers.

Among monophosphine ligands, *tris*(*t*-butyl)phosphane **L5** was also particularly efficient for the cross-coupling of aryl chlorides with aryl Grignard at ambient temperature [17]. The sulfur analog $(t\text{-Bu})_2P(S)H$ **L6** [18] and its oxygen analog $(t\text{-Bu})_2P(O)H$ **L8**, which revealed to be the best ligand [19], were also used successfully at room temperature in THF, with the advantage of being air stable and particularly low cost and accessible (Fig. 2.8).

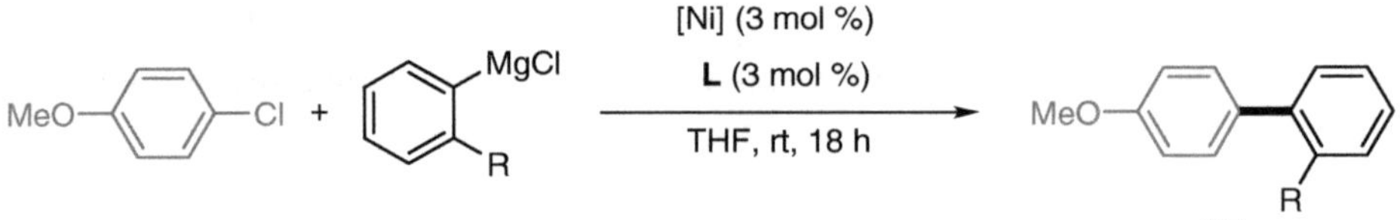

6-7

R	L		yield (%)	
H	**L5**	$P(t\text{-Bu})_3$	**6**	71
H	**L6**	$(t\text{-Bu})_2P(S)H$	**6**	96
CH_3	**L7**	$(t\text{-Bu})_2P(S)H$	**7**	79
CH_3	**L8**	$(t\text{-Bu})_2P(O)H$	**7**	93

Fig. 2.8: Monophosphine ligands and Kumada-Corriu coupling reaction.

Unactivated, deactivated and functionalized aryl chlorides reacted also with aryl Grignard reagents when monophosphine based catalysts **C1a-b**, in connection with additives such as LiCl or $ZnCl_2$ were used (Fig. 2.9). This significantly expanded the scope of Kumada cross-coupling reaction [20].

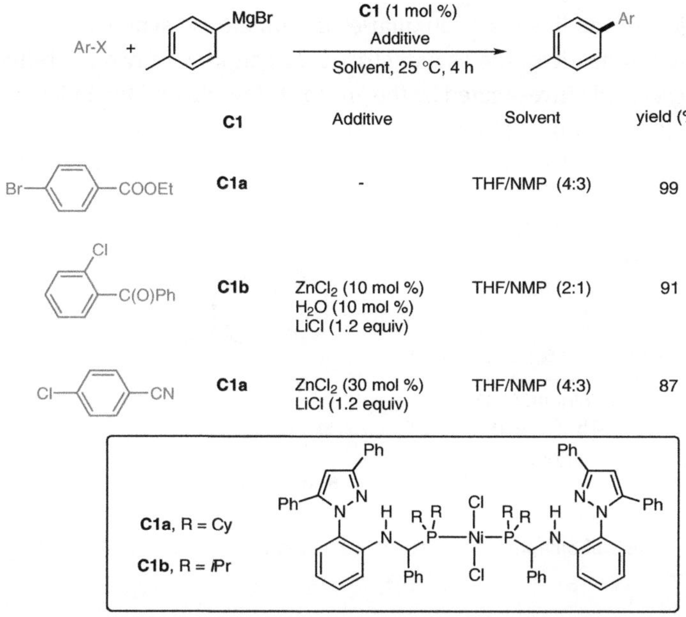

Fig. 2.9: Influence of additives on the coupling of aryl halides with tolyl Grignard reagent.

Phosphine oxides turned out to be excellent ligands for the activation of C–Cl bonds. Thus, the particularly congested structure **L9** showed one of the highest activity among the already described ligands [21]. Its sulfur analog, the diaminophosphine sulfide ligand **L10**, turned out to be less active and selective (Fig. 2.10) [22].

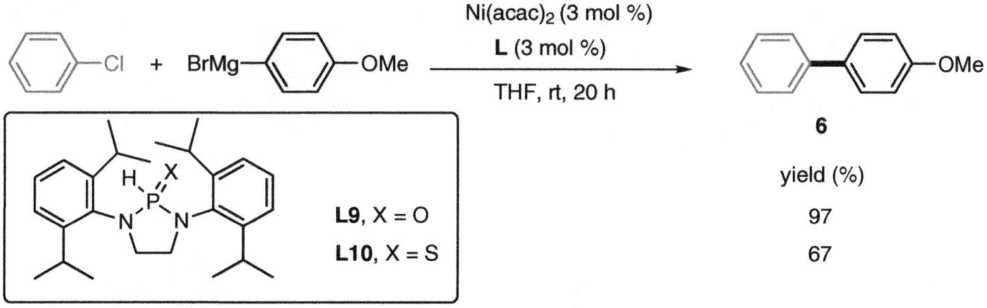

Fig. 2.10: Application of phosphine oxide **L9** in the Kumada-Corriu reaction.

Binaphthyls play a major role as ligands or auxiliaries in numerous asymmetric reactions. To access to related derivatives, the asymmetric cross-coupling of two naphthyl derivatives has been successfully investigated by the group of Hayashi and Ito, in Japan. To significantly favor the formation of one of the two atropoisomers, chiral ferrocenylphosphine ligand **L11** was used and gave excellent stereoselectivities (Fig. 2.11) [23].

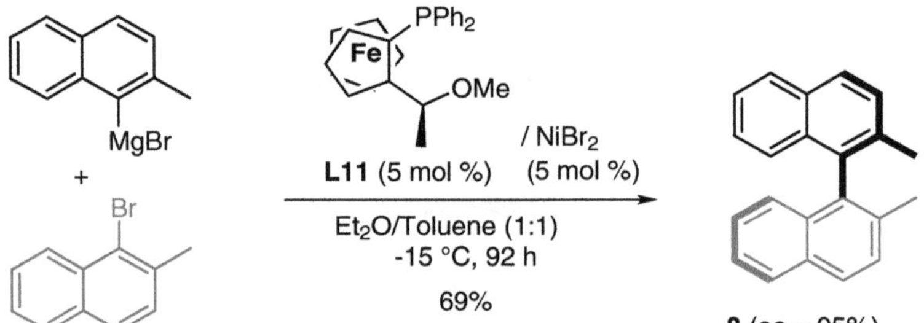

8 (ee = 95%)

Fig. 2.11: Stereoselective synthesis of binaphthyl derivatives *via* an asymmetric Kumada-Corriu cross-coupling.

Use of bidentate P,N Ligands

Bidentate diarylamido phosphine nickel chelates as **C2** turned out to be involved in very active catalytic systems (Fig. 2.12) [24].

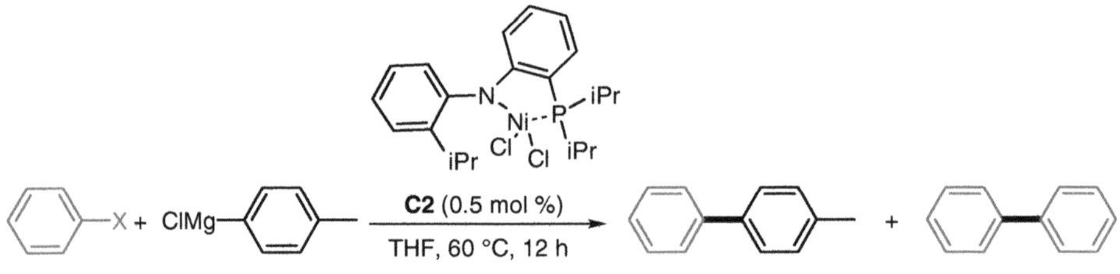

9 **10**

X	Conversion (%)	Ratio **9**/**10**
Cl	89	93/7
Br	89	93/7
I	94	95/5

Fig. 2.12: Bidentate diarylamido phosphine **C2** as ligand for the Kumada-Corriu reaction of aryl halides.

Very recently, the Sommer's group [25] used the hybrid P,N ligand **C3** for Kumada-nickel catalyst transfer polycondensation of sterically hindered thiophenes (Fig. 2.13). They showed that such a system is highly suitable to control the polymerization of thiophene-based monomers with sterically very demanding side chain, much more efficient than the commercially available $NiCl_2$(dppe) and $NiCl_2$(dppp) catalysts.

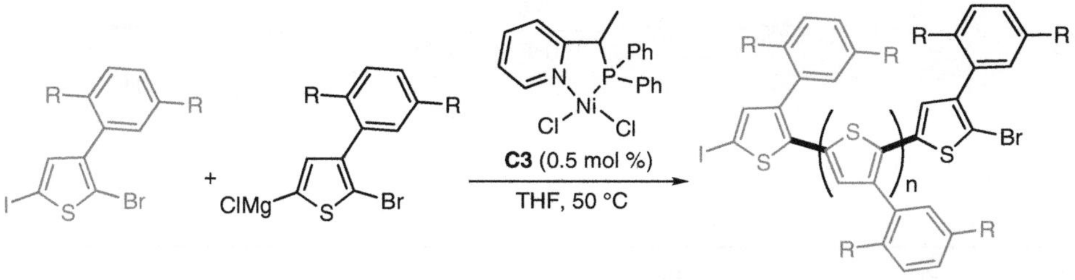

Fig. 2.13: Polycondensation of sterically hindered thiophenes.

Use of tridentate pincer ligands

Tridentate pincer ligands containing phosphorus were also considered (Fig. 2.14). Thus, P,N,P ligands as **C4**, [26] **C5**, [27] or **C6**, P,C,N ligands as **C7**, **C8** [28] and P,C,P ligand such as **C9** [29] have been tested successfully in the aryl–aryl coupling, even with aryl chlorides.

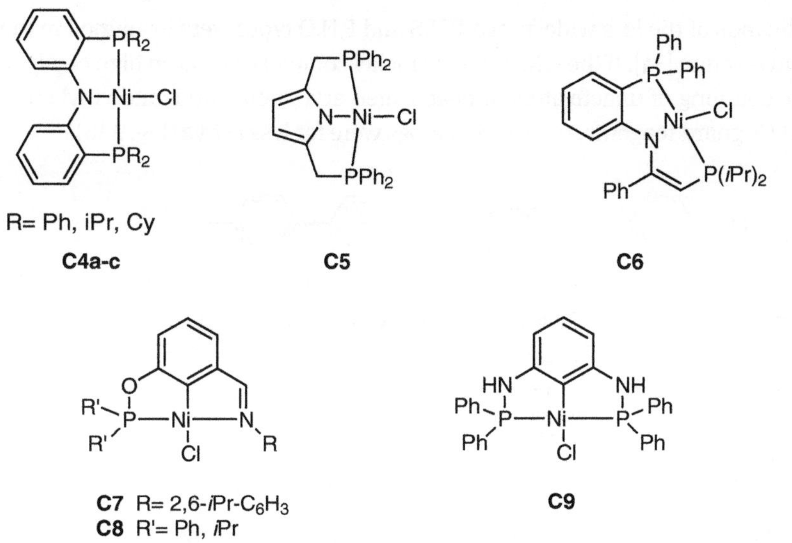

Fig. 2.14: Various P,N,P ligands, P,C,N and P,C,P ligands used for the aryl–aryl coupling.

As depicted in Fig. 2.15, **C6** and other catalysts such as **C10**, **C11** or **C12** are particularly relevant for the formation of biphenyl derivatives [30–32]. It has been assumed that (P,N,P) pincers in **C6** and **C11** are less labile than (P,N,N) ligands as in **C10** and **C12**. This difference directly impacts the oxidative addition step and favors the C–C bond formation [33].

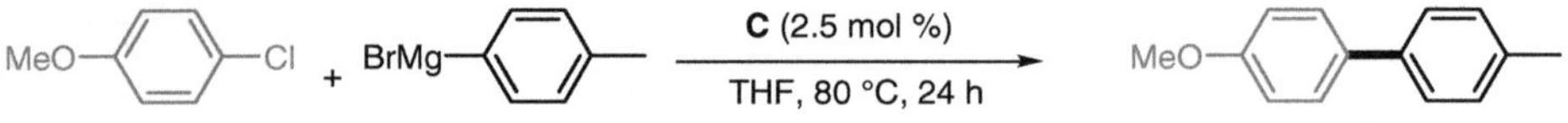

	11
C6	46 %
C10	92 %
C11	98 %
C12	98.6 %

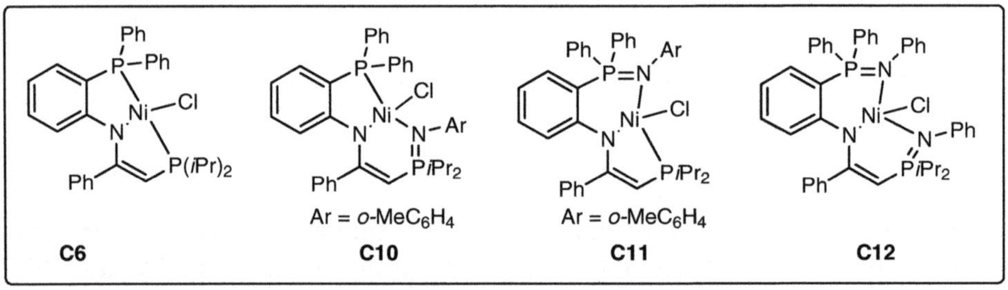

Fig. 2.15: Comparison of (P,N,P) and (P,N,N) pincer ligands for the formation of biphenyl derivatives.

Other pincer ligands of the less widespread P,N,S and P,N,O types were examined in the Kumada-Corriu reaction [34]. If the P,N,O chelate nickel complexes showed high catalytic activity for the coupling of unactivated, or deactivated aryl-, heteroaryl- and vinyl chlorides with aryl Grignard reagents, the P,N,S analogs were far less active (Fig. 2.16).

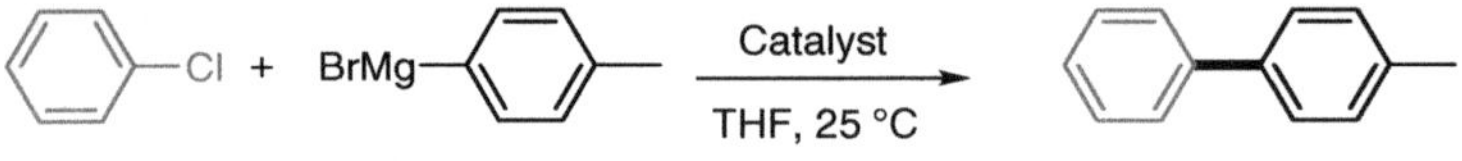

Catalyst (mol %)		X	Time (h)	Isolated yield (%)
C13a	1	S	24	51
C13b	0.5	O	12	93
C14a	1	S	24	69
C14b	0.5	O	12	85

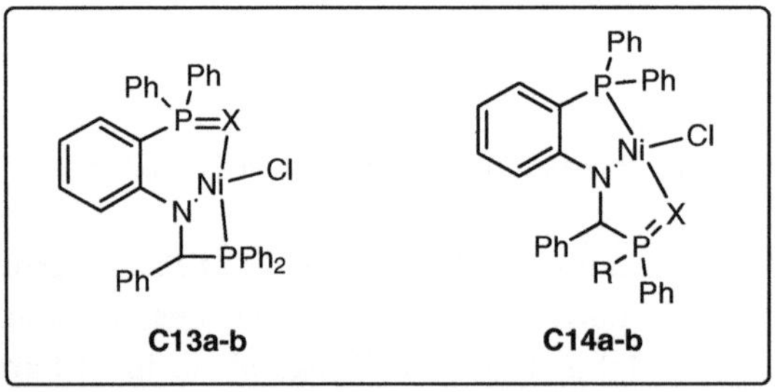

Fig. 2.16: Comparison of (P,N,S) and (P,N,O) pincer ligands for the formation of biphenyl derivatives.

Use of NHC carbenes as ligands

N-Heterocyclic carbenes are strong σ-donor ligands especially effective when the oxidative addition of an aryl halide to the catalyst is the rate-determining step. The main interest of *N*-heterocyclic carbene ligands resides in the robustness of the associated catalysts in comparison to the phosphorus ones, avoiding decomposition or deactivation. The first described air-stable efficient carbene ligand **L12** (Fig. 2.17) [17] combined with $Ni(acac)_2$, allowed the cross-coupling of aryl chlorides with aryl Grignard at room temperature, in THF with a catalyst loading of 3 mol %.

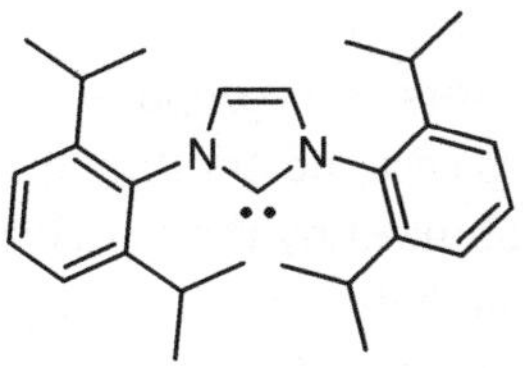

L12 (IPr)

Fig. 2.17: First efficient air-stable NHC ligand for the Kumada-Corriu reaction.

Nakamura *et al.* successfully used NHC carbenes such as **L12** for the coupling of aryl halides with aryl Grignard reagents [35]. The formation of homocoupling byproducts was mainly suppressed by using a nickel halide salt (Fig. 2.18).

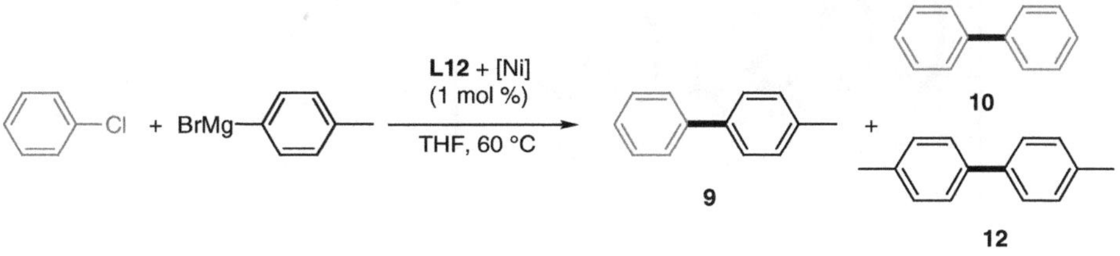

[Ni]	Time (h)	9	10	12
NiF_2	24	98	Trace	3
$NiF_2 .4H_2O$	24	96	Trace	1
$NiCl_2$	15	64	18	23

Fig. 2.18: Inhibition of the homocoupling process induced by **L12** and nickel(II) fluoride.

Only one or two NHC ligands could be coordinated to the metal, depending on the steric hindrance on the carbene ring. With ligands **L13** and **L14** (Fig. 2.19), the metal can be coordinated to two NHC subunits leading to a stabilized nickel complex which is efficient for the cross-coupling between an aryl Grignard reagent with an aryl halide [36–37].

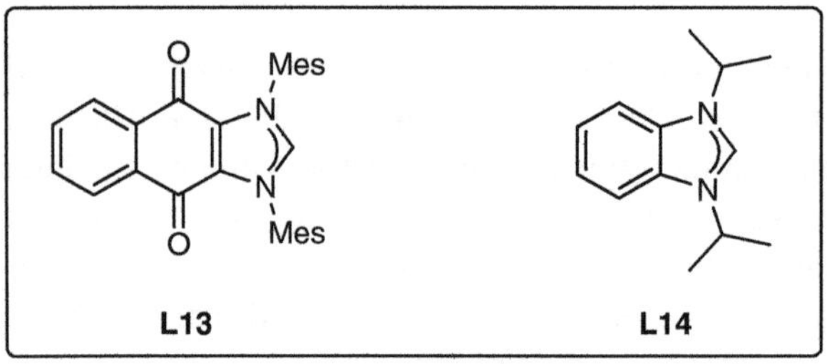

Fig. 2.19: Other NHC ligands used in Kumada-Corriu reaction.

Heterogeneous version of *N*-heterocyclic carbene precursors was proposed using copolymer-embedded nickel nanoparticles **C15** [38]. In this case, no additional coordination site was necessary, the nickel particles being stabilized by the structure of the polymer. After 10 recyclings of the catalyst, no loss of activity and selectivity was observed in the coupling. This catalyst is efficient for the coupling of aryl halides with aryl Grignard reagents (Fig. 2.20).

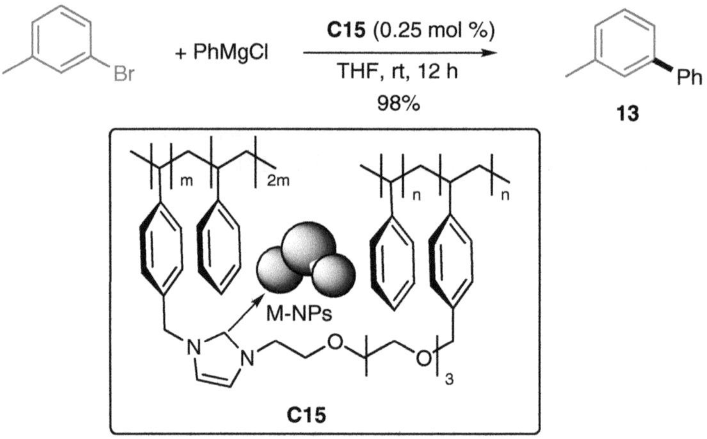

Fig. 2.20: Heterogeneous NHC ligand.

Other NHC-ligands containing additional coordinating functions which should provide a good catalyst efficiency have been successfully developed for the Kumada-Tamao-Corriu reaction involving aryl chlorides and aryl Grignard reagents. Thus, at room temperature in THF with 3 mol% catalyst loading, the ligand **L15** with the phosphane-NHC bidentate ligand [39–40] turned out to be far more active than **L12**, and remained the best nickel-NHC catalyst to date for the coupling of 4-chloroanisole with phenyl magnesium chloride (Fig. 2.21).

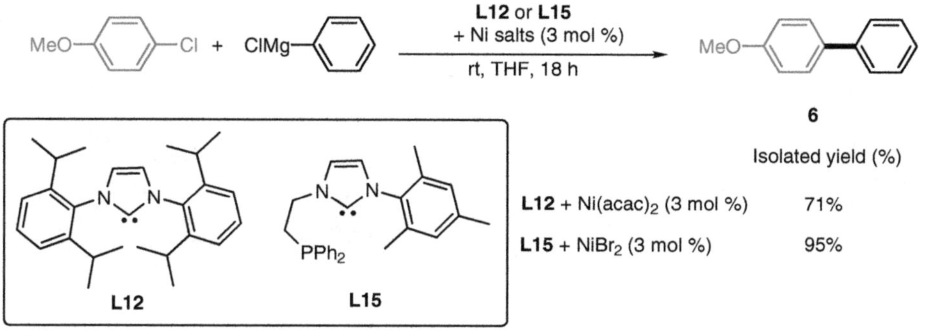

Fig. 2.21: Cross-coupling catalyzed by a NHC-ligand containing an additional phosphane group.

Nickel catalyst possessing six-, seven- or eight-membered ring *N*-heterocyclic carbenes **C16–18** (Fig. 2.22) have also been tested [41]. At room temperature in THF, the best results were obtained in the coupling of aryl chlorides in the presence of **C16** as catalyst.

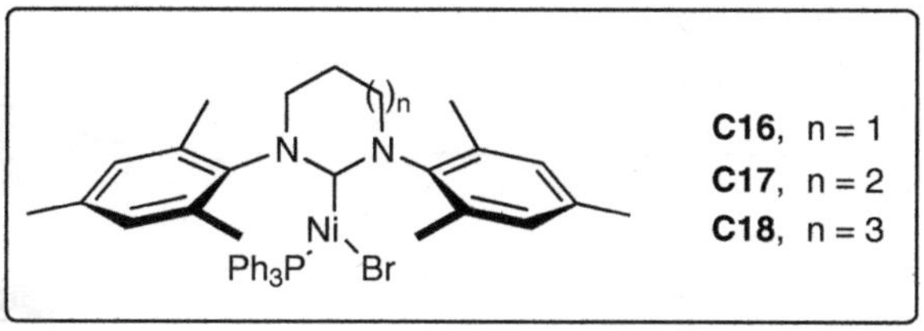

Fig. 2.22: Six-, seven- and eight-membered ring NHC ligands.

NHC-carbene ligand **L16** combined with $Ni(acac)_2$ [42] or a diaza function [CNN]-pincer nickel complexes such as **C19** [43], **C20** [44], and **C21** [45], were synthesized and successfully involved in Kumada-Corriu coupling between phenyl chloride and tolyl Grignard reagent (Fig. 2.23).

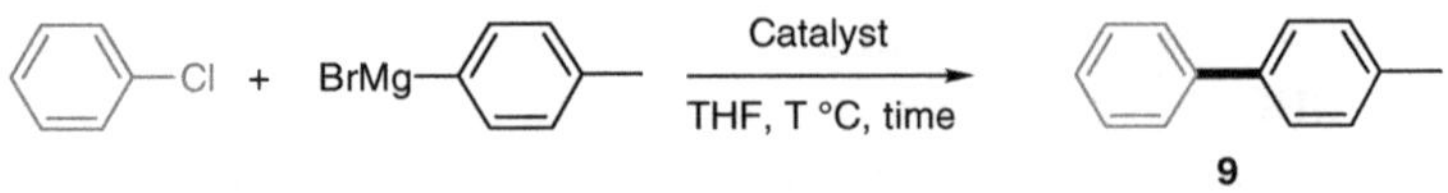

Catalyst	T (°C)	time	Yield
[**L16** + $Ni(acac)_2$] (1 mol %)	60	20 min	94 %
C19 (1 mol %)	rt	16 h	91 %
C20 (1 mol %)	rt	12 h	98 %
C21(1 mol %)	rt	24 h	92 %

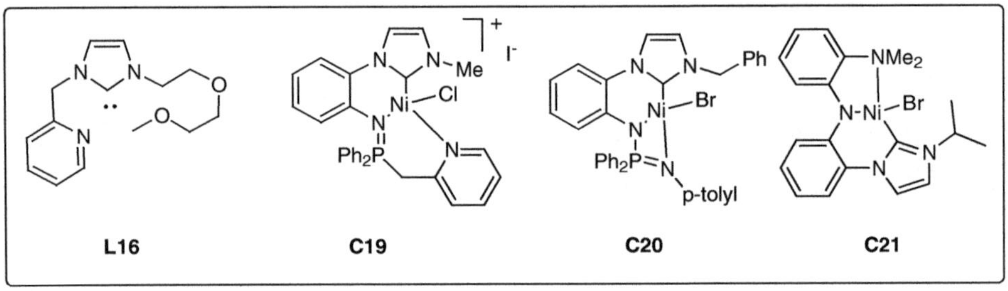

Fig. 2.23: NHC ligands with additional coordinating sites used in the Kumada-Corriu reaction.

Bridged di-NHC complexes, **C22a-c and C23,** were also developed and tested for the synthesis of **6** as a model reaction [46–48]. The bridge length influences the formation of monochelate complex *versus* less reactive dichelate species. The 1,3-propanediyl-bridged dicarbene ligand **C22c** turned out to be the most active of the series for the coupling of aryl Grignard reagents with aryl halides (Fig. 2.24).

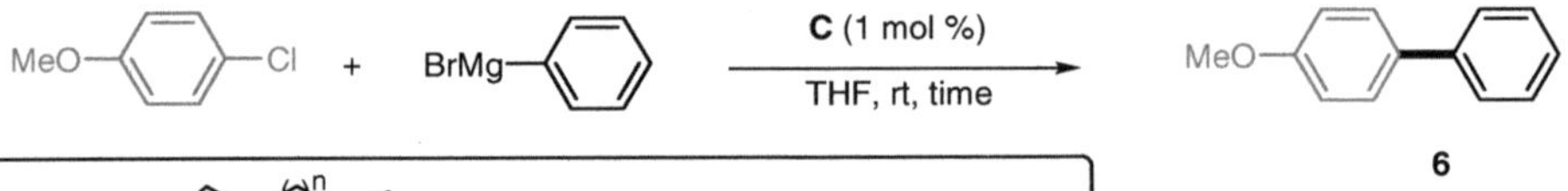

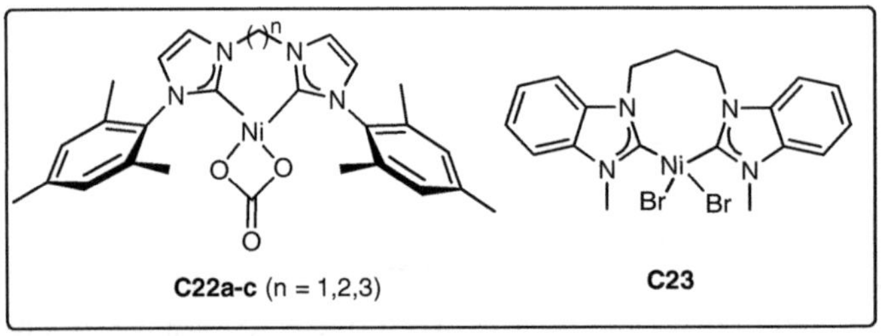

C	Time	Yield
C22a, n = 1	40 h	23 %
C22b, n = 2	40 h	47 %
C22c, n = 3	40 h	99 %
C23	12 h	85 %

Fig. 2.24: Bridged di-NHC complexes.

Tridentate bis(carbene)-derived nickel(II)-pincer complexes were also developed. Thus, Inamoto *et al.* tested successfully catalyst **C24** on a wide range of aryl halides as electrophiles [49]. The reaction was performed in THF, at room temperature, with a catalyst loading of 5 mol % (Fig. 2.25).

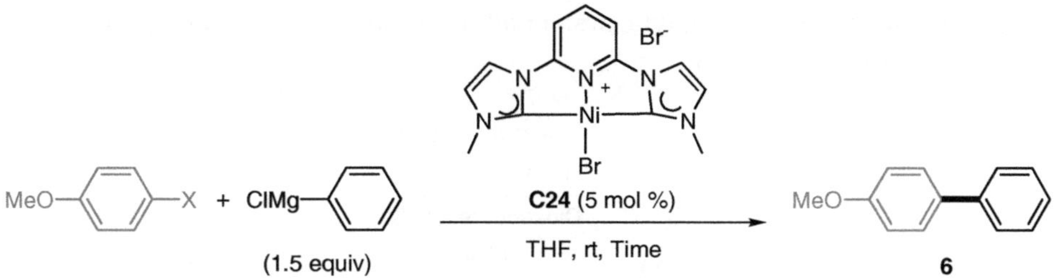

X	Time	Yield (%)
Cl	16 h	64
Br	24 h	67
F	96 h	44

Fig. 2.25: Tridentate bis(carbene)-derived nickel (II)-pincer complexes.

Heterogeneous catalysis

Several heterogeneous systems have been developed for the Kumada-Corriu reaction using Salen ligands bound either on a Merrifield resin, such as **C25** [50–51] or aminomethylpyrrolidine framework attached on silica, such as **C26** [52] for the aryl–aryl coupling (Fig. 2.26). Isolated yields of cross-coupling reaction were quiet good (around 70 % e.g. for the reaction of 4-bromoanisole and phenyl magnesium bromide). In addition, in both cases, the catalytic systems remained active even after 5 runs with almost no leaching (<1 %), which could be interesting for the development of green industrial processes.

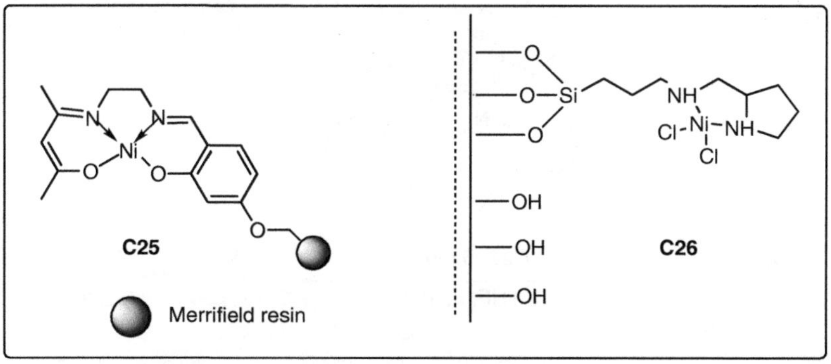

Fig. 2.26: Heterogeneous systems using N containing ligands for the Kumada-Corriu reaction.

As a source of nickel, Ni/C can be used for the Kumada-Corriu reaction of aryl chlorides and aryl Grignard reagents (Fig. 2.27). The best conditions required refluxing THF, the presence of a phosphine (generally PPh_3) to minimize the homocoupling side reaction. While this catalytic system allowed the scale-up, no recycling of the catalyst was reported [53–55].

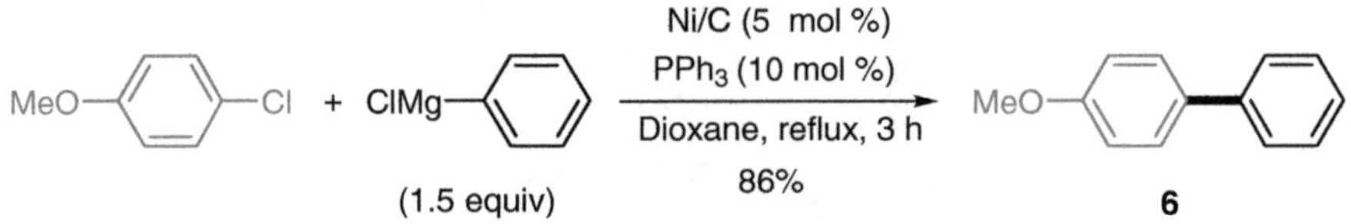

Fig. 2.27: Kumada-Corriu reaction of aryl chlorides and aryl Grignard reagents catalyzed by Ni/C.

2.2.1.2 With aryl fluorides

For a long time, fluorinated compounds have been considered as inert materials in cross-coupling reactions. Due to the design of new catalysts NHC-Ni(0) **C27**, the Kumada-Corriu reaction has been successfully achieved between fluoroarenes and aryl magnesium bromides. The reaction was efficient in a large number of cases even with electron-deficient substrates. Steric hindrance induced by substitution on the *ortho* position considerably reduced the yields of the couplings. Mechanistic studies revealed that the process occurred probably *via* a polar process, while reaction promoted by $NiCl_2$ took place *via* a radical process [17]. Air stable phosphine oxide ligands such as **L17** [56] and P,N,N-pincer nickel complexes such as **C28** [57] have also been tested to be efficient even for the coupling of electron-enriched substrates which are usually more reluctant to cross-couplings (Figs. 2.28 and 2.29).

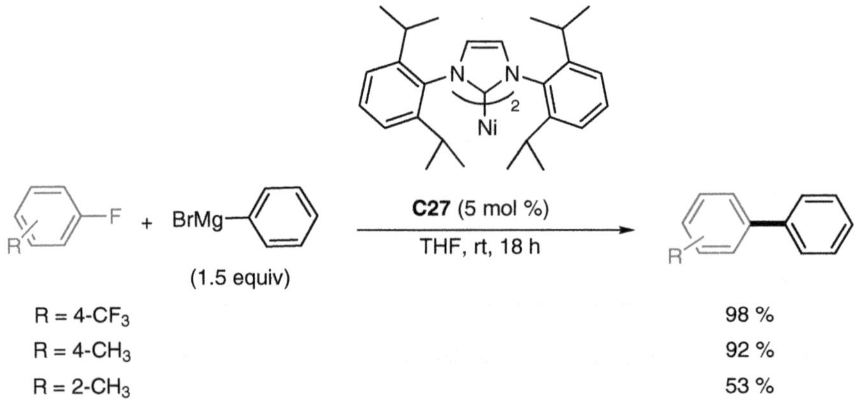

Fig. 2.28: Kumada-Corriu reaction between fluoroarenes and aryl magnesium bromides.

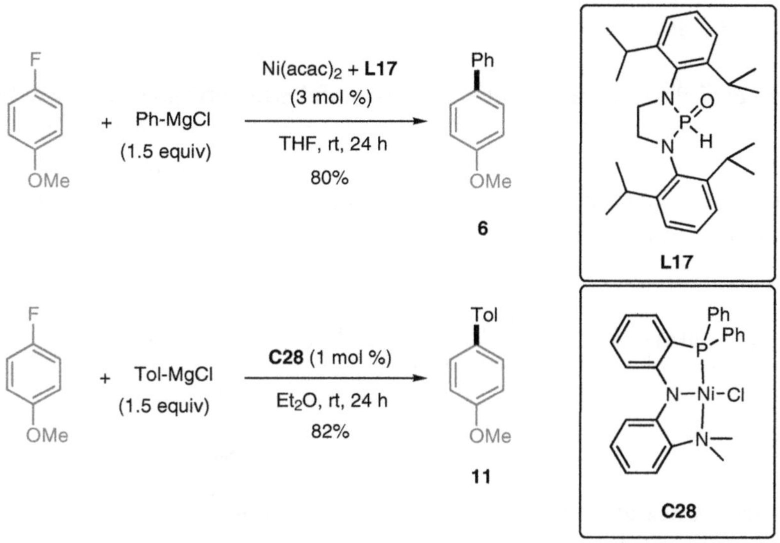

Fig. 2.29: Coupling between fluoroanisoale and aryl magnesium bromides.

In order to reach mono- or difluorinated compounds, the activation of di- and trifluoroarenes has been deeply investigated [58–60]. However, no significant selectivities were observed when reactions were realized in the presence of nickel salts. In contrast, palladium catalysis employing $PdCl_2$(dppf) gave only the monoarylated adduct **14** in 91 % yield (Fig. 2.30).

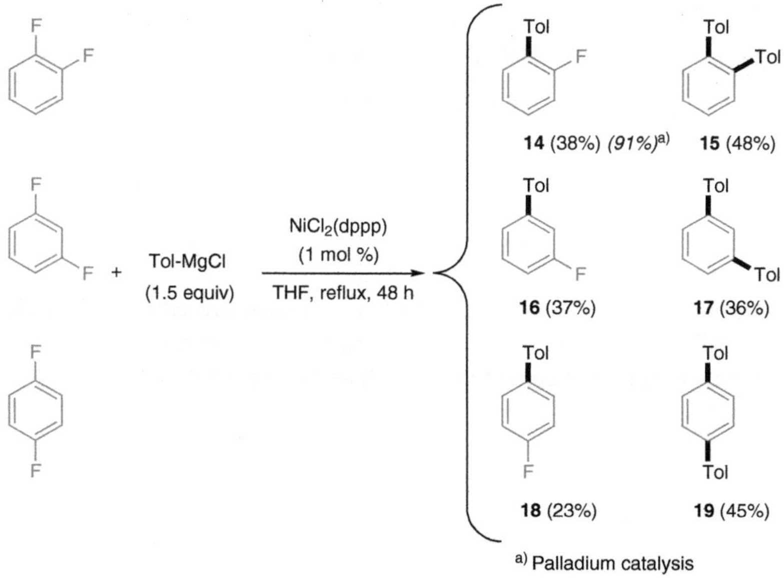

Fig. 2.30: Selectivities of nickel-catalyzed Kumada-Corriu reactions starting from difluoroarenes.

2.2.1.3 With aryl thioethers

In 1979, Wenkert *et al.* demonstrated that arylthiols, sulfides underwent coupling with phenyl- or *p*-tolyl-magnesium bromides in the presence of 10 mol % of $NiCl_2(PPh_3)_2$ catalyst (Fig. 2.31) [61]. The process required higher temperatures and longer reaction times than with halides as the carbon-sulfide bond is stronger and therefore more reluctant to oxidative addition [62].

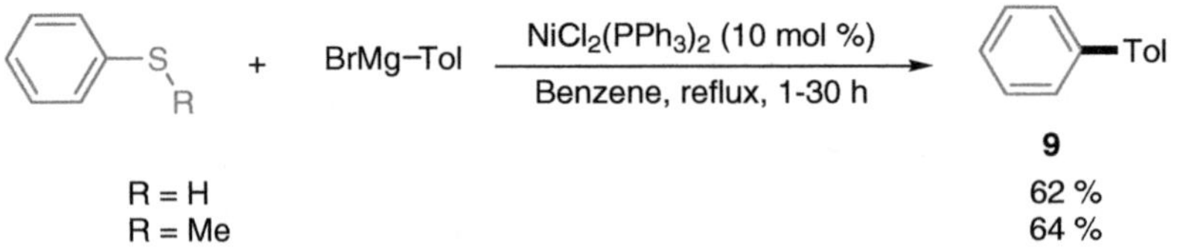

Fig. 2.31: Nickel-catalyzed Kumada-Corriu reaction with arylthiol derivatives.

The scope was expanded to a great variety of heterocyclic aryl sulfides (Fig. 2.32) [63].

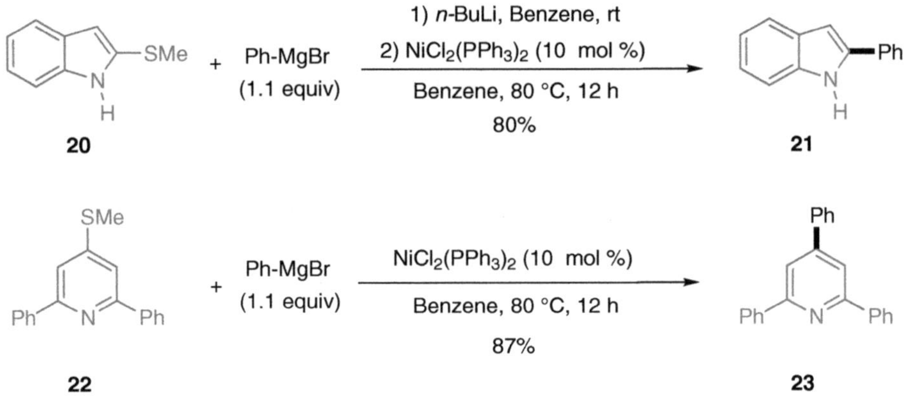

Fig. 2.32: Nickel-catalyzed Kumada-Corriu reaction with heterocyclic aryl sulfides.

The electron-rich $NiCl_2(PPh_3)(IPr)$ catalyst **C29** was successfully employed in the cross-coupling of bulky 2-methylthio-substituted benzofurans with aryl Grignard reagents, yielding highly fluorescent benzofurans such as **25** (Fig. 2.33) [64].

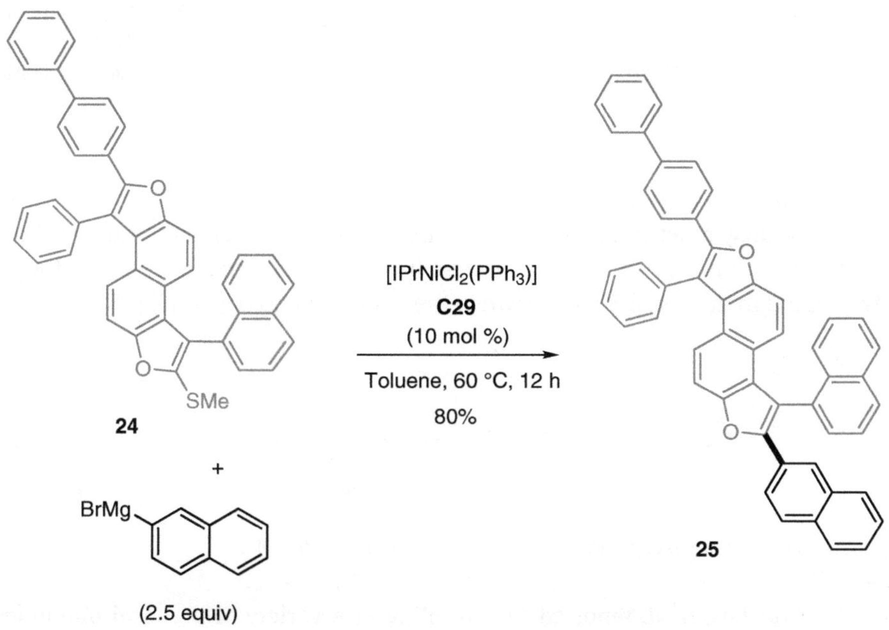

Fig. 2.33: Access to highly fluorescent benzofurans *via* the nickel-catalyzed Kumada-Corriu reaction.

Taking advantage of this coupling, a flexible access to eupomatenoids, a class of natural products has been designed [64]. A Kumada-Corriu coupling between 2-methylthiobenzofuran **26** and the Grignard reagent **27**, prepared from 5-bromo-1,3-benzodioxole, led the expected compound **28** in high yield (Fig. 2.34).

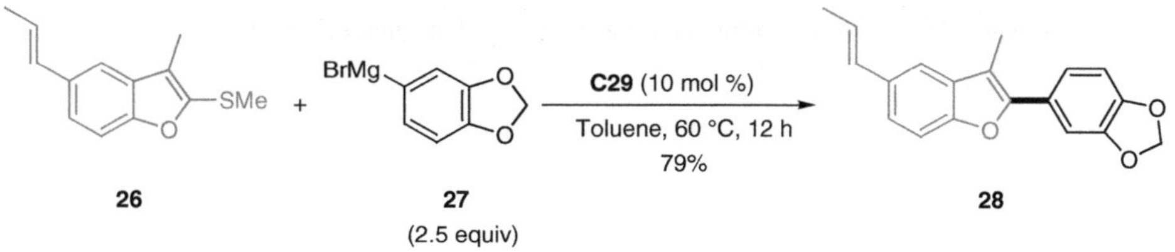

Fig. 2.34: Access to eupomatenoids using the Kumada-Corriu reaction.

2.2.1.4 With aryl sulfones, sulfonates, sulfonamides

The use of aryl *tert*-butyl sulfones as electrophilic coupling partners was described by the group of Julia. The ability of the *tert*-butyl sulfonyl group to direct *ortho*-lithiation was later applied to the synthesis of *ortho*-substituted unsymmetrical biaryls (Fig. 2.35) [65].

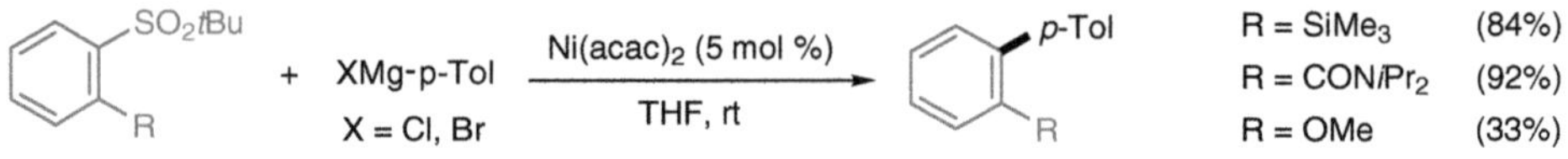

Fig. 2.35: Aryl *tert*-butyl sulfones as electrophilic coupling partners.

Arenesulfonates can undergo nickel-catalyzed cross-coupling with aryl Grignard nucleophiles to give unsymmetrical biaryls. Park *et al.* studied the C–C bond formation from neopentyl arylsulfonates with arylmagnesium halides and showed that the $NiCl_2$(dppf) catalyst inserts in the C(sp_2)-S rather than in the C(sp_3)-O bond (Fig. 2.36) [66].

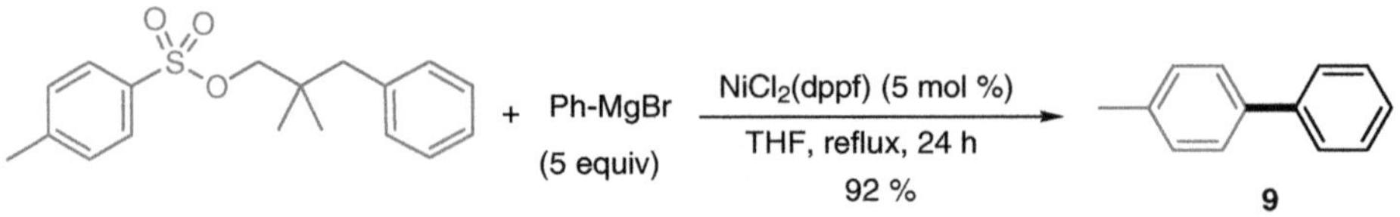

9

Fig. 2.36: Arenesulfonate used for the synthesis of unsymmetrical biaryls.

Similarly, Snieckus *et al.* reported the coupling of a variety of arenesulfonamides with phenylmagnesium halides in refluxing toluene under classical nickel catalysis (Fig. 2.37) [67].

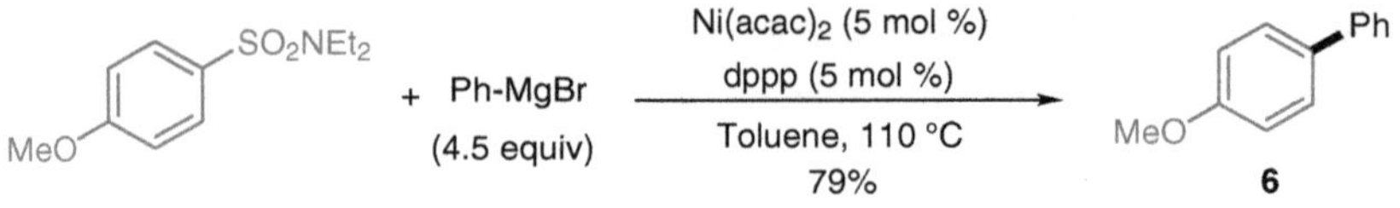

6

Fig. 2.37: Arenesulfonamides as partners for the synthesis of unsymmetrical biaryls.

2.2.1.5 With aryl ethers

Cross-couplings with oxygenated derivatives represent new perspectives in Kumada-Corriu reactions. Due to their large availability from natural biosources, they represent an interesting pool of starting materials. Their use has been recently reviewed [68] and will probably increase in the near future.

2.2.1.5.1 Aryl alkylethers and aryl silylethers

Aryl ethers constitute more attractive electrophiles than sulfonates as they are more readily available and more environmentally benign as they generate less waste. However, the C-OMe bond is more reluctant towards oxidative addition and, thus, less reactive in cross-coupling reactions due to the higher activation energy for its cleavage. In a pioneering work, Wenkert and co-workers described the cross-coupling of aryl methyl ethers with phenyl magnesium bromide in the presence of 10 mol% of $NiCl_2(PPh_3)_2$ in refluxing benzene [69]. Two decades later, Dankwardt found that nickel complexes with electron-rich phosphine ligands possessing a larger cone angle (e.g., PCy_3 or $PhPCy_2$) were efficient at 60 °C for the arylation of diversely substituted methoxyarenes (Fig. 2.38) [70].

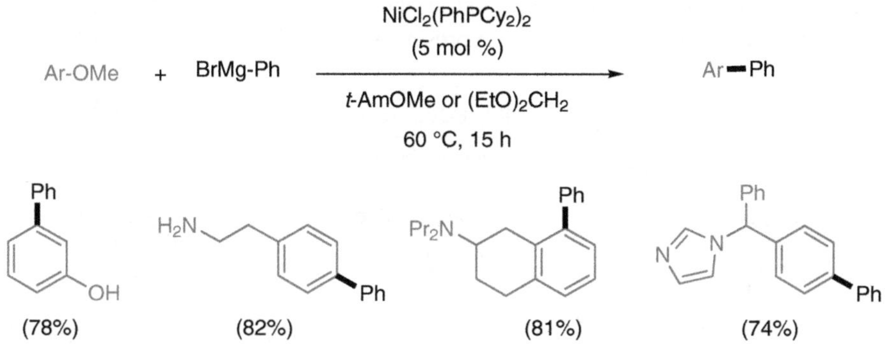

Fig. 2.38: Fonctionalized arylethers as efficient electrophiles for the Kumada-Corriu reaction.

More recently, several contributions reported the design of new ligands to improve the efficiency and the selectivity of nickel-catalyzed Kumada-Corriu coupling reactions of aryl ethers. In 2011, Wang and Xie reported that electron-rich and sterically hindered pyrazolyl amino phosphines **C1b** (cf. Fig. 2.9) allowed for the coupling of π-extended aromatic and alkenyl ethers under mild conditions (THF, room temperature or 50 °C) [71]. The more challenging aryl methyl ethers still required high temperatures to react (toluene, 120 °C). In 2012, and for the first time, Nicasio *et al.* reported the use of *N*-heterocyclic carbene ligands as an alternative to phosphine ligands in the nickel-catalyzed Kumada-Tamao-Corriu coupling of π-extended and regular aryl methyl ethers under relatively mild conditions (THF, 60 °C) [72]. The scope was again limited to non-functionalized coupling partners due to the high reactivity of Grignard reagents (Fig. 2.39).

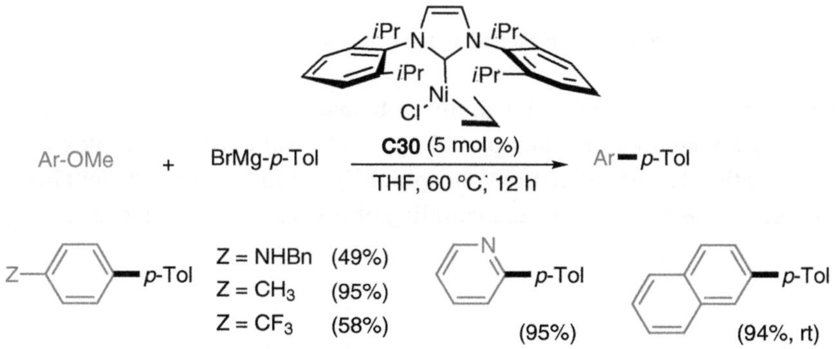

Fig. 2.39: NHC ligand for the coupling of arylethers.

An initial study on the nickel-catalyzed alkenyl C-OSi bond activation for C-C bond formation with aryl and alkyl Grignard reagents was reported in 1980 by Kumada *et al.* [73]. Although a few aromatic silyl ethers were mentioned to react under the conditions optimized for the Kumada-Corriu coupling of aromatic alkyl ethers, no systematic study had been carried out on these substrates until 2011. Different silyloxy groups were tested and led to the cross-coupling products in similar yields compared to the OTMS derivatives whatever the steric hindrance of the silyl protecting group. It is worth to note that the C-OSi bond can be activated in the presence of a methoxy group (Fig. 2.40) [74].

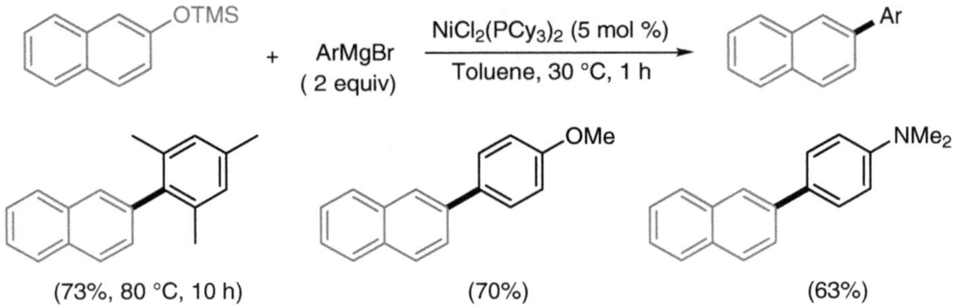

Fig. 2.40: C–C bond formation using silyloxyaryl derivatives.

2.2.1.5.2 Aryl and alkenyl carbamates and phosphates

In 1992, the group of Snieckus showed that aryl carbamates and triflates could be coupled to aryl Grignard in similar conditions [75].

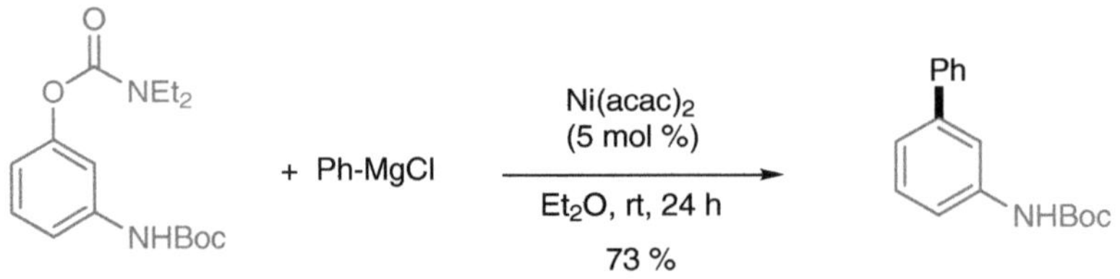

Fig. 2.41: Carbamates as electrophiles in nickel-catalyzed Kumada-Corriu reaction.

Compared to other *O*-based electrophiles, the carbamate moiety offers the ability to effect directed *ortho*-metallation, thus providing a potential entry to complex polysubstituted aromatics. The use of hydroxyphosphine ligand **L18** proved efficient in the rate acceleration of the nickel-catalyzed coupling of aryl carbamates (Fig. 2.42) [76].

The intermediate nickel phosphine/magnesium alkoxide bimetallic catalytic species induces the C–O bond cleavage by a synergetic push-pull action. These conditions were also effective with aryl phosphates, which were initially considered in 1981 by Kumada *et al.* using $Ni(acac)_2$ as the catalyst [77–78].

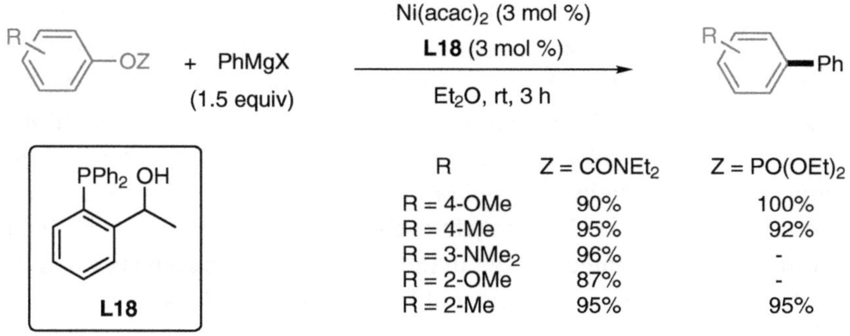

R	Z = $CONEt_2$	Z = $PO(OEt)_2$
R = 4-OMe	90%	100%
R = 4-Me	95%	92%
R = 3-NMe_2	96%	-
R = 2-OMe	87%	-
R = 2-Me	95%	95%

Fig. 2.42: Hydroxyphosphine **L18** as an efficient ligand for the nickel-catalyzed coupling of aryl carbamates.

2.2.1.5.3 Diaryl sulfates and aryl sulfamates

In the search of environmentally benign cross-coupling transformations, diaryl sulfates were found by Shi *et al.* to be efficient green organic electrophiles [79]. Both of the aryl groups can be inserted into the product, while the process generates harmless inorganic salts (e.g. $MgSO_4$ and MgX_2) as by-products. Interestingly, the reaction conditions are mild enough to leave C-F and C-OMe bonds unreactive. The proposed mechanism involves a monosulfate magnesium salt which proved to be a reactive intermediate in the catalytic cycle (Fig. 2.43). In 2005, Snieckus *et al.* described the use of more electron-rich IMesNHC ligands for the nickel-catalyzed cross-coupling of *O*-sulfamates electrophiles in moderate to good yields [80].

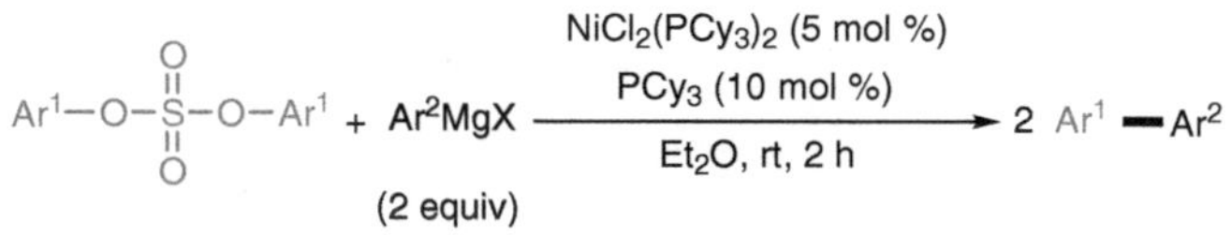

Ar^1	Ar^2	Yield
2-naphthyl	2-tolyl	93%
2-naphthyl	4-FC_6H_4	72%
2-tolyl	Ph	88%
4-$MeOC_6H_4$	2-tolyl	75%
4-$Me_2NC_6H_4$	Ph	83%

Fig. 2.43: Diaryl sulfates as efficient green organic electrophiles for the Kumada-Corriu reaction.

2.2.1.5.4 Magnesium aryloates

The direct transformation of readily available phenol derivatives would represent a major breakthrough in nickel-catalyzed cross-coupling reactions as the use of free alcohols would be step- and atom-economic as well as environmentally benign. The process would indeed generate less organic waste and would avoid additional transformations of the alcohol to good leaving groups (e.g., phosphonates or sulfonates) or the use of toxic organic halides as electrophiles. The first achievement towards this goal was reported by Shi *et al.* using magnesium naphtholates [81]. The NiF_2/PCy_3-catalyzed cross-coupling of aryl Grignard reagents with different naphthols was performed in non-polar solvents after *in situ*-preformation of the corresponding magnesium salts with methylmagnesium bromide. The process was believed to proceed through the Ni(0)/Ni(II) catalytic cycle and the C-O naphthoxide bond cleavage was supposedly activated by Mg^{2+} in the oxidative addition step to the Ni(0) species. Unfortunately, the parent phenols could not be converted to biphenyls under these conditions (Fig. 2.44).

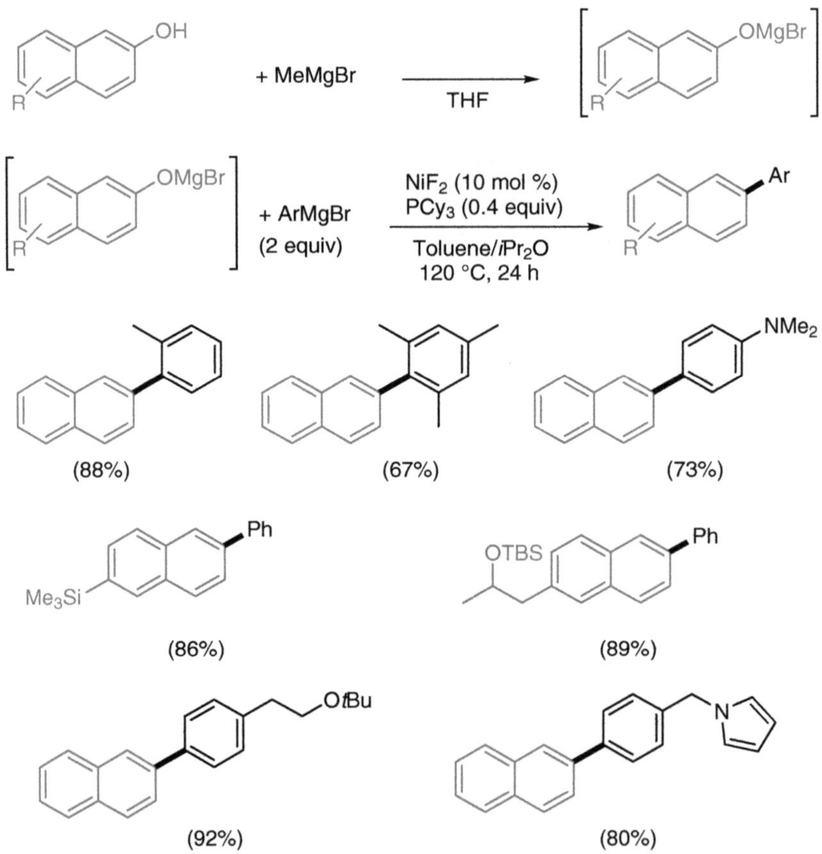

Fig. 2.44: Cross-coupling of aryl Grignard reagents with functionalized 2-naphthols.

2.2.2 With vinyl derivatives

2.2.2.1 Vinyl halides

Vinyl chloride is one of the most reactive C(sp_2)-halides and has been widely used in the industry for the synthesis of styrene derivatives in multi-tons scale. Chemists at Hokko Chemical industry in Japan have used inexpensive $NiCl_2$(dppp) as catalyst. For security reasons, they have replaced highly flammable ether by a mixture of benzene/THF or toluene/THF (Fig. 2.45) [82].

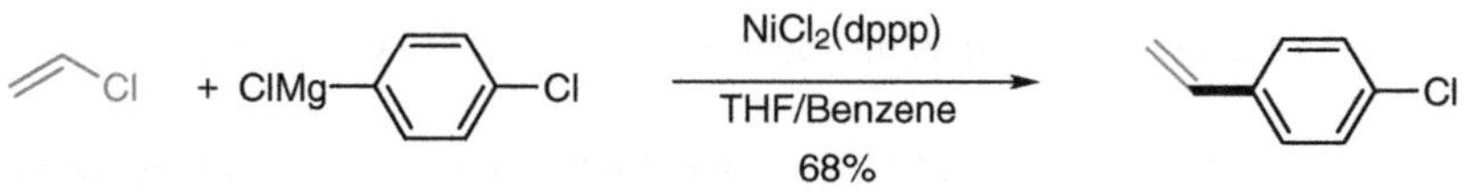

Fig. 2.45: Cross-coupling of vinyl chloride with aryl Grignard reagents.

Dppp can be replaced by 4,6-bis(3-di*iso*propylphosphinophenyl) dibenzofuran **L19**, a related disphosphine ligand, which coordinate to Ni(I) to produce the [(*iPr*DPDB-Fphos)NiCl] complex and allowed the synthesis of styrene from vinyl chloride and phenyl Grignard reagent in nearly 85 % isolated yield (Fig. 2.46) [83].

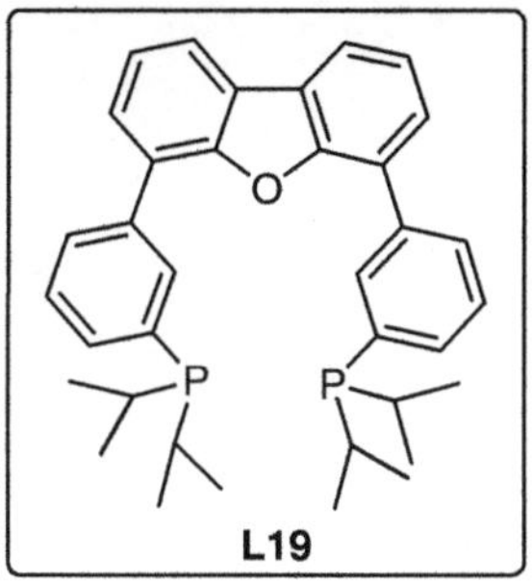

Fig. 2.46: Diphosphine ligand **L19** designed for the synthesis of styrenes from vinyl chloride.

Tetradentate nickel catalyst **C31** was also particularly efficient for the coupling of unactivated vinyl chlorides (Fig. 2.47) [84].

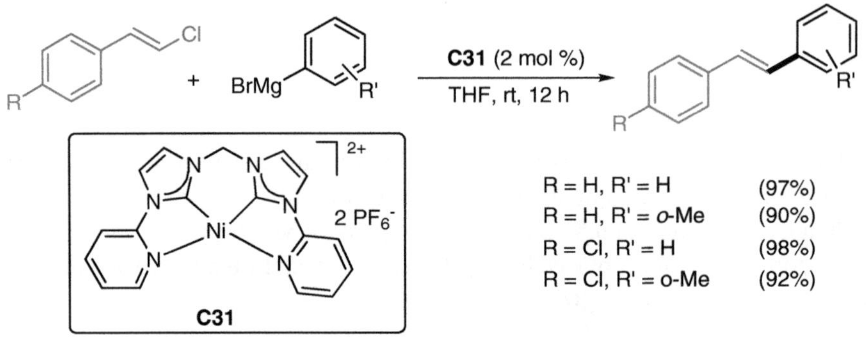

Fig. 2.47: Tetradentate nickel complex **C31** for the coupling of unactivated vinyl chlorides.

In 2014, Pugh *et al.* used the same $NiCl_2$(dppp) system for the synthesis of a polymerizable 1-ethoxyvinylbenzocyclobutenes. These latter could undergo radical polymerization at 100–150 °C leading to cross-linked polymers (Fig. 2.48) [85].

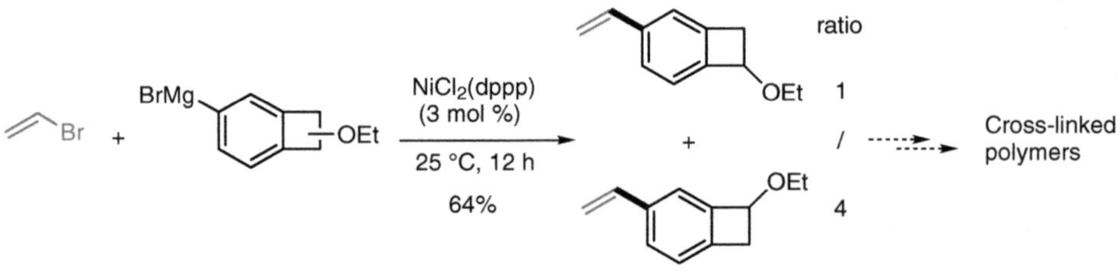

Fig. 2.48: Access to polymerizable cyclobutene from vinylbromide.

2.2.2.2 With vinyl thioethers

Following the early results obtained by Wenkert *et al.*, Takei *et al.* reported the reaction between alkenyl and aryl sulfides in nickel-catalyzed cross-coupling reactions with phenyl Grignard reagents [86]. The olefinic products were obtained with retention of configuration suggesting a high stereospecificity for each step of the process (Fig. 2.49).

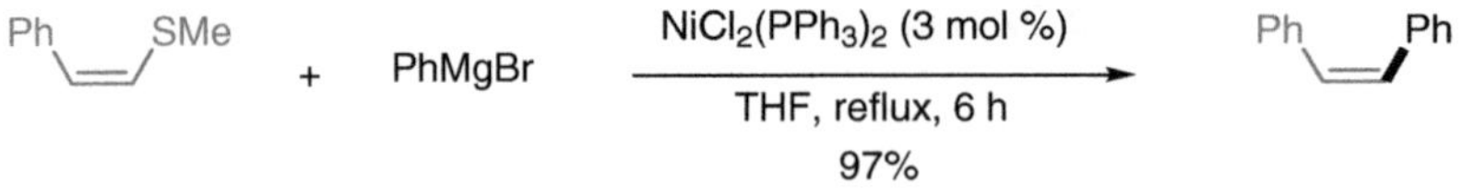

Fig. 2.49: Stereoselective cross-coupling of alkenyl sulfides.

2.2.2.3 With cyclic enol ethers

The group of Martin found that a combination of $Ni(cod)_2$ with **L20.** HCl, a NHC precursor, and LiCl as the additive allowed the cross-coupling of cyclic vinyl ethers when performed at low temperatures [87]. Under these conditions, the use of functionalized Grignard reagents furnished the corresponding homoallylic alcohols with complete *Z*-stereoselectivity (Fig. 2.50).

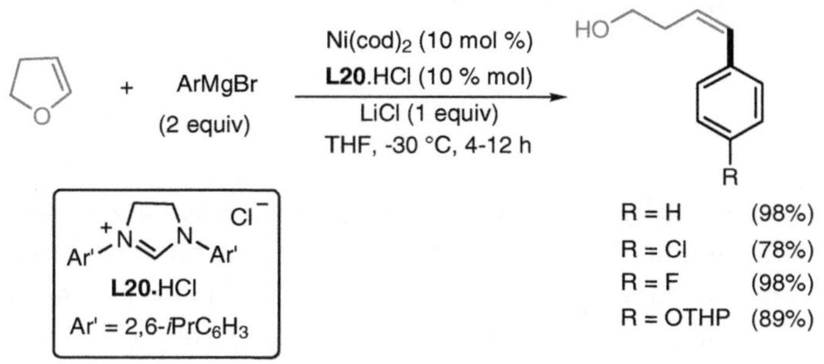

Fig. 2.50: Stereoselective access to homoallylic alcohols from dihydrofuran.

2.2.3 With alkynyl halides

1-Bromoalkynes have also been used as electrophiles to achieve Ni-catalyzed cross-coupling reactions with aryl magnesium bromides [88]. Among the different catalysts tested, the combination of $NiCl_2$ with a triarylphosphine appeared as the most promising system to obtain high yields in the coupling product (Fig. 2.51).

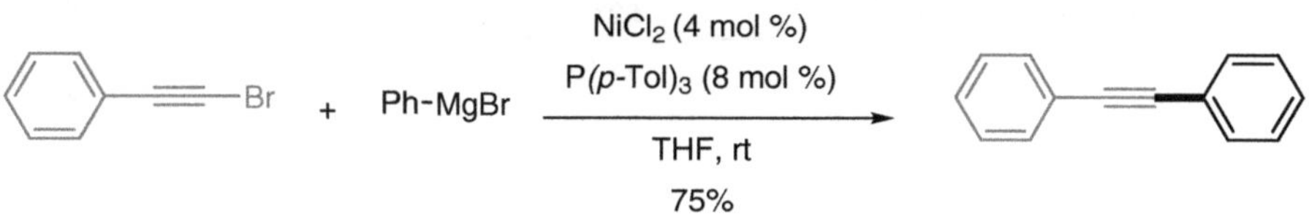

Fig. 2.51: 1,2-Disubstituted acetylenes from bromo(ethynyl)benzene.

2.2.4 Miscellaneous

2.2.4.1 With propargyl bromide

The coupling of aryl Grignard with propargyl bromide led to an arylallene derivative **29** contaminated by small amounts of the corresponding propargyl arene **30** (Fig. 2.52) [89].

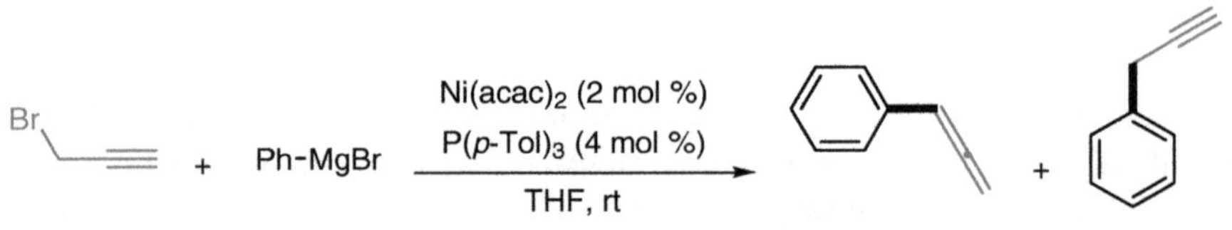

29 (80%) **30** (16%)

Fig. 2.52: Allene derivatives from propargyl bromide.

2.2.4.2 With α-bromoketones

α-Bromopropiophenones can be easily substituted by an aryl group when treated with an aryl Grignard and a small amount of a nickel catalyst. This reaction has been extensively studied in the presence of chiral ligands. Of interest, the reaction is stereoconvergent: both enantiomers of the starting material led to the final chiral product with the same configuration and a similar level of induction. The best results were obtained when dimethyl bis-oxazoline ligand **L21**, derived from arylglycine, was used. To reach similar ee with dialkylketones, the process needed however to be conducted in the presence of the less common and more hindered ligand **L22** and with this ligand, the ee values culminate to the range of 85–90 % (Fig. 2.53) [90].

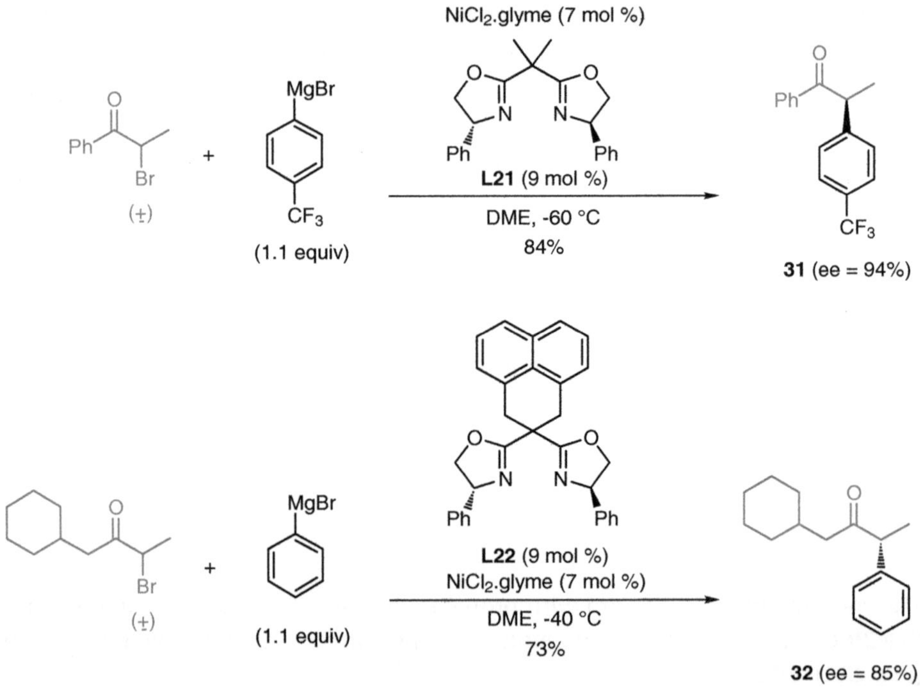

Fig. 2.53: Enantioselective formation of α-arylketones.

We have to point out that, recently, the group of Walsh has shown that the Kumada-Corriu reaction gave even better results when performed in the presence of cobalt salts associated to the same bis-oxazolidine ligands [91].

2.2.4.3 Three component reaction

To extend the scope of the Kumada-Corriu coupling reaction, nickel-catalyzed three component process has been designed [92]. The reaction was carried out in the presence of the two classical partners, an alkyl halide and a phenyl Grignard reagent, and also in the presence of 2,3-dialkyl-1,3-butadienes. Interestingly, functionalizations occurred selectively on the two opposite terminal positions of the diene and delivered polysubstituted alkenes in reasonable yields but with a poor *E*/*Z* ratio (Fig. 2.54). It was assumed that alkyl radical species were generated by a Single Electron Transfer (SET) process from a nickelate complex intermediate.

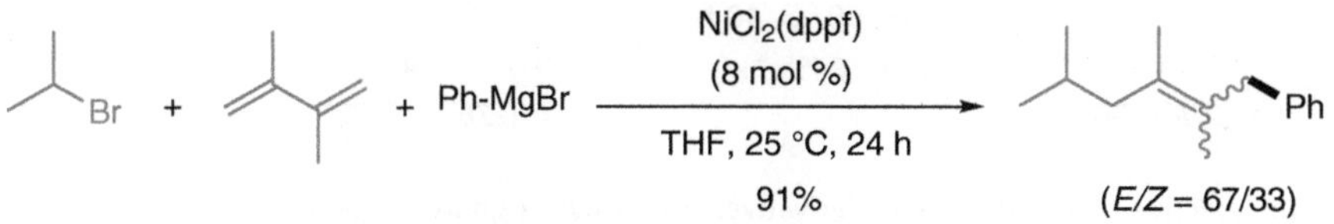

Fig. 2.54: Cross-coupling of alkyl halides, 1,3-butadienes and aryl Grignard reagents.

A three-component reaction leading to tetrasubstituted alkenes has been devised by Zhao, Hayashi *et al.* [93]. By adding a very small amount of a nickel salt to a mixture of an internal alkyl(phenyl)acetylene, an aryl Grignard reagent and an aryl iodide, an arylmagnesiation occured rapidly, followed by an efficient Kumada-Corriu coupling. The *Z*-selectivity observed for the first step was essential for the stereoselective synthesis of pharmaceutical compounds such as tamoxifen **33** (Fig. 2.55).

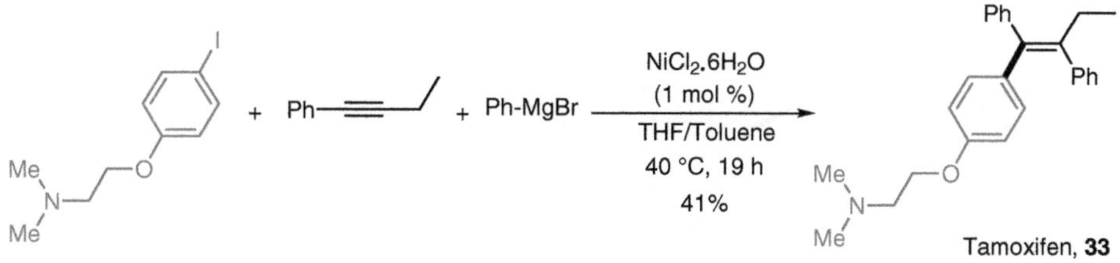

Tamoxifen, **33**

Fig. 2.55: Tetrasubstituted alkenes by a three-component reaction.

2.3 Coupling reactions of alkenyl Grignard reagents

2.3.1 With aryl derivatives

2.3.1.1 With aryl halides

In contrast to the Kumada-Tamao-Corriu reaction with aryl Grignard reagents, the nickel-catalyzed coupling of alkenyl magnesium reagents received much less attention since a first report in 1975 by the group of Kumada [94]. The cross-coupling of alkenyl Grignard reagents with aryl or vinyl halides was found to occur with *cis/trans* isomerization of the olefinic bond depending on the reactivity of the halide, the bulkyness of the Grignard reagent and the ArX/R'MgX ratio (Fig. 2.56). The vinyl Grignard reagents appeared far less reactive than their aryl counterparts [95].

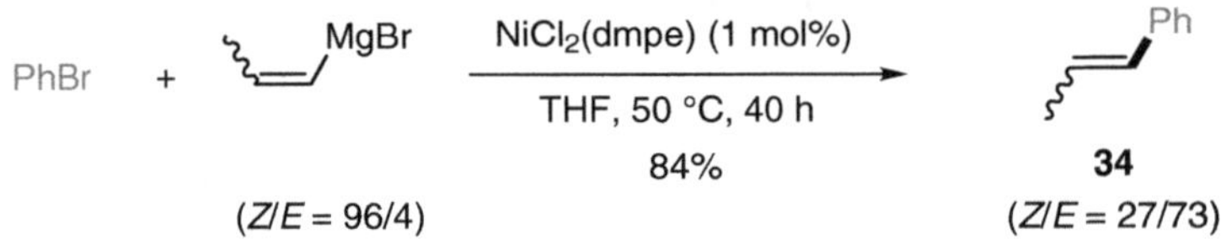

Fig. 2.56: Unselective access to styrene derivatives from vinyl Grignard reagents.

The group of Bach studied the functionalization of dibromobenzofurans **35** and observed a selective alkenylation at the C-5 position in high yields under Kumada-Corriu conditions. A further Negishi coupling reaction led to different eupomatenoids **37** in high yields (Fig. 2.57) [96].

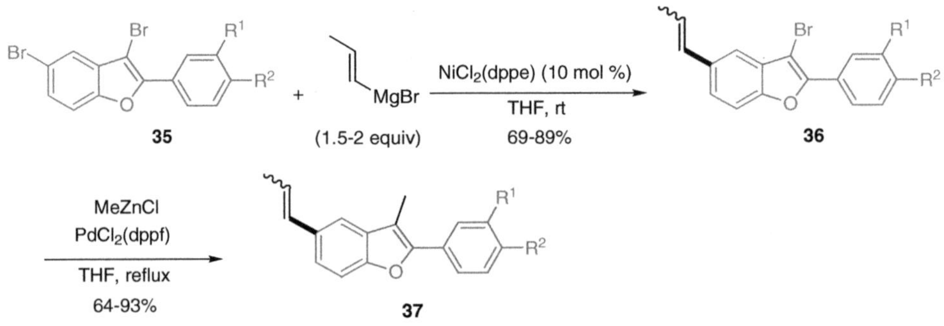

Fig. 2.57: Selective alkenylation of dibromobenzofurans with vinyl Grignard reagents.

2.3.1.2 With aryl sulfides

The group of Wenkert reported the selective cross-coupling of methyl aryl sulfides over *iso*-propyl aryl sulfides with alkenyl Grignard reagents, albeit in moderate yields [97]. The coupling conditions using $NiCl_2(PPh_3)_2$ as catalyst also led to the reduction product of the thiomethyl substituent (Fig. 2.58).

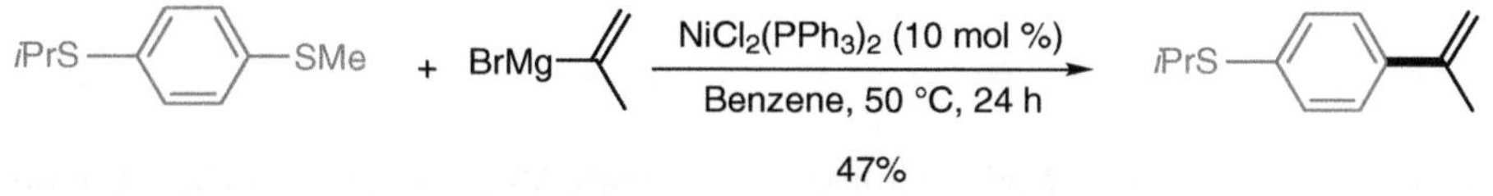

Fig. 2.58: Regioselective coupling of aryl sulfides with vinyl Grignard reagents.

2.3.1.3 With aryl carbamates

The coupling of a phenanthrenyl (*N,N*-diethyl)carbamate was performed with vinyl magnesium bromide and $Ni(acac)_2$ as the catalyst to deliver a styrene derivative however, in moderate yield [75].

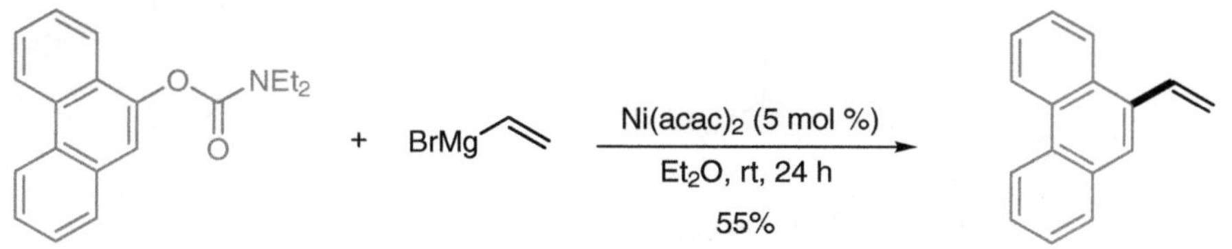

Fig. 2.59: Coupling of arylcarbamates with vinyl magnesium bromide.

2.3.2 With vinyl derivatives

2.3.2.1 With vinyl sulfones

Conjugated dienes can be obtained with complete retention of configuration by the Ni-catalyzed coupling of vinyl sulfones with vinylmagnesium bromide (Fig. 2.60) [98]. The outcome of the reaction is highly dependent on the nickel catalyst (40 % yield with $Ni(acac)_2$ and 23 % yield with $NiCl_2/2PPh_3$) leading, in all cases, to the formation of the reduction product along with the coupling product.

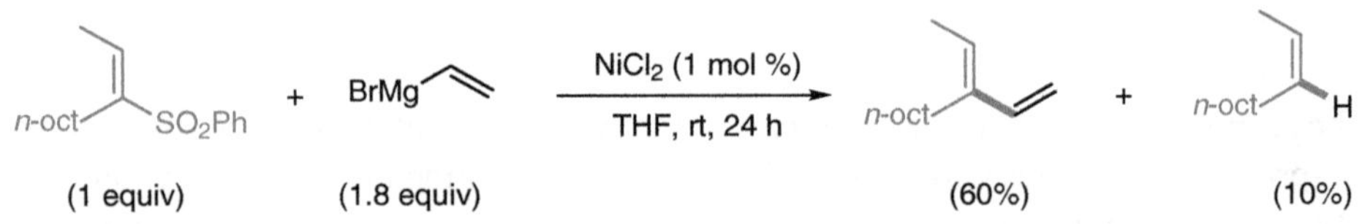

Fig. 2.60: Synthesis of 1,3-dienes from vinylsulfones and vinyl Grignard reagent.

2.3.2.2 With vinyl carbamates

The coupling of (*Z*)-vinyl (*N*,*N*-di*iso*propyl)carbamate **38** with different vinyl magnesium bromides in the presence of 10 mol % of Ni(acac)$_2$ led to (*Z*)-alkenyl derivatives with high stereoselectivity (Fig. 2.61) [99].

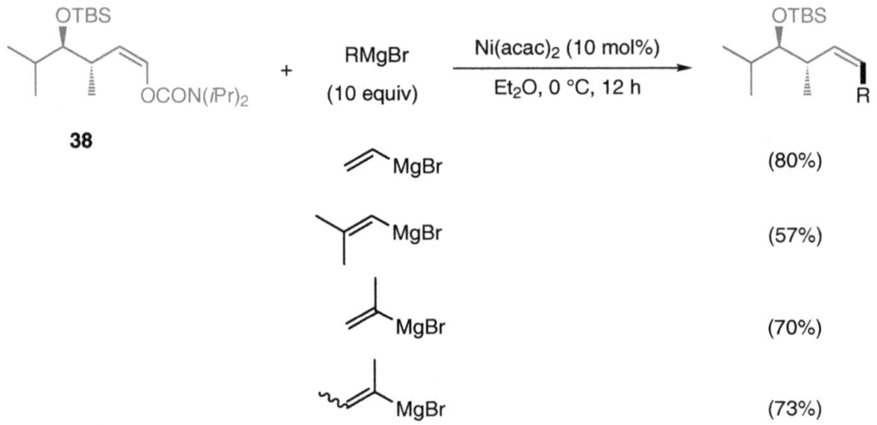

Fig. 2.61: Coupling of (*Z*)-vinyl carbamates with vinyl magnesium bromides.

2.3.3 With alkyl derivatives

The coupling of vinyl magnesium chloride with alkyl fluorides in the presence of a nickel catalyst gave rise to an alkylative dimerization of the vinyl Grignard reagent to form a 2-alkyl-3-butenyl Grignard intermediate which is hydrolyzed to the corresponding butene (Fig. 2.62) [100]. The nickelate complex **C32** was suggested by Kambe *et al.* to be a reactive intermediate in the different plausible reaction pathways.

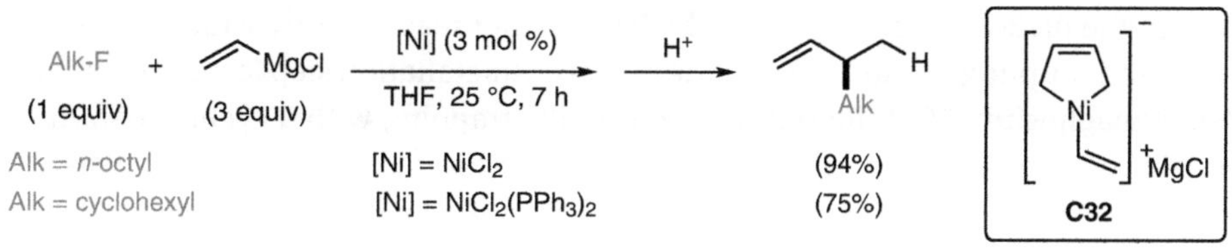

Fig. 2.62: Coupling of alkylfluorides with vinyl Grignard reagent.

2.3.4 Miscellaneous

2.3.4.1 Three component reactions

A novel nickel-catalyzed coupling reaction between alkenyl magnesium reagents and methylenecyclopropanes in the presence of a chlorosilane, as the terminal electrophile, was developed by Kambe *et al.* [101]. The corresponding carbomagnesation products were obtained by the selective cleavage of the distal C–C bond of methylenecyclopropanes through nickel oxidative addition (Fig. 2.63).

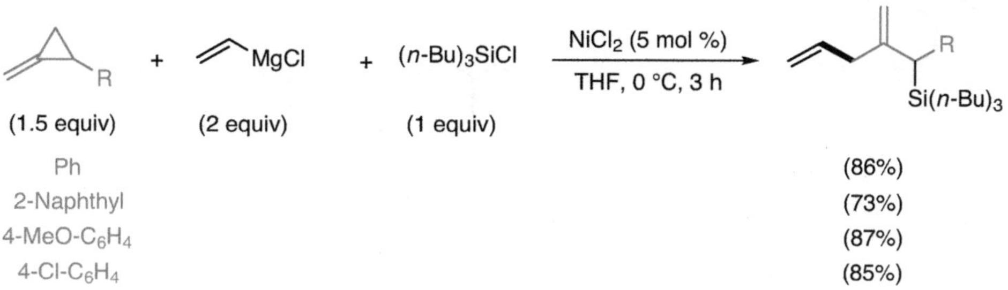

Fig. 2.63: Selective cleavage of methylenecyclopropanes with vinyl Grignard reagents.

2.3.4.2 Ring-opening of cyclic ethers

Kambe *et al.* have explored the nickel-catalyzed coupling of vinyl magnesium chloride with 3- or 4-membered cyclic ethers [102]. As for alkyl fluorides, the vinyl Grignard reagent underwent a dimerization prior to the coupling with the electrophile giving access to pentenol or hexenol derivatives **40** (Fig. 2.62). The postulated mechanism involves the nickelate complex **C47** which can either react directly with cyclic ethers to form the coupling product **40** or undergo a transmetallation step to yield (2-butene-1,4-diyl)magnesium **39**, followed by electrophilic trapping with a cyclic ether. Interestingly, no reaction took place when using substituted vinyl Grignard reagents (Fig. 2.64). Chlorosilanes as electrophilic partners led to 1,4-disilyl-2-butenes under the same reaction conditions.

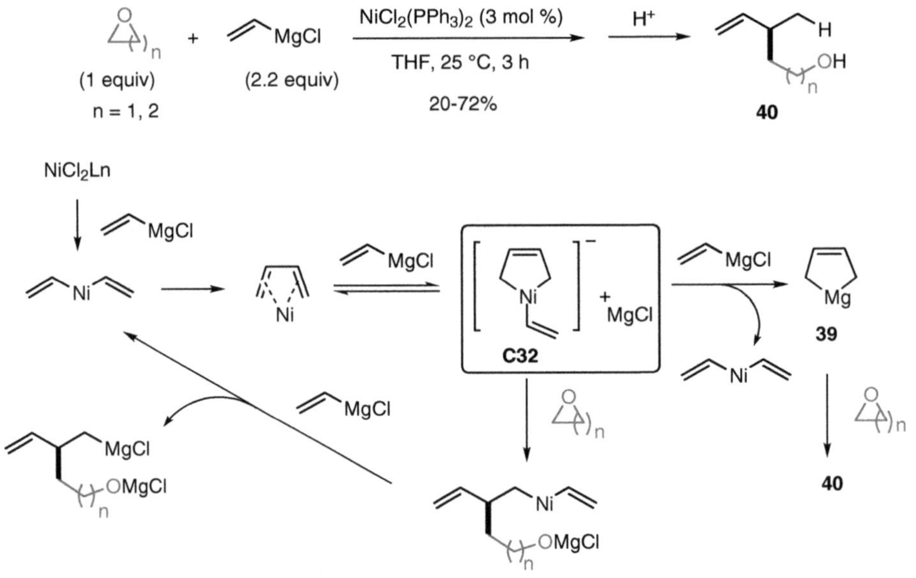

Fig. 2.64: Catalyzed dimerization of vinyl Grignard with 3- or 4-membered cyclic ethers.

2.4 Coupling reactions of Alkynyl Grignard Reagents

2.4.1 With alkyl halides

The functionalization of terminal alkynes is usually achieved by well-established and high yielding Sonogashira reaction conditions. In order to avoid the use of expensive palladium salts, the development of reactions between alkynyl Grignard reagents and primary alkyl halides are still challenging, especially due to the lower nucleophilic character of metal alkynylides and their propensity to dimerize. Very recently, successful processes have been published (Fig. 2.65). For example coupling between acetylenic Grignard reagents and primary alkyl iodides or bromides have been reported in the presence of Nickamine **C33** and diamine *O*-TMEDA as the additive [103]. From a mechanistic point a view, this additive plays an important role: it could coordinate to the magnesium salt favouring the transmetallation and the formation of an alkynyl complex with Nickamine, but it could also minimize the homocoupling side-reaction. Moreover, radical intermediates are probably involved as suggested by the ring-opening observed with cyclopropylmethyl bromide used in this reaction as the cross-coupling partner.

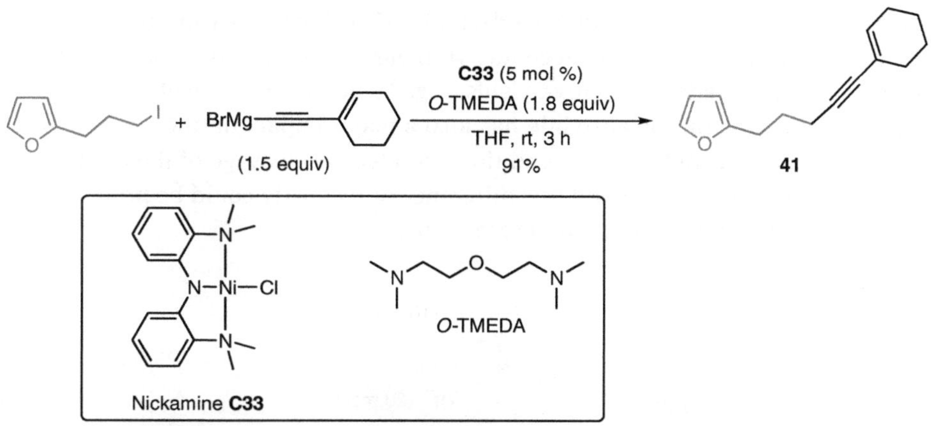

Fig. 2.65: Disubstituted alkynes from alkynyl Grignard reagents using Nickamine complex **C33**.

Interestingly, the reaction is compatible with sulfur-containing groups. Attempts to perform the same process with secondary halides are however unsuccessful.

2.4.2 With alkynyl sulfones

Diyne derivatives have also been prepared at room temperature by condensation of readily available alkynyl Grignard reagents onto alkynyl sulfones (Fig. 2.66). The process is an extension of the Ni-catalyzed cross-coupling of aryl sulfones with Grignard reagents previously discovered by Julia *et al.* [65], in which the catalyst Ni(acac)$_2$ was reduced *in situ* to Ni(0) by the acetylide present in excess. The yields are in the range from 42% to 83%. The best results were observed for the formation of bis-aryldiynes (R = Ar) [104].

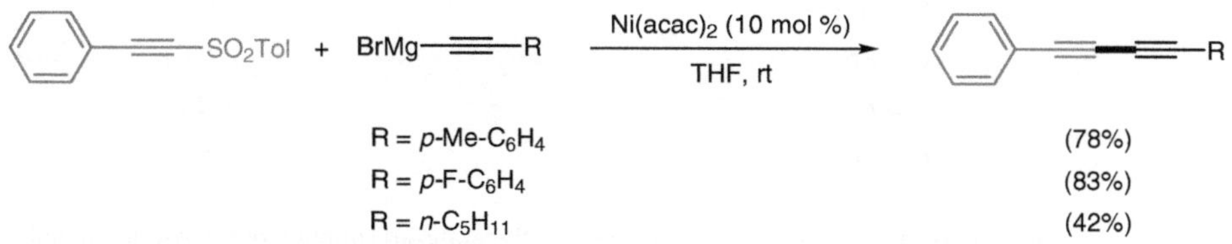

Fig. 2.66: 1,3-Diyne derivatives from alkynyl sulfones.

2.4.3 With ethers

2.4.3.1 Aryl ethers

The functionalization of terminal alkynes has also been extended to anisole derivatives [105]. In the presence of Ni(cod)$_2$ and a NHC ligand bearing bulky groups, the reaction of tri*iso*propylsilylacetylide with electron-enriched anisoles delivered the cross-coupling adducts in high yields (Fig. 2.67). The presence of sterically hindered substituents on the acetylide has also a major impact on the success of the process (Fig. 2.68). Under these conditions, a selective cleavage of the C-F bond of fluoroanisole was observed and this difference of reactivity could be promising for multi- and sequential cross-coupling reactions.

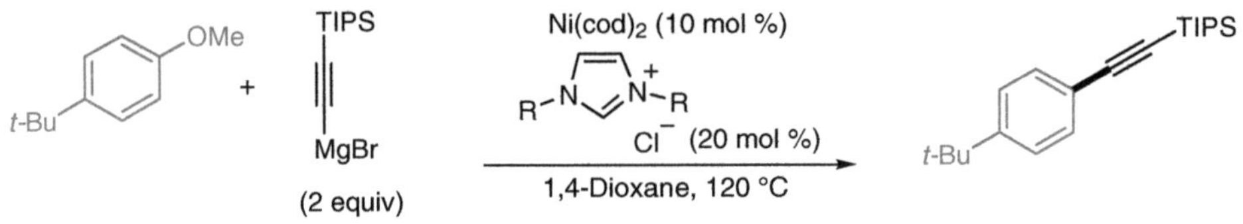

	GC Yield
R = Me	(7%)
R = *i*Pr	(38%)
R = *c*-Hex	(76%)
R = *t*-Bu	(45%)

Fig. 2.67: Influence of the NHC ligand on the coupling of TIPS-alkynyl-MgBr with *p*-*tert*-butylanisole.

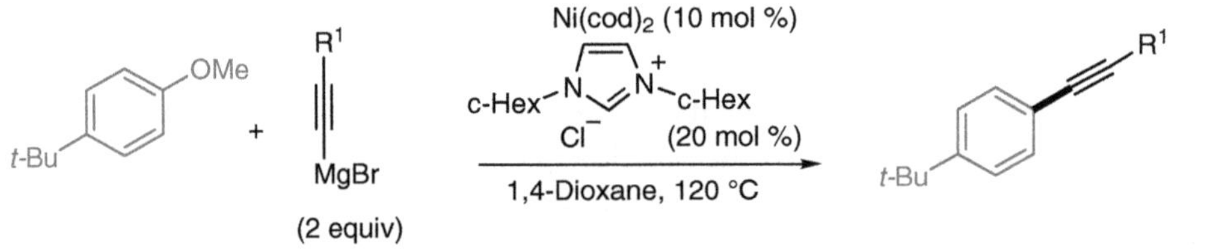

	GC Yield
R^1 = *n*-Hex	(0%)
R^1 = Ph	(0%)
R^1 = *t*-Bu	(22%)
R^1 = TES	(28%)

Fig. 2.68: Influence of the alkynyl nucleophile on the NHC-Ni-catalyzed coupling of p-*tert*-butylanisole.

This coupling reaction found application for the ethynylation of a large number of methoxyaryl derivatives and tolerate numerous functionalities (alcohol, phenol, amine, etc.) without significant decrease of the yields.

2.4.3.2 Enol derivatives

Enol ethers

Silyl enol ether **42** underwent a smooth coupling with an alkynyl Grignard reagent leading to enyne **43**, while dihydrofuran **44** underwent a ring-opening leading to compound **45** (Fig. 2.69) [105].

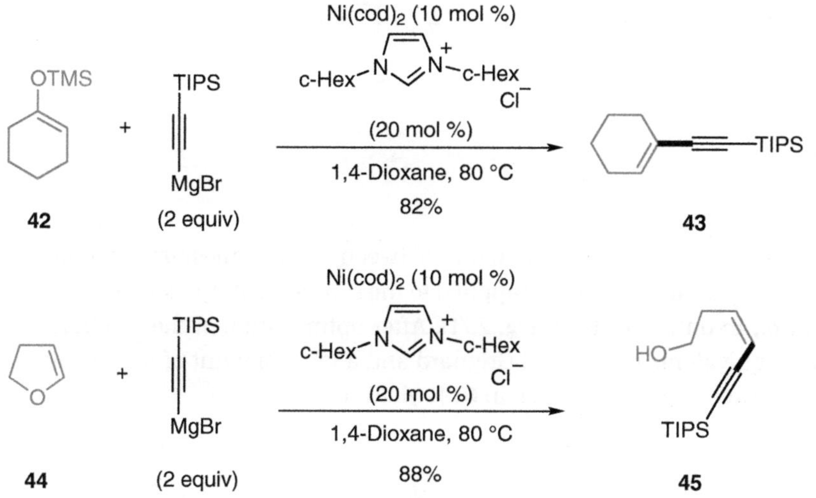

Fig. 2.69: 1,3-Enynes by Ni-catalyzed alkynylation of TMS enol ethers and dihydrofurans.

Vinyl carbamates

We have to point out that *(Z)*-vinylcarbamates **46** were also implicated in reaction with alkynyl Grignard reagents to deliver enynes **47** which could find application in the total synthesis of complex molecules such as avermectins [106]. The *Z/E* ratio was affected by both the nature of the catalyst but also by the substitution of the oxygen atom present at the homoallylic position (Fig. 2.70).

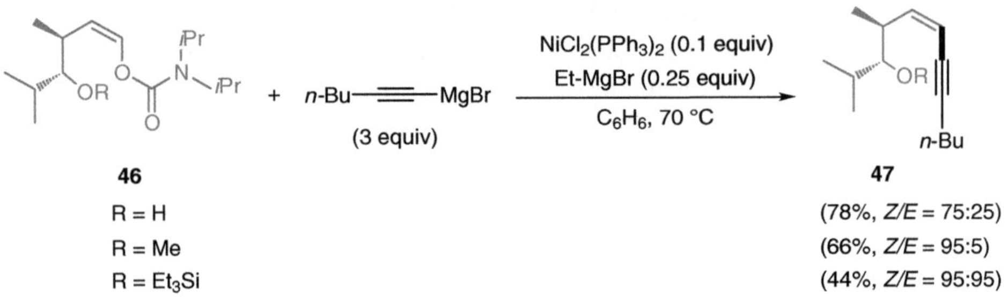

Fig. 2.70: *(Z)*-Enynes from *(Z)*-vinyl carbamates and hexynyl magnesium bromide.

2.5 Coupling reactions of Alkyl Grignard Reagents

2.5.1 With aryl derivatives

2.5.1.1 Aryl halides

As an example, the Kumada-Corriu reaction between alkylmagnesium reagents and aryl halides has been advantageously applied to the synthesis of *(S)*-macrostomine **48** by Comins *et al.* in only five steps (Fig. 2.71). After optimization of the reaction conditions, three equivalents of a benzyl Grignard and a large amount of nickel catalyst were required to achieve the final step in 63 % yield [107].

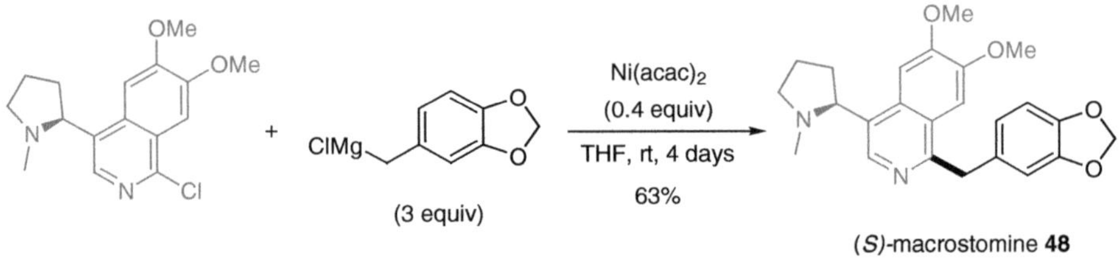

Fig. 2.71: Key Kumada cross-coupling of a benzyl Grignard fragment in the synthesis of *(S)*-macrostomine.

Similarly, as reported for aryl Grignard reagents in Section 2.2, the coupling between aryl fluorides and alkyl magnesium reagents can be also efficiently achieved. As depicted in Fig. 2.72, the reaction can be performed in the presence of catalyst **C34**, with alkyl Grignard without observation of β-hydride elimination. It is anticipated that magnesium phenolate could be the key element for the success of the reaction. Moreover, cleavage of C-F bond, under these conditions, was observed in the presence of weaker carbon-halogen bonds [108].

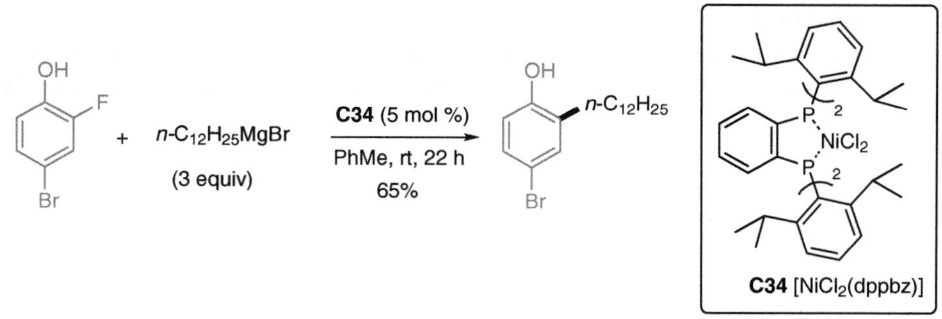

Fig. 2.72: Selective coupling of aryl fluorides with alkyl-MgX using catalyst **C34**.

2.5.1.2 Aryl sulfides

As already pointed out with aryl Grignard, efficient coupling reactions were noticed between methyl Grignard and aryl thiols, thioethers, sulfoxides or sulfones when performed in the presence of 10 mol % of $NiCl_2(PPh_3)_2$ (Fig. 2.73) [61].

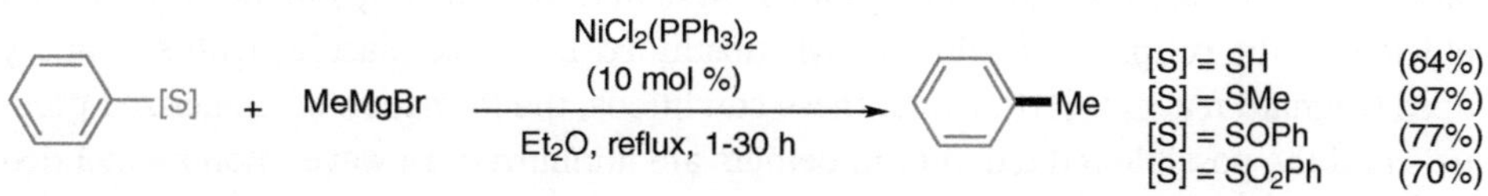

Fig. 2.73: Coupling of methyl magnesium bromide with different thioaryl derivatives.

In the presence of this additive, the reductive cleavage of $C(sp_2)$-sulfide bonds by secondary alkyl Grignard reagents has been reported. Fortunately, the use of bidentate phosphine ligands allowed a higher selectivity for the cross-coupling product [109]. For example, the synthesis of 4-cyclohexyl-2,6-diphenylpyridine *via* the coupling of cyclohexylmagnesium bromide and the corresponding 4-thiomethyl derivative occurred when $NiCl_2$(dppp) was present in the reaction media (Fig. 2.74) [63].

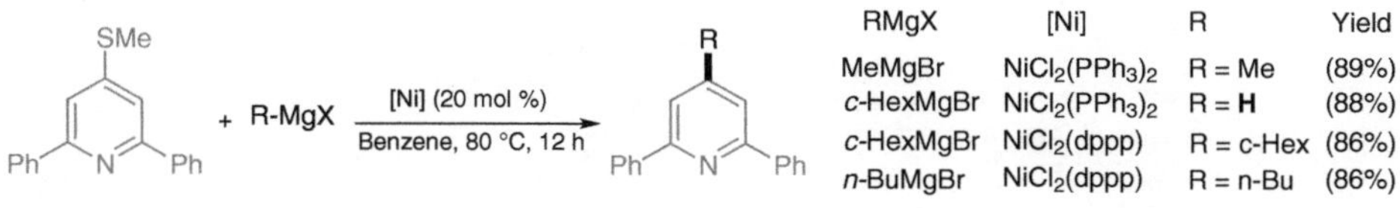

Fig. 2.74: Coupling of aryl thioethers with primary and secondary alkyl magnesium reagents.

A combination of $Ni(acac)_2$ and *(Z)*-1,2-diphosphinoalkene **L23** was also described for the cross-coupling of arylsulfides with primary and secondary alkyl Grignard reagents (Fig. 2.75) [110]. Nevertheless, the coupling reactions with *i*PrMgX led to partial isomerization of the alkyl nucleophile.

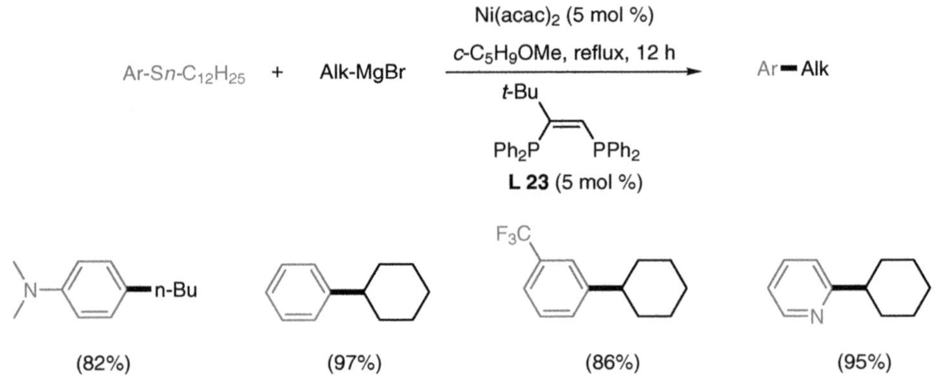

Fig. 2.75: Coupling of aryl thioethers with $Ni(acac)_2$ and (*Z*)-1,2-diphosphinoalkene **L23** as an additive.

More recently, the $NiCl_2$/butadiene catalytic system developped by Kambe *et al.* proved efficient in the coupling of 2-thiomethyl-substituted 1,3-benzothiazoles with secondary alkyl Grignard reagents [111]. Under these conditions, the formation of isomerized products and the possible reduction of thioethers are minimized. However, non π-extended systems such as thiazoles and thiazolines are not reactive. In the case of arylsulfonates, alkylmagnesium reagents bearing no β-hydrogen reacted efficiently at room temperature, when the more electron-rich NHC ligand $IPr.HBF_4$ was used as the ligand (Fig. 2.76) [112].

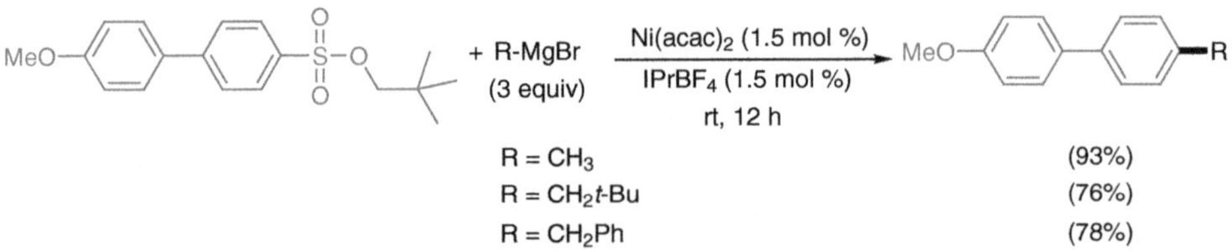

Fig. 2.76: Coupling of aryl sulfonates with alkyl Grignard reagents using the electron-rich NHC ligand.

2.5.1.3 Aryl methyl ethers

In 2008, Shi *et al.* reported the methylation of different alkoxy naphthalenes including alkyl-, aryl- and silyl ethers by using 5 mol% of $NiCl_2(PCy_3)_2$ in toluene [113]. Comparative experiments showed that π-extended aryl methyl ethers reacted faster than phenyl methyl ethers (Fig. 2.77).

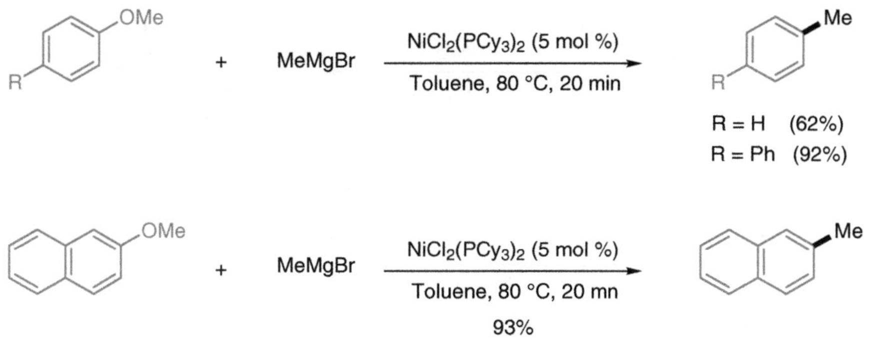

Fig. 2.77: Ni-catalyzed methylation of aryl or naphthyl methyl ethers with MeMgBr.

2.5.1.4 Aryl phosphonates

Functionalization of **49** by using a similar coupling allowed the introduction of a prenylated chain onto 1,5-naphthalenediol derivatives (Fig. 2.78) [114]. The resulting compound was transformed in few steps into (±)-cryptotanshinone **51**.

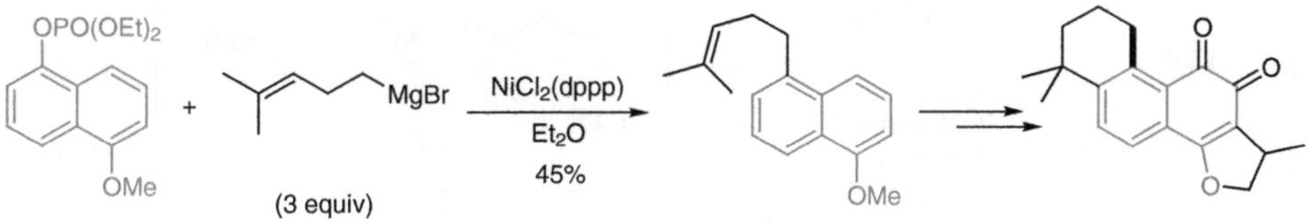

49 **50** (+)-cryptotanshinone, **51**

Fig. 2.78: Selective coupling of aryl phosphate with an alkyl Grignard in the synthesis of **51**.

2.5.2 With alkenyl derivatives

2.5.2.1 Alkenyl halides

The access to chiral 2-aryl propanoic acids is of great interest due to their pharmaceutical properties. In this context, the synthesis of precursors such as 3-aryl-1-butenes, has been intensively studied. Enantioselective Kumada-Tamao-Corriu reaction between benzyl Grignard reagents and vinyl bromide has been carried out in the presence of various chiral phosphine ligands [115–119]. Enantiomeric excesses up to 88 % were recorded with the readily available ligands **L27** and **L28** (Fig. 2.79) [118–119]. However, the process cannot actually compete in terms of selectivities with other procedures to access α-aryl propionic acids as hydrogenation or hydroformylation of parent styrenes are extremely efficient.

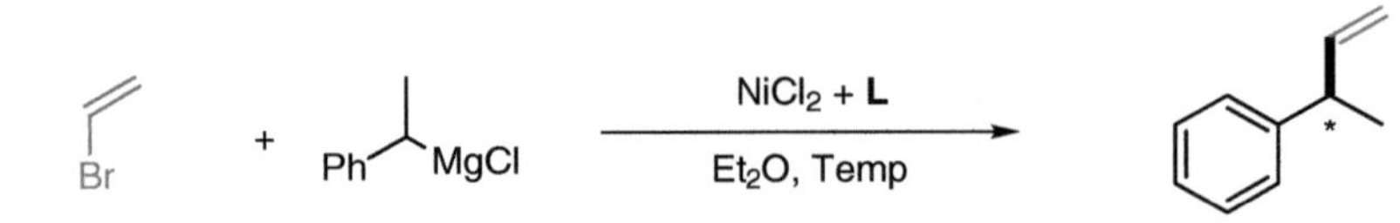

Ligand	$NiCl_2$/**L**	Temp	Yield	ee	Configuration
L24	1:1	0 °C	40%	26%	(*R*)
L25a	1:1 (0.5 mol%)	0 °C	55%	9%	(*R*)
L25b	1:1 (0.5 mol%)	0 °C	67%	46%	(*R*)
L26	1:2.2 (3 mol %)	-78 °C	60%	68%	(*R*)
L27	1:1 (0.4 mol%)	-50 °C	90%	88%	(*R*)
L28	1:1 (0.5 mol %)	0 °C	50%	85%	(*R*)

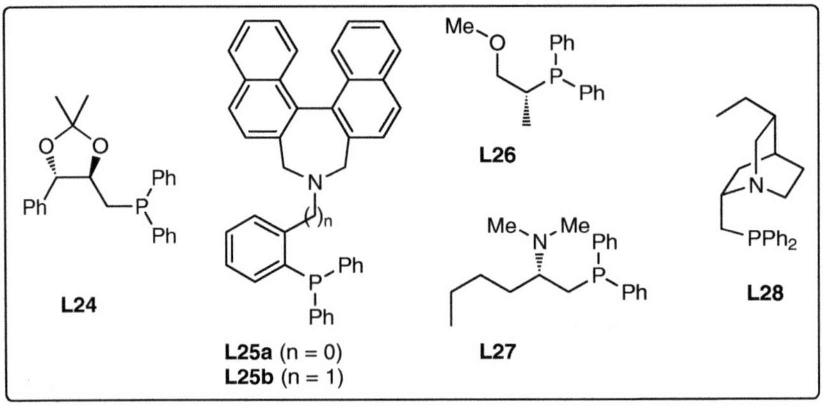

Fig. 2.79: Enantioselective coupling of secondary benzyl Grignard reagents with vinyl bromide.

The coupling of *gem*-difluoroalkenes has also been considered (Fig. 2.80). The Ni-catalyzed reaction was attractive with alkyl Grignard reagents compared to the process performed in the presence of palladium salts for which lower yields were recorded, probably due to side processes such as β-elimination [120].

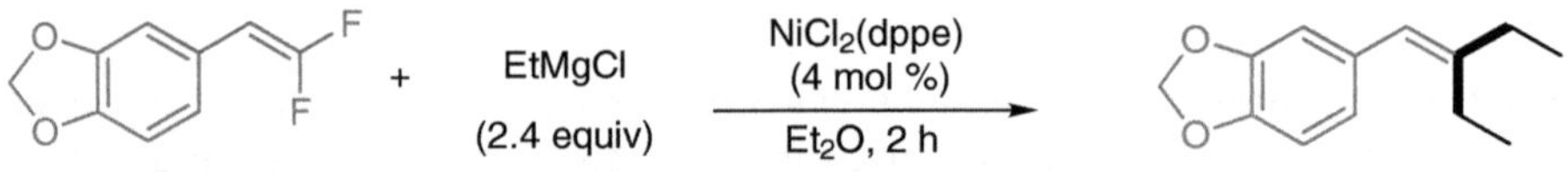

52 84% 53

Fig. 2.80: Trisubstituted alkenes from *gem*-difluoroalkenes and ethyl magnesium chloride.

2.5.2.2 Enol carbamates and phosphates

The reaction of enol carbamates **54** with alkyl Grignard reagents initially reported by Kocienski *et al.* occurred with retention of configuration (Fig. 2.81) [121].

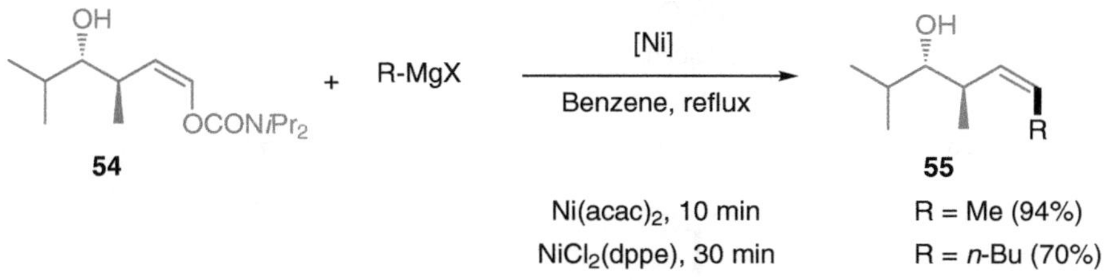

Fig. 2.81: Coupling of (*Z*)-enol carbamates and alkyl Grignard nucleophiles.

Kumada-Corriu reactions involving the cleavage of a C–O bond has been advantageously achieved to prepare different natural products. For example, phosphate enol ether **56** was easily generated from the corresponding bicyclic ketone. When this phosphate enol ether was involved in a nickel-catalyzed coupling with methyl Grignard reagent, it was transformed to the corresponding alkene in 73 % yield and was further transformed into the marine diterpene fuscol (Fig. 2.82) [122].

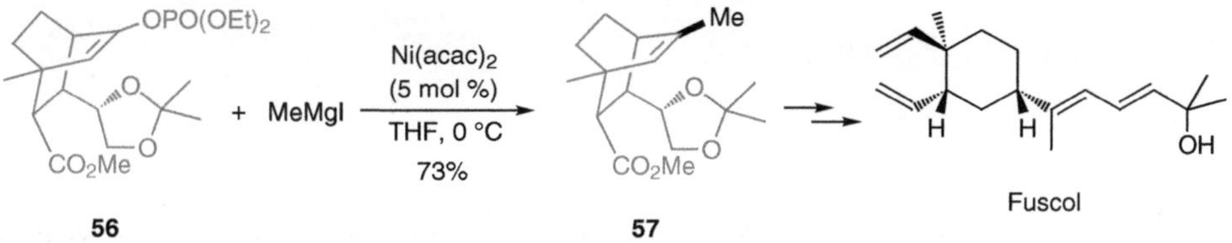

Fig. 2.82: Synthesis of fuscol involving the coupling of intermediate enol phosphate **56**.

A key intermediate in the total synthesis of Δ^1-*trans*-tetrahydrocannabinol **58** was also prepared in excellent yield by a 1,4-addition/regioselective phosphorylation/ Kumada-Corriu coupling sequence (Fig. 2.83) [123].

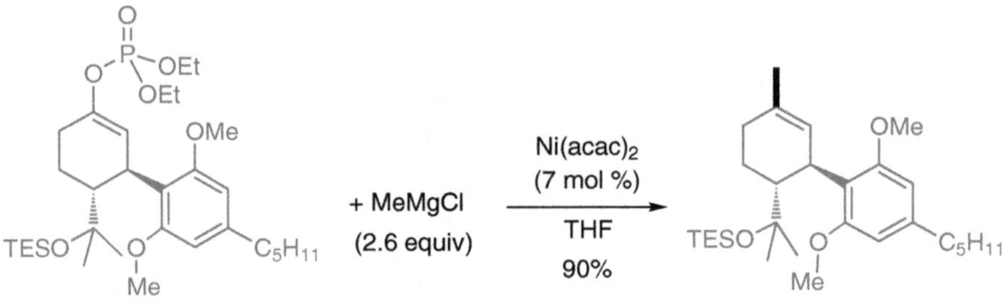

56 **57**

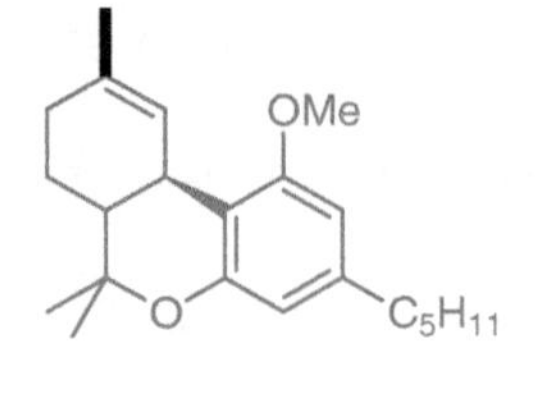

Δ^1-*trans*-tetrahydrocannabinol, **58**

Fig. 2.83: Methylation of enol phosphates **56** in the total synthesis of **58**.

2.5.3 With alkyl derivatives

2.5.3.1 Alkyl halides

The major challenge in the alkyl–alkyl Kumada cross-coupling reactions is to suppress the β-hydride elimination step that may occur from the intermediate alkyl–nickel complex, leading to olefins as the side products. Therefore, the reductive elimination step has to occur rapidly to avoid this competing process. The first Ni-catalyzed alkyl–alkyl cross-coupling between primary alkyl halides (Br, Cl) and tosylates and primary or secondary Grignard reagents was reported by Kambe *et al.* in 2002 using $NiCl_2$/1,3-butadiene as a unique catalytic system [4, 124]. Interestingly, the Csp$_3$-Br bond reacted faster than the Csp$_2$-Br bond yielding *p*-bromohexylbenzene in quantitative GC yield even if Csp$_3$-X bonds are more electron-rich than Csp$_2$-X bonds and thus more reluctant to oxidative addition. Moreover, (bromomethyl)cyclopropane led to the corresponding alkylcyclopropane in 87 % yield which seems to preclude the formation of an alkyl radical as intermediate (Fig. 2.84).

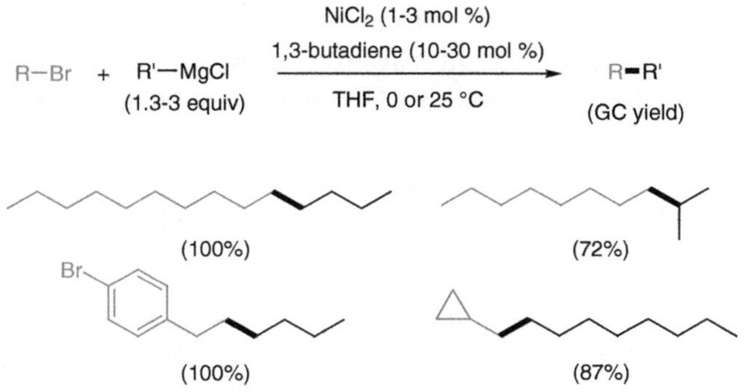

Fig. 2.84: Alkyl–alkyl coupling in the presence of $Ni(acac)_2$ and 1,3-butadiene.

Based on these results, a non-radical mechanism involving Ni(II)/Ni(IV) species in the catalytic cycle was proposed by Kambe *et al.* (Fig. 2.85). The 1,3-butadiene additive converts Ni(0) to a bis-π-allyl nickel(II) complex **E** which reacts first with R'-MgX to form an anionic complex **F** with enhanced nucleophilicity as a key intermediate. The oxidative addition of alkyl halides to complex **F** then yields Ni(IV) complex **G** which undergoes reductive elimination to the coupled product, regenerating the bis-π-allyl nickel complex **E**. The activation of the alkyl halide may also be due to a strong interaction with the magnesium salt.

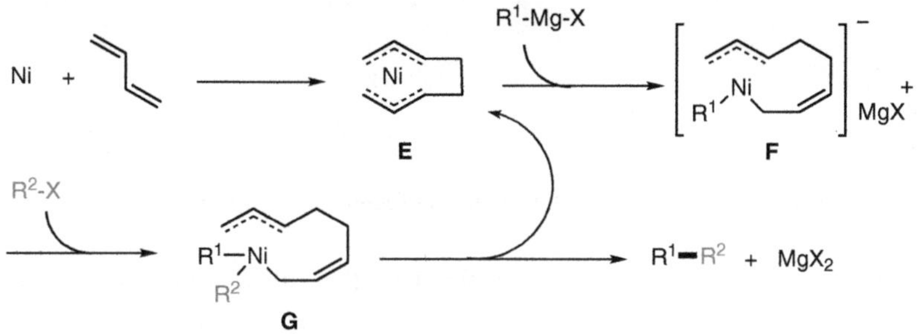

Fig. 2.85: Postulated mechanism for the alkyl–alkyl coupling using Ni/1,3-butadiene system.

This catalytic system also proved to be efficient for the coupling of primary alkyl fluorides albeit in lower yields due to the formation of elimination products [125]. In this case, $CuCl_2$ was found to be superior to $NiCl_2$ for the coupling. It was later shown that the combination of $Ni(acac)_2$ (0.6 mol %) and a 1,3,8,10-tetraene as an additive improved the efficiency of the coupling of alkyl fluorides [126]. Functionalized alkyl halides and tosylates could also be coupled by using these conditions [127].

The use of 3 mol % of a pincer (NN_2)-Ni catalyst **C33** synthesized from an amido-bis(amine) tridentate ligand was described by Hu *et al.* for the coupling of functionalized primary and secondary alkyl iodides and bromides with alkylmagnesium nucleophiles [128–129]. At low temperature (–35 °C) and, using dimethylacetamide as solvent, a slow addition of an alkyl Grignard reagent (1 equiv) on a large range of unactivated alkyl halides yielded the coupled product in moderate to high yields (Fig. 2.86). Notably, alkyl triflates and alkyl tosylates were inert. A structure-activity study demonstrated that bidentate nickel catalysts **C35** and **C36** gave better yields for the coupling of secondary alkyl halides, which was attributed to a higher structural flexibility around the nickel center [130].

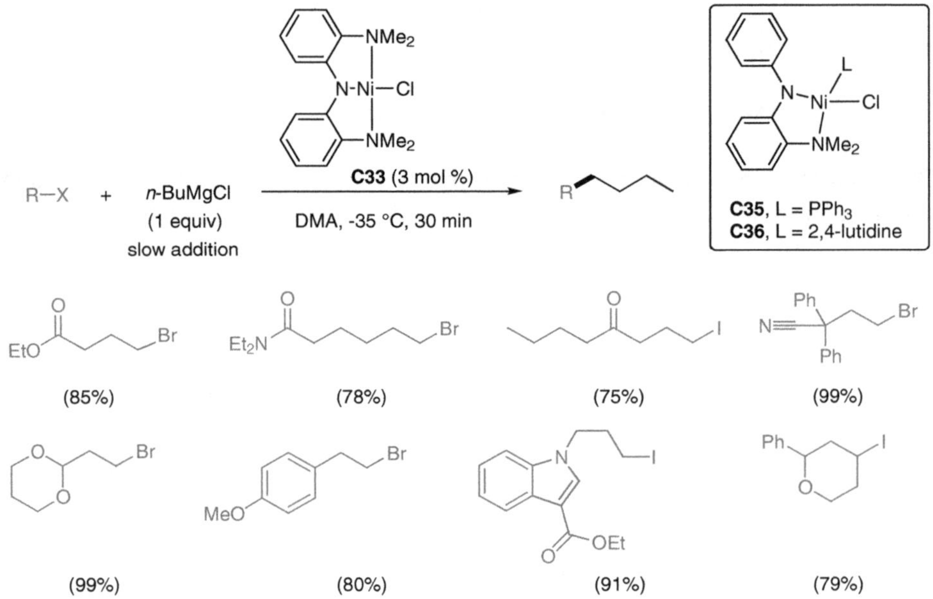

Fig. 2.86: Alkyl–alkyl Kumada-Corriu coupling by using pincer **C33**-nickel complex.

Under these catalytic conditions, a diastereoselective alkyl–alkyl Kumada coupling was also reported by Hu *et al.* for the formation of the conformationally most stable 1,4-*trans*-dialkyl cyclohexanes and 1,3-*cis*-dialkyl cyclohexanes and tetrahydropyrans, whatever the diastereomeric ratio of the corresponding cyclic halide substrates [131]. The alkyl–alkyl cross-coupling was believed to occur *via* the transmetallation of **C33** by RMgX in a first step to form a Ni–alkyl intermediate, followed by oxidative addition of the alkyl halide and reductive elimination step to produce the coupled product. An extensive mechanistic investigation revealed that the key intermediate for the activation of the alkyl halide was a bimetallic [(N_2N)Ni-R](RMgCl) complex and that the oxidative addition step involved radical alkyl intermediates [132].

More recently, the group of Cardenas reported the use of tetramethylethylenediamine (TMEDA) as additive in the Ni(acac)$_2$-catalyzed coupling of non-activated primary and secondary iodides with alkylmagnesium reagents [133]. On the basis of experimental and computational studies evidencing the intermediacy of an alkyl radical from the alkyl iodide, a cascade cyclization-coupling was applied to a variety of iodoalkene substrates for the sequential formation of two alkyl–alkyl bonds (Fig. 2.87).

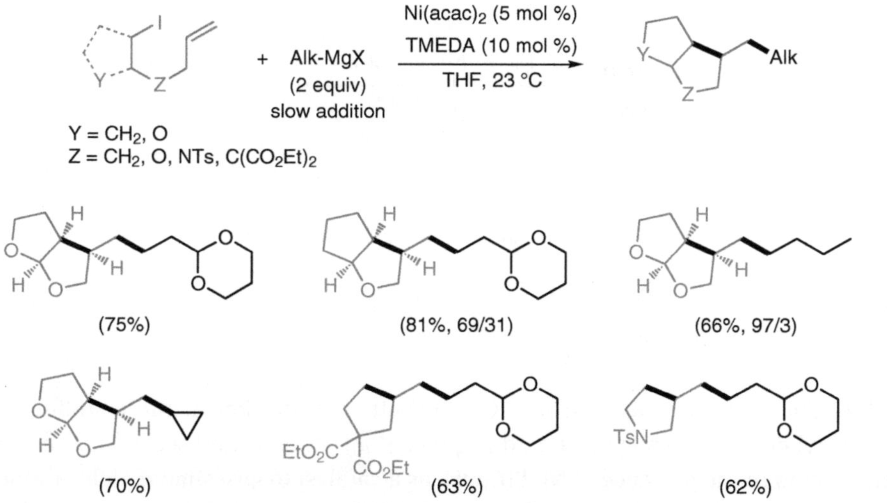

Fig. 2.87: Ni-catalyzed sequential radical cyclization Kumada-Tamao-Corriu coupling of iodoalkenes.

The catalytic system was also efficient in the cross-coupling of functionalized benzyl chlorides with alkyl Grignard reagents [134].

2.5.3.2 Sulfones

Li and co-workers reported the Kumada coupling of activated alkyl sulfones with Grignard reagents with the $NiI_2(PPh_3)_2/PCy_3$ system (Fig. 2.88) [135]. A range of aryl ketones were synthesized in good yields (61–77 %) from 1-aryl-2-tosylethanones without affecting the carbonyl group. Interestingly, the outcome of the coupling of 1-(1-tosylethyl)arenes was completely reversed by changing the solvent: the coupled product **62** was obtained in toluene or cyclohexane whereas the use of THF yielded the methylstyrene derivatives **64**.

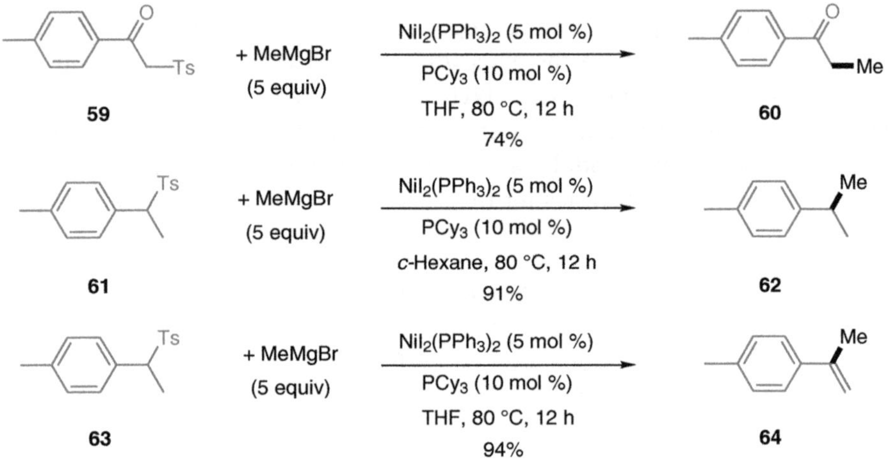

Fig. 2.88: Ni(II)-catalyzed reactions of ketosulfones and sulfones with methyl Grignard reagent.

2.5.3.3 Alkyl ethers

In 2008, Shi *et al.* reported the first C(sp$_3$)-O(ether) activation in the coupling of benzyl alkyl ethers with alkyl Grignard reagents [136]. Primary and secondary benzyl ethers reacted in the presence of $NiCl_2(dppf)_2$ as a catalyst to give similar yields in the cross-coupling products (Fig. 2.89). The use of *i*PrMgX and *n*-BuMgBr led to lower yields in the coupled products, probably due to a competitive α-β-H elimination pathway.

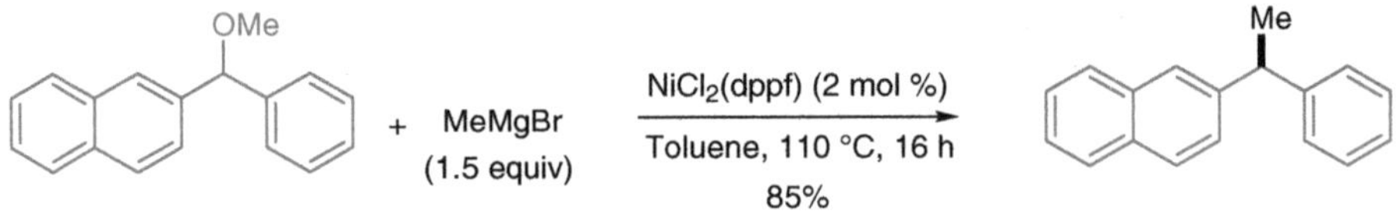

Fig. 2.89: Methylation of benzyl methyl ether using $NiCl_2(dppf)_2$ as catalyst.

Cross-coupling of non racemic bis-aryl ethers **65** with methyl Grignard reagent has been applied to prepare enantio-enriched diarylethanes **66**, precursors of anti-insomnia drugs (Fig. 2.90) [137–138] It is worth noting that the transfer of chirality is sizeable.

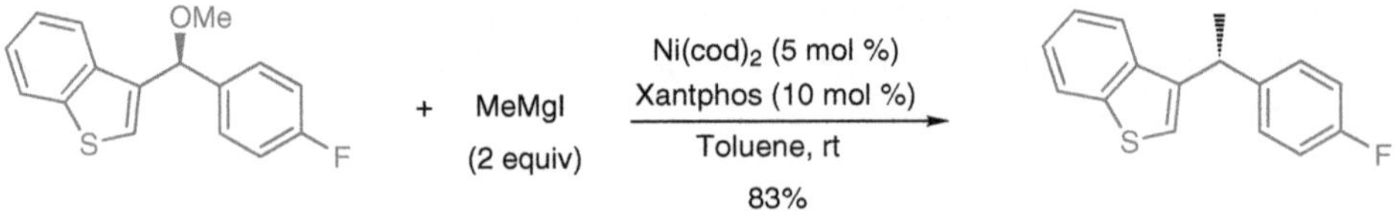

65 (ee = 90%)
66 (ee = 85%)

Fig. 2.90: Diastereoselective Ni-catalyzed methylation of chiral bis-aryl ether **65**.

A similar coupling has been also realized starting from both racemic chiral cyclic ethers **67** and **69**. Introduction of a methyl group occured in a stereospecific manner with a complete inversion of configuration at the electrophilic center (Fig. 2.91). It is noteworthy that the same authors reported the ring-opening of the corresponding lactones with organozinc reagents (Negishi coupling) which also occured nicely and with very high stereoselectivities [139]. The reactions were conducted in the presence of DPEphos **L29** and of *(R)*- or *(S)*-BINAP, **L30**. No match/mismatch effects were observed in the later case demonstrating that the chirality of the ligand has no influence on the stereoselectivity of the reaction.

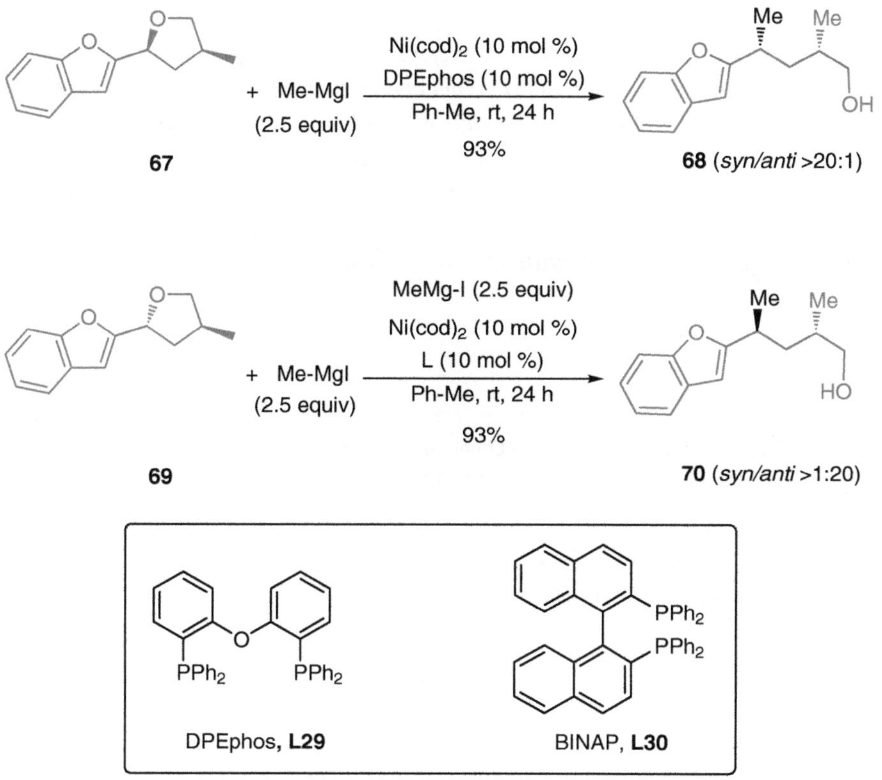

Fig. 2.91: Diastereoselective ring-opening of chiral cyclic benzyl ethers by Kumada-Corriu reaction.

2.5.3.4 Benzyl alcohols

The cross-coupling of benzylic C(sp$_3$)-OH bond was also successfully achieved with methyl Grignard reagent (Fig. 2.92) [140]. Starting from naphthylmethanol, 2-ethylnaphthalene was obtained in 68% yield by using bidentate 1, 2-bis(dicylohexylphosphino)ethane (DCyPE) **L31** as ligand. Unfortunately, other alkyl Grignard reagents led mainly to reductive products.

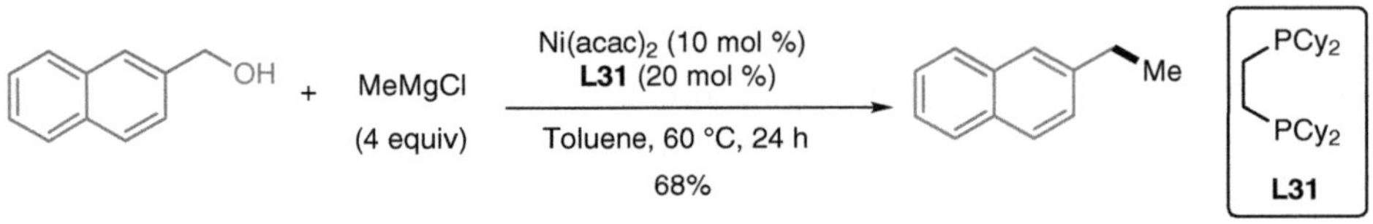

Fig. 2.92: Methylation of benzyl alcohols with $Ni(acac)_2$ and bidentate phosphine **L31**.

2.6 Conclusion

In parallel to the well-established reactivity of Grignard reagents towards electrophiles, a complementary set of reactions has been designed when these organomagnesium derivatives are associated with appropriate nickel catalysts. Due to the lower cost of nickel salts, the coupling reactions depicted in this Chapter seems of particular interest when compared to procedures using palladium salts. The major drawbacks are still the use of substoechiometric amounts of Grignard reagents and the β-hydride elimination frequently observed with alkyl derivatives.

References

[1] Tamao K, Sumitami M, Kumada M, J Am Chem Soc 1972, 94, 4374–76.
[2] Corriu RJP, Masse J, J Chem Soc, Chem Commun 1972, 144.
[3] Hölzer B, Hoffmann RW, Chem Commun 2003, 732–3.
[4] Terao J, Kambe N, Acc Chem Res 2008, 41, 1545–54.
[5] Tamao K, J Organomet Chem 2002, 653, 23–6.
[6] Kiso Y, Tamao K, Kumada M, J Organomet Chem 1973, 50, C12–C14.
[7] Picquet M, Platinum Metals Rev 2013, 57, 272–80.
[8] Magano J, Dunetz JR, Chem Rev 2011, 111, 2177–250.
[9] Manley PW, Acemoglu M, Marterer W, Pachinger W, Org Process Res Dev 2003, 7, 436–45.
[10] Bold G, Fässler A, Capraro H-G, Cozens R, Klimkait T, Lazdins J, Mestan J, Poncioni B, Rösel J, Stover D, Tintelnot-Blomley M, Acemoglu F, Beck W, Boss E, Eschbach M, Hürlimann T, Masso E, Roussel S, Ucci-Stoll K, Wyss D, Lang M, J Med Chem 1998, 41, 3387–401.
[11] Fan X, Song Y-L, Long Y-Q, Org Process Res Dev 2008, 7, 69–75.
[12] Lau SYW, Hughes G, O'Shea PD, Davies IW, Org Lett 2007, 9, 2239–42.
[13] Ghosh R, Sarkar A, J Org Chem 2010, 75, 8283–6.
[14] Monnereau L, Semeril D, Matt D, Toupet L, Mota AJ, Adv Synth Catal 2009, 351, 1383–9.
[15] Monnereau L, Semeril D, Matt D, Chem Commun 2011, 47, 6626–8.
[16] Monnereau L, Semeril D, Matt D, Adv Synth Catal 2013, 355, 1351–60.
[17] Bohm VPW, Weskamp T, Gstöttmayr CWK, Herrmann WA, Angew Chem Int Ed 2000, 39, 1602–4.
[18] Li GY, Marshall WJ, Organometallics 2002, 21, 590–1.
[19] Li GY, Angew Chem Int Ed 2001, 40, 1513–1516
[20] Xie L-G, Wang Z-X, Chem Eur J 2010, 16, 10332–6.
[21] Ackermann L, Born R, Spatz JH, Meyer D, Angew Chem Int Ed 2005, 44, 7216–9.

[22] Ackermann L, Wechsler C, Kapdi AR, Althammer A, Synlett 2010, 294–8.
[23] Hayashi T, Hayashizaki K, Kiyoi T, Ito Y, J Am Chem Soc 1988, 110, 8153–6.
[24] Chen M-T, Lee W-Y, Tsai T-L, Liang L-C, Organometallics 2014, 33, 5852–62.
[25] Schiefer D, Wen T, Wang Y, Goursot P, Komber H, Hanselmann R, Braunstein P, Reiter G, Sommer M, ACS Macro Lett 2014, 3, 617–21.
[26] Liang L-C, Chien P-S, Lin J-M, Huang M-H, Huang Y-L, Liao J-H, Organometallics 2006, 25, 1399–411.
[27] Venkanna GT, Tammineni S, Arman HD, Tonzetich ZJ, Organometallics 2013, 32, 4656–63.
[28] Sanford J, Dent C, Masuda JD, Xia A, Polyhedron 2011, 30, 1091–4.
[29] Sun Y, Li X, Sun H, Inorg Chim Acta 2014, 415, 95–7.
[30] Sun K, Wang L, Wang Z-X, Organometallics 2008, 27, 5649–56.
[31] Liang L-C, Lee W-Y, Hung Y-T, Hsiao Y-C, Cheng L-C, Chen W-C, Dalton Trans 2012, 41, 1381–8.
[32] Wang Z-X, Wang L, Chem Commun 2007, 2423–5.
[33] Zhang X-Q, Wang Z-X, Synlett 2013, 24, 2081–4.
[34] Liu N, Wang Z-X, J Org Chem 2011, 76, 10031–8.
[35] Hatakeyama T, Hashimoto S, Ishizuka K, Nakamura M, J Am Chem Soc 2009, 131, 11949–63.
[36] Tennyson AG, Lynch VM, Bielawski CW, J Am Chem Soc 2010, 132, 9420–9.
[37] Jothibasu R, Huang K-W, Huynh HV, Organometallics 2010, 29, 3746–52.
[38] Soulé J-F, Miyamura H, Kobayashi S, J Am Chem Soc 2013, 135, 10602–5.
[39] Wolf J, Labande A, Daran J-C, Poli R, J Organomet Chem 2006, 691, 433–43.
[40] Wolf J, Labande A, Natella M, Daran J-C, Poli R, J Mol Catal A: Chem 2006, 259, 205–12.
[41] Page MJ, Lu WY, Poulten RC, Carter E, Algarra AG, Kariuki BM, MacGregor SA, Mahon MF, Cavell KJ, Murphy DM, Whittlesey MK, Chem Eur J 2013, 19, 2158–67.
[42] Mitsudo K, Doi Y, Sakamoto S, Murakami H, Mandai H, Suga S, Chem Lett 2011, 40, 936–8.
[43] Zhang C, Wang Z-X, Organometallics 2009, 28, 6507–14.
[44] Guo W-J, Wang Z-X, J Org Chem 2013, 78, 1054–61.
[45] Sun Y, Li X, Sun H, Dalton Trans 2014, 43, 9410–3.
[46] Guo J, Lv L, Wang X, Cao C, Pang G, Shi Y, Inorg Chem Commun 2013, 31, 74–8.
[47] Berding J, Lutz M, Spek AL, Bouwman E, Organometallics 2009, 28, 1845–54.
[48] Vinh Huynh H, Jothibasu R, Eur J Inorg Chem 2009, 1926–31.
[49] Inamoto K, Kuroda J-i, Sakamoto T, Hiroya K, Synthesis 2007, 2853–61.
[50] Styring P, Grindon C, Fisher CM, Catal Lett 2001, 77, 219–25.
[51] Phan NTS, Brown DH, Adams H, Spey SE, Styring P, Dalton Trans 2004, 33, 1348–57.
[52] Chen X, Wang L, Liu J, Synthesis 2009, 2408–12.
[53] Lipshutz BH, Tomioka T, Blomgren PA, Sclafani JA, Inorg Chim Acta 1999, 296, 164–9.
[54] Tasler S, Lipshutz BH, J Org Chem 2003, 68, 1190–9.
[55] Frieman BA, Taft BR, Lee C-T, Butler T, Lipshutz BH, Synthesis 2005, 2989–93.
[56] Ackermann L, Born R, Spatz JH, Meyer D, Angew Chem, Int Ed 2005, 44, 7216–9.
[57] Wu D, Wang Z-X, Org Biomol Chem 2014, 12, 6414–24.
[58] Ahrens T, Kohlmann J, Ahrens M, Braun T, Chem Rev 2015, 115, 931–72.
[59] Saeki T, Takashima T, Tamao K, Synlett 2005, 1771–4.
[60] Guo W-J, Wang Z-X, J Org Chem 2013, 78, 1054–61.
[61] Wenkert E, Ferreira TW, Michelotti EL, J Chem Soc, Chem Commun 1979, 637–8.
[62] Dubbaka SR, Vogel P, Angew Chem Int Ed 2005, 44, 7674–84.
[63] Wenkert E, Hanna JM Jr, Leftin MH, Michelotti EL, Potts KT, Usifer D, J Org Chem 1985, 50, 1125–6.
[64] Murakami K, Yorimitsu H, Osuka A, Angew Chem Int Ed 2014, 53, 7510–3.
[65] Clayden J, Cooney JJA, Julia M, J Chem Soc, Perkin Trans 1, 1995, 7–14.
[66] Cho C-H, Yun H-S, Park K, J Org Chem 2003, 68, 3017–25.

[67] Milburn RR, Snieckus V, Angew Chem Int Ed 2004, 43, 888–91.
[68] Li W-N, Wang Z-L, RSC Advances 2013, 3, 25565–75.
[69] Wenkert E, Michelotti EL, Swindell CS, J Am Chem Soc 1979, 101, 2246–7.
[70] Dankwardt JW, Angew Chem Int Ed 2004, 43, 2428–32.
[71] Xie L-G, Wang Z-X, Chem Eur J 2011, 17, 4972–5.
[72] Iglesias MJ, Prieto A, Nicasio C, Org Lett 2012, 14, 4318–21.
[73] Hayashi T, Katsuro Y, Kumada H, Tetrahedron Lett 1980, 21, 3915–18.
[74] Zhao F, Yu D-G, Zhu R-Y, Xi Z, Shi Z-J, Chem Lett 2011, 40, 1001–3.
[75] Sengupta S, Leite M, Raslan DS, Quesnelle C, Snieckus V J Org Chem 1992, 57, 4066–8.
[76] Yoshikai N, Matsuda H, Nakamura E, J Am Chem Soc 2009, 131, 9590–9.
[77] Hayashi T, Fujiwa T, Okamoto Y, Katsuro Y, Kumada M, Synthesis 1981, 1001–3.
[78] Hayashi T, Katsuro Y, Okamoto Y, Kumada M, Tetrahedron Lett 1981, 22, 4449–52.
[79] Guan B-T, Lu X-Y, Zheng Y, Yu D-G, Wu T, Li K-L, Li B-J, Shi Z-J, Org Lett 2010, 12, 396–9.
[80] Macklin TK, Snieckus V, Org Lett 2005, 7, 2519–22.
[81] Yu D-G, Li B-J, Zheng S-F, Guan B-T, Wang B-Q, Shi Z-J, Angew Chem Int Ed 2010, 49, 4566–70.
[82] Banno T, Hayakawa Y, Umeno M J Organomet Chem 2002, 653, 288–291.
[83] Marlier EE, Tereniak SJ, Ding K, Milliken JE, Lu CC, Inorg Chem 2011, 50, 9290–9.
[84] Xi Z, Liu B, Chen W, J Org Chem 2008, 73, 3954–7.
[85] Pugh C, Baker J, Storms WK, Synlett 2014, 25, 148–52.
[86] Okumara H, Miura M, Takei H, Tetrahedron Lett 1979, 43–6.
[87] Cornella J, Martin R, Org Lett 2013, 6298–301.
[88] Li Q-H, Ding Y, Yang X-J, Chin Chem Lett 2014, 25, 1296–1300.
[89] Li Q, Gau H, Synlett 2012, 747–50.
[90] Lou S, Fu GC, J Am Chem Soc 2010, 132, 1264–6.
[91] Mao J, Liu F, Wang M, Wu L, Zheng B, Liu S, Zhong J, Bian Q, Walsh PJ, J Am Chem Soc 2014, 136, 17662–8.
[92] Terao J, Nii S, Chowdhury FA, Nakamura A, Kambe N, Adv Synth Catal 2004, 346, 905–8.
[93] Xue F, Zhao J, Hor TSA, Hayashi T, J Am Chem Soc 2015, 137, 3189–92.
[94] Zembayashi M, Tamao K, Kumada M, Tetrahedron Lett 1975, 21, 1719–22.
[95] Tamao K, Sumitani K, Kiso Y, Zembayashi M, Fujioka A, Kodama S, Nakajima I, Minato A, Kumada M, Bull Chem Soc Jpn 1976, 49, 1958–69.
[96] Bach T, Bartels M, Synthesis 2003, 925–939.
[97] Tiecco M, Testaferri L, Tingoli M, Wenkert E, Tetrahedron 1983, 39, 2289–94.
[98] Fabre J-L, Julia M, Verpeaux J-N, Bull Soc Chim Fr 1985, 762–71.
[99] Porée F-H, Clavel A, Betzer J-F, Pancrazi A, Ardisson J, Tetrahedron Lett 2003, 44, 7553–6.
[100] Terao J, Watabe H, Kambe N, J Am Chem Soc 2005, 127, 3656–7.
[101] Terao J, Tomita M, Singh SP, Kambe N, Angew Chem Int Ed 2010, 49, 144–7.
[102] Fujii Y, Terao J, Watabe H, Watanabe H, Kambe N, Tetrahedron 2007, 63, 6635–41.
[103] Vechorkin O, Godinat A, Scopelliti R, Hu X, Angew Chem Int Ed 2011, 50, 11777–81.
[104] Fang K, Xie M, Zhang Z, Ning P, Shu G, Tetrahedron Lett 2013, 54, 3819–21.
[105] Tobisu M, Takahira T, Ohtsuki A, Chatani N, Org Lett 2015, 17, 680–3.
[106] Madec D, Pujol S, Henryon V, Férézou JP, Synlett 1995, 435–8.
[107] Enamorado MF, Ondacchi PW, Comins DL, Org Lett 2010, 12, 4513–5.
[108] Wang J-R, Manabe K, Org Lett 2009, 11, 741–4.
[109] Luh T-Y, Ni Z-J, Synthesis 1990, 89–103.
[110] Kanemura S, Kondoh A, Yorimitsu H, Oshima K, Synthesis 2008, 2659–64.
[111] Ghaderi A, Iwasaki T, Fukuoka A, Terao J, Kambe N, Chem Eur J 2013, 19, 2951–5.
[112] Kim C-B, Jo H, Ahn B-K, Kim CK, Park K, J Org Chem 2009, 74, 9566–9.
[113] Guan B-T, Xiang S-K, Wu T, Sun Z-P, Wang B-Q, Zhao K-Q, Shi Z-J, Chem Comm 2008, 1437–9.

[114] Jiang Y-Y, Li Q, Lu W, Cai J-C, Tetrahedron Lett 2003, 44, 2073–5.
[115] Chelucci G, Cabras MA, Botteghi C, Marchetti M, Tetrahedron: Asymmetry 1994, 5, 299–302.
[116] Wimmer P, Widhaim M, Tetrahedron: Asymmetry 1995, 6, 657–60.
[117] Terfort A, Brunner H, J Chem Soc, Perkin Trans 1, 1996, 1467–79.
[118] Cross G, Vriesema BK, Boven G, Kellogg RM, van Bolhuis F, J. Organomet. Chem. 1989, 370, 357–81.
[119] Pellet-Rostaing S, Saluzzo C, Ter Halle R, Breuzard J, Vial L, Le Guyader F, Lemaire M, Tetrahedron: Asymmetry 2001, 12, 1983–5.
[120] Dai W, Xiao J, Jin G, Wu J, Cao S, J Org Chem 2014, 79, 10537–46.
[121] Kocienski P, Dixon NJ, Synlett 1989, 52–4.
[122] Iwashima M, Nagaoka H, Kobayashi K, Yamada Y, Tetrahedron Lett 1992, 33, 81–2.
[123] William AD, Kobayashi Y, Org Lett 2001, 3, 2017–20.
[124] Terao J, Watanabe H, Ikumi A, Kuniyasu H, Kambe N, J Am Chem Soc 2002, 124, 4222–3.
[125] Terao J, Ikumi A, Kuniyasu H, Kambe N, J Am Chem Soc 2003, 125, 5646–7.
[126] Terao J, Todo H, Watanabe H, Ikumi A, Kambe N, Angew Chem Int Ed 2004, 43, 6180–3.
[127] Singh SP, Terao J, Kambe N, Tetrahedron Lett 2009, 50, 5644–6.
[128] Vechorkin O, Csok Z, Scopelliti R, Hu X, Chem Eur J 2009, 15, 3889–99.
[129] Vechorkin O, Hu X, Angew Chem Int Ed 2009, 48, 2937–40.
[130] Ren P, Vechorkin O, von Allmen K, Scopelliti R, Hu X, J Am Chem Soc 2011, 133, 7084–95.
[131] Perez Garcia PM, Di Franco T, Orsino A, Ren P, Hu X, Org Lett 2012, 14, 4286–9.
[132] Breitenfeld J, Ruiz J, Wodrich MD, Hu X, J Am Chem Soc 2013, 135, 12004–12.
[133] Guisan-Ceinos M, Soler-Yanes R, Collado-Sanz D, Phapale VB, Bunuel E, Cardenas DJ, Chem Eur J 2013, 19, 8405–10.
[134] Soler-Yanes R, Guisan-Ceinos M, Bunuel E, Cardenas DJ, Eur J Org Chem 2014, 6625–9.
[135] Wu J-C, Gong L-B, Xia Y, Song R-J, Xie Y-X, Li J-H, Angew Chem Int Ed 2012, 51, 9909–13.
[136] Guan B-T, Xiang S-K, Wang B-Q, Sun Z-P, Wang Y, Zhao K-Q, Shi Z-J, J Am Chem Soc 2008, 130, 3268–9.
[137] Taylor BLH, Swift EC, Waetzig JD, Jarvo ER, J Am Chem Soc 2011, 133, 389–91.
[138] Taylor BLH, Jarvo ER, Synlett 2011, 2761–5.
[139] Tollefson EJ, Dawson DD, Osborne CA, Jarvo ER, J Am Chem Soc 2014, 136, 14951–8.
[140] Yu D-G, Wang X, Zhu R-Y, Luo S, Zhang X-B, Wang B-Q, Wang L, Shi Z-J, J Am Chem Soc 2012, 134, 14638–41.

Julien Legros, Bruno Figadère

3 Grignard Reagents and Iron

3.1 Introduction

Iron exhibits a decisive advantage over most of the other metals: it is truly inexpensive and environmentally benign. Therefore, synthetic methods involving iron salts in organic chemistry have much expanded over the last decades, especially in coupling reactions. In 1972, Kumada and Corriu independently reported the cross-coupling reaction between Grignard reagents and organic halides catalyzed by nickel salts, later extended to palladium complexes. However, one year before, Tamura and Kochi already reported some successful examples of formal nucleophilic substitution between alkyl magnesium bromides and vinyl bromides in presence of a catalytic amount of $FeCl_3$ in THF (Fig. 3.1, eq 1) [1, 2]. Moreover, they early demonstrated the stereoselectivity of the coupling (Fig. 3.1, eq 2).

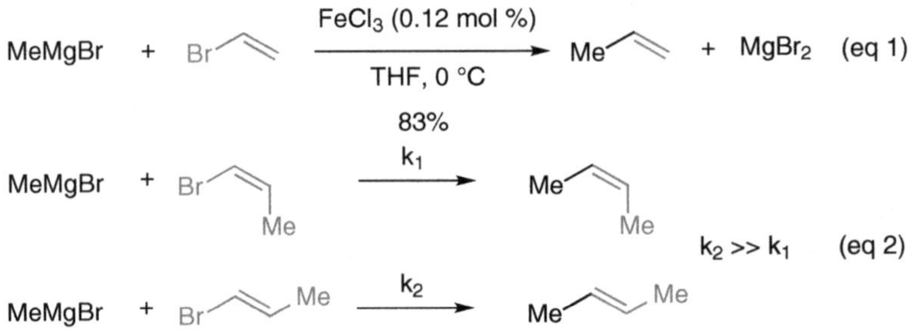

Fig. 3.1: Fe-catalyzed cross-couplings with Grignard reagents by Tamura and Kochi.

While the potential of Fe-catalyzed cross-coupling remained in the shadow of palladium catalysis for years and did not undergo much studies, Cahiez *et al.* reported in 1998 an improvement of the coupling reaction of alkyl Grignard reagents with poorly reactive substrates, such as alkenyl chlorides (Fig. 3.2); by using $Fe(acac)_3$ as catalyst and *N*-methylpyrrolidinone (NMP) as co-solvent, the yield jumped from 5% in THF to 85% in a mixture of solvents constituted by THF/NMP [3]. Even if recent improvements with more sophisticated catalysts can prevent the use of NMP as additive, these convenient conditions are still often used [4].

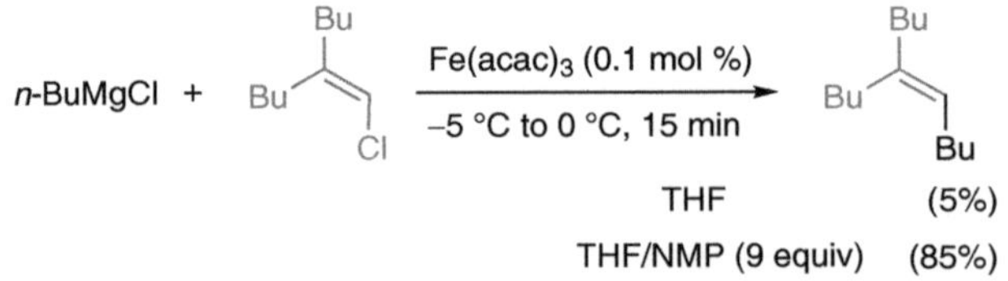

Fig. 3.2: Cahiez *et al.* conditions for Fe-catalyzed cross-coupling.

Thus, these early works have paved the way for numerous developments of iron catalysts [mostly $Fe(acac)_3$ as the precatalyst] and organomagnesium reagents, which have been partially reported in some general reviews [5–14].

3.2 Coupling reactions of alkyl Grignard reagents

Alkyl Grignard reagents react with a wide variety of electrophiles through Fe-catalyzed cross-coupling reactions.

3.2.1 With aryl derivatives

Aryl halides

In 2002, Fürstner *et al.* reported in a seminal article [15] that the conditions implemented by Cahiez *et al.*, in the 1990s, could be successfully applied to the coupling between alkyl magnesium halides and aromatic halides (Fig. 3.3) [15–19].

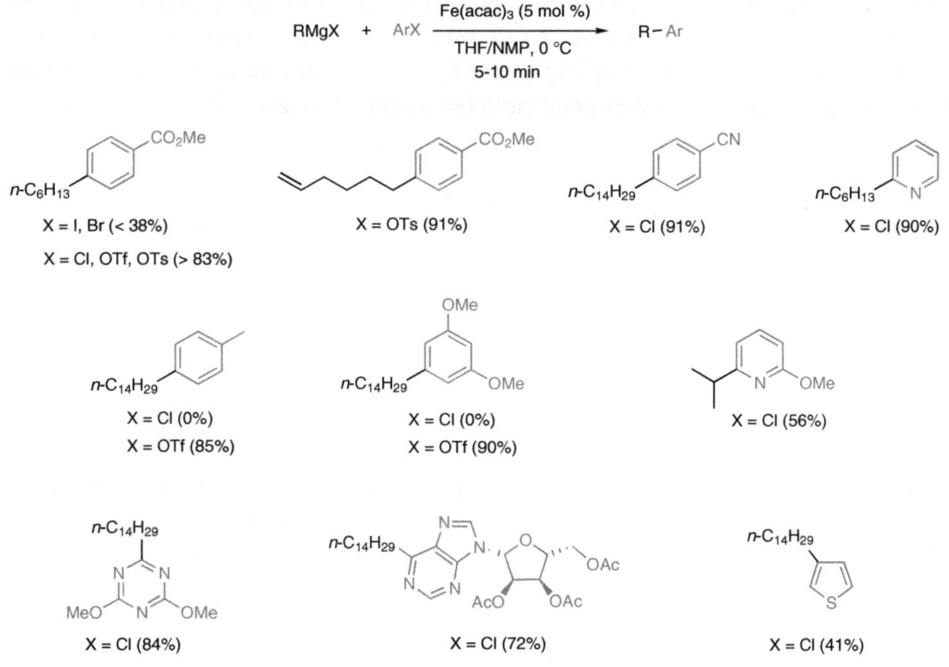

Fig. 3.3: Cross-coupling between alkyl Grignard reagents and aryl (pseudo)halides.

The reaction in presence of $Fe(acac)_3$ in a THF/NMP mixture at 0 °C, between *n*-butyl magnesium bromide and benzoates with a (pseudo)halide in the *para* position gives excellent results provided that the latter is a chloride atom, a triflate or a tosylate group (<83 % yield). In contrast with iodo- or bromo-aryl derivatives, only low yields in the coupling product were obtained along with significant amounts of reduced arene. In all cases, the Grignard reagent does not affect the carboxylic moiety. The reaction scope is broad with electron-poor arenes (X = Cl, OTf, OTs) but limited to triflates and tosylates for aromatics bearing electron-donating groups. However, Nakamura *et al.* showed in 2015 that this problem could be overcome to couple chloroanisoles with alkyl Grignard reagents by using the FeF_3/*N*-heterocyclic carbene catalytic system, previously assessed in the coupling with aryl Grignard reagents [20].

Besides linear alkyl chains, Fürstner *et al.* also used *i*PrMgX and the best results were obtained with iron(salen) salts rather than with the classical $Fe(acac)_3$ pre-catalyst. Impressively, a broad range of heteroaromatic chlorides can be coupled with alkyl Grignard reagents.

Malhotra *et al.* observed an interesting selectivity in the cross-coupling of dihalogenated heteroarenes where alkyl Grignard reagents reacted with one halogen with very high selectivity. For example, MeMgBr reacted selectively with 1,3-dichloroisoquinoline at the C1 position and the subsequent addition of EtMgBr allowed the cross-coupling at the C3 position, affording the corresponding bis-alkylated product in a very good yield (84 %) (Fig. 3.4) [21].

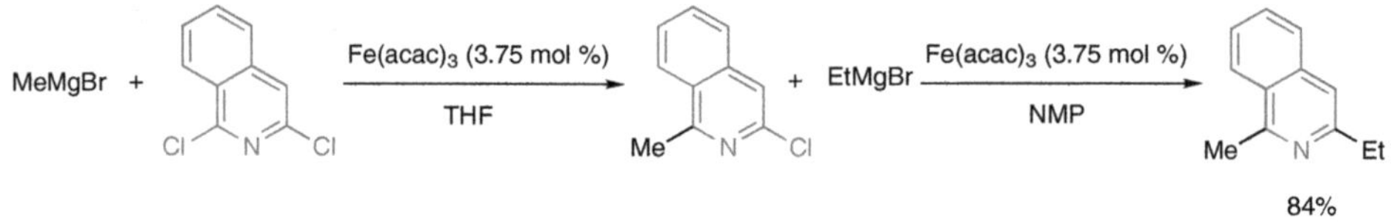

Fig. 3.4: Selective coupling with dichloroisoquinoline.

As an application in medicinal chemistry, Risatti *et al.* performed a $Fe(acac)_3$ catalyzed coupling, in high yield, between a chloropyridine and cyclopropylmagnesium bromide in their route to the synthesis of a dual NK-1/serotonin receptor antagonist (Fig. 3.5) [22].

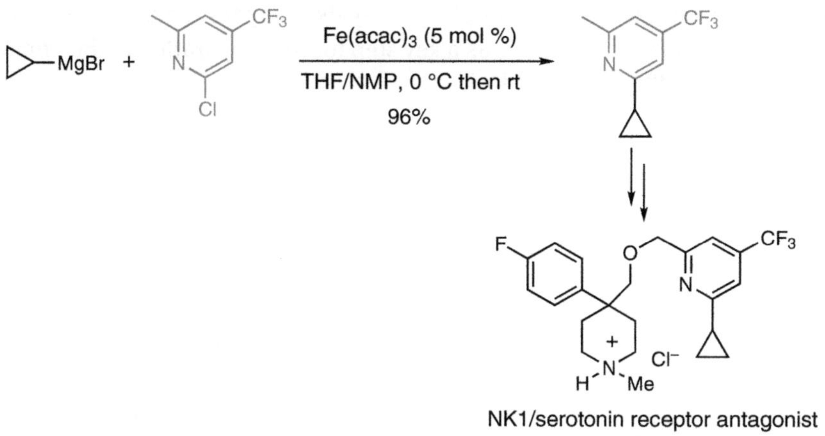

Fig. 3.5: Fe-catalyzed coupling of a cyclopropyl magnesium bromide with a chloropyridine triflate for the synthesis of NK-1/serotonin receptor antagonist.

Aryl pseudohalides

Fürstner *et al.* have shown that iron-catalyzed couplings of alkyl Grignard reagents with aryl chlorides and triflates might be useful tools for an efficient synthesis of (*R*)-muscopyridine, an odoriferous alkaloid from animal origin. Indeed, mixing the chloro-triflate pyridine with a first alkyl Grignard in the presence of an iron-salen salt (5 mol %) in a THF/NMP mixture, followed by a second alkyl Grignard, afforded the expected double adduct, which underwent further a ring-closing metathesis, which after hydrogenation, gave rise to the expected natural product in 57 % overall yield (Fig. 3.6) [23].

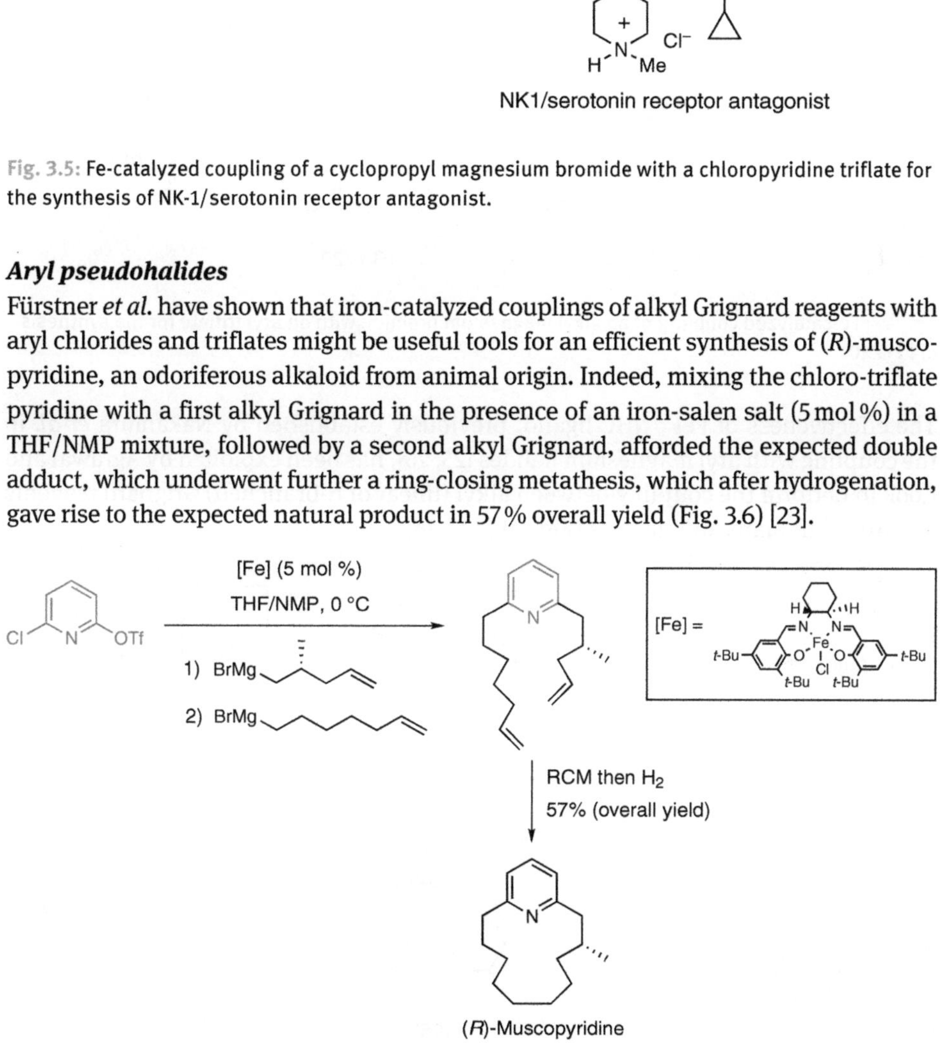

Fig. 3.6: Fe-catalyzed coupling of alkyl magnesium bromides with a 6-chloro-2-pyridine triflate for the synthesis of (*R*)-muscopyridine.

The same authors have also used the coupling of an alkylmagnesium bromide with a phenyl triflate under $Fe(acac)_3$ catalysis as a key step for the preparation of the immunosuppressive agent FTY270 (Fig. 3.7) [24].

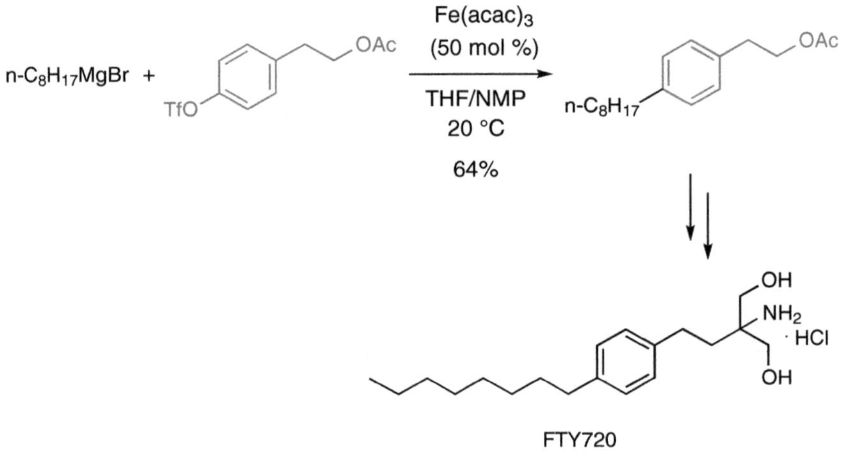

Fig. 3.7: Fe-catalyzed coupling of an alkyl magnesium bromide with an aryl triflate for the synthesis of FTY720.

The effectiveness of FeF_3/NHC ligand, previously established by Nakamura *et al.* in the coupling with aryl magnesium halides [25, 26], has been exploited by Agrawal and Cook to perform the coupling between alkyl (linear or α-branched) Grignard reagents and aryl sulfamates or tosylates (Fig. 3.8) [27, 28].

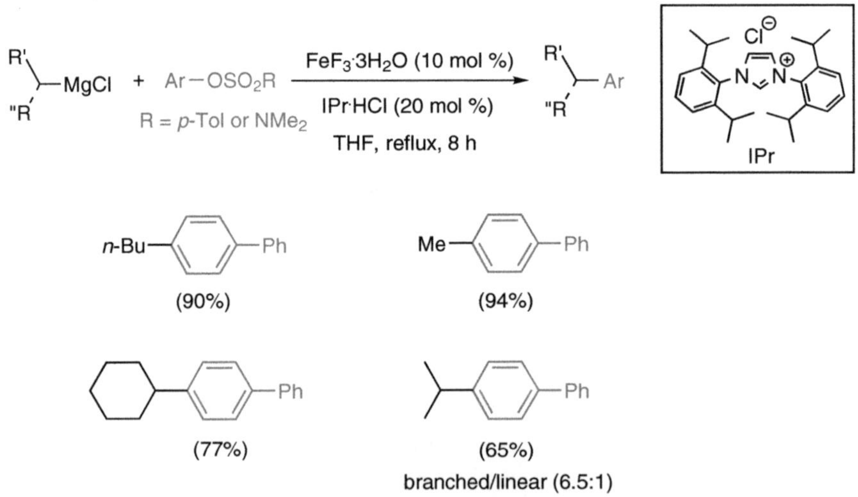

Fig. 3.8: Aryl sulfamates or tosylates as partners in the cross-coupling.

3.2.2 With alkyl derivatives

Alkyl halides

Nucleophilic substitution between Grignard reagents and sp^3-electrophiles in the presence of catalytic loadings of iron salts has remained a difficult task in C–C bond formation. Thus, in their first report Kochi *et al.* [29] noted that when alkyl Grignard reagents reacted, in THF, with alkyl bromides (primary and secondary) in the presence of $FeCl_3$, yields in the cross-coupled products remained close to zero, due to disproportionation [29, 30]. In 2007, Chai *et al.* reported the iron-catalyzed cross-coupling reaction of alkyl Grignard reagents with alkyl bromides, catalyzed by iron acetate in the presence of XantPhos in Et_2O. The cross-coupled products were obtained in modest to good yields (Fig. 3.9) [31].

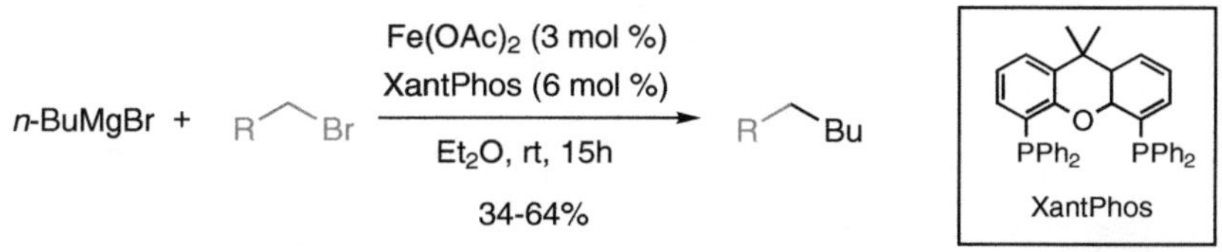

Fig. 3.9: Iron catalyzed cross-coupling between alkyl Grignard reagents and alkyl bromides.

In 2015, Rueping *et al.* reported the cross-coupling between 3-iodoazetidines and MeMgBr catalyzed by $Fe(acac)_3$ (Fig. 3.10) [32].

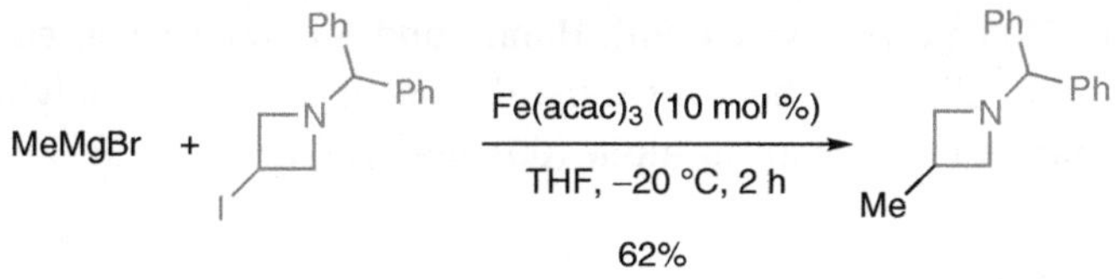

Fig. 3.10: Iron-catalyzed cross-coupling between MeMgBr and iodoazetidine.

Alkyl ethers

Shi *et al.* described the cross-coupling of homobenzylic methyl ethers with an excess of primary and secondary alkyl magnesium chlorides, in the presence of FeF_2 and tricyclohexylphosphine (PCy_3) in *o*-xylene under reflux (Fig. 3.11) [33]. The authors proposed that the reaction proceeded through a first dehydro-alkoxylation and then a carbomagnesation, catalyzed by iron salts.

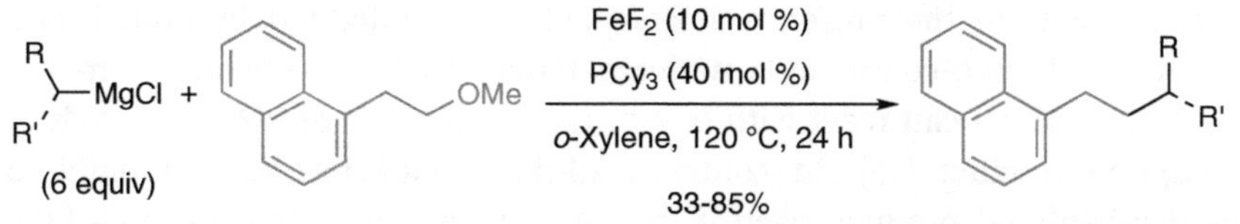

Fig. 3.11: Iron-catalyzed cross-coupling between alkyl Grignard reagents and homobenzylic methyl ethers.

3.2.3 With vinyl derivatives

Vinyl halides

Alkenyl halides are also very good partners for the coupling with alkyl Grignard reagents. Interestingly, in the presence of a leaving group on an alkenyl halide substrate, alkyl Grignard reagents react only with the halide connected to a sp^2 carbon. Thus, *i*PrMgBr reacts with a vinyl chloride without affecting an acetyl moiety (Fig. 3.12, eq 1), whereas *n*-BuMgBr reacted with a vinyl bromide, without any coupling with a primary chloride (Fig. 3.12, eq 2) [3].

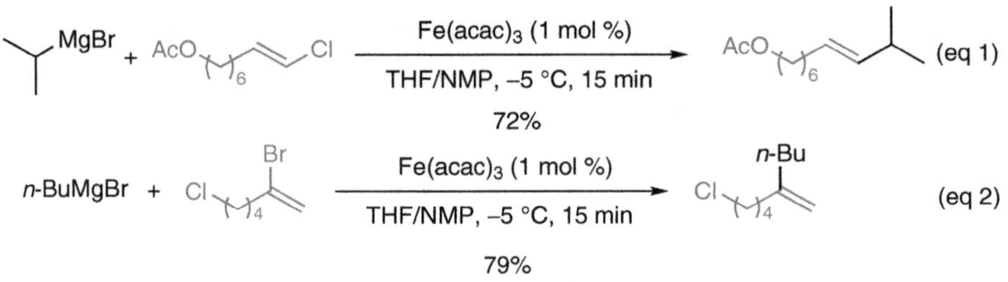

Fig. 3.12: Selective Fe-catalyzed cross-coupling between alkyl Grignard reagents and vinyl halides bearing a leaving group.

During their study on the synthesis of isocombretastatins (1,1-diarylethylenes) from aryl Grignard reagents and α-halogenostyrenes using iron(III) chloride and copper(I) thiophene carboxylate (CuTC) catalysts (*vide infra*), Hamze and Alami have successfully extended the method to alkyl and benzyl Grignard reagents. This co-catalytic system is more efficient than Fe or Cu catalysts alone (60–98 % yield) (Fig. 3.13) [34].

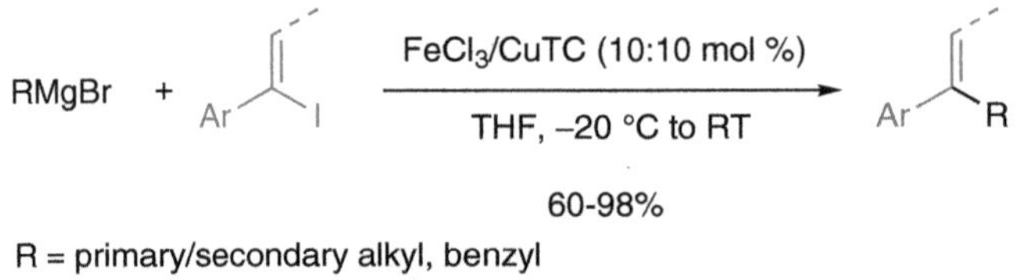

Fig. 3.13: Fe/Cu-catalyzed cross-coupling between alkyl or benzyl Grignard reagents and vinyl halides.

Aryl-*gem*-dihalogenoalkenes exhibit a very interesting reactivity towards organomagnesium reagents in the presence of Fe(acac)$_3$, as highlighted by Franck and Figadère. Thus, dibromo-compounds undergo reduction to selectively afford the (*E*)-bromoalkene, which can react with one further equivalent of RMgX to produce a cross-coupling product [35]. In contrast, 1,1-dichloroalkenes react rapidly to afford the disubstituted product, even if less than 2 equiv of RMgX are used [36]. For example, when *gem*-dichlorovinyl quinolines were engaged in a cross-coupling with *bis*-Grignard reagents, obtained from 1,5-dibromopentane, an intermolecular cross-coupling followed by an intramolecular cross-coupling led to an original cyclohexylidene moiety (Fig. 3.13) [36]. It is worth noting that the same authors reported the cross-coupling between chloroenynes and alkyl Grignard reagents (Fig. 3.14) [37].

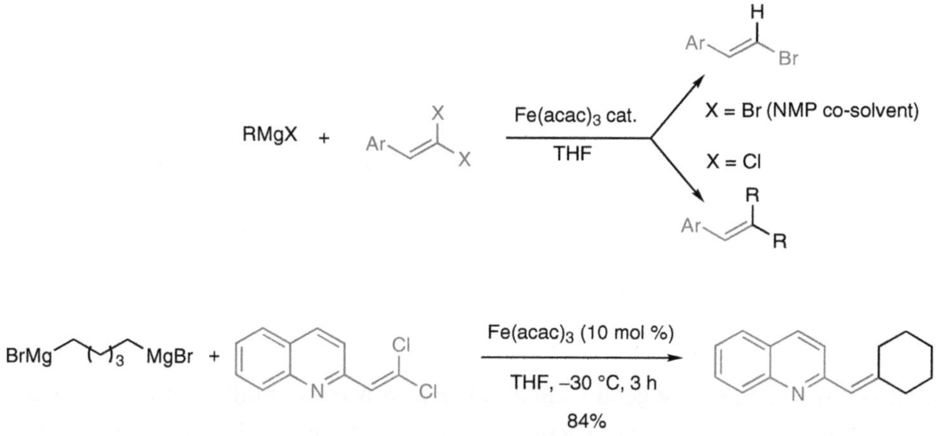

Fig. 3.14: Fe-catalyzed cross-coupling between Grignard reagents and *gem*-dihalogenoalkanes.

Thomas *et al.* reported an interesting iron-catalyzed reductive cross-coupling for the preparation of alkanes from vinyl halides and Grignard reagents. This coupling is performed *via* a sequential cross-coupling between β-bromostyrenes and Grignard reagents/*in situ* reduction affording a product that corresponds to a formal sp^3–sp^3 cross-coupling reaction. The reaction afforded the coupling products in very good yields from primary and secondary alkyl Grignard reagents while phenyl and allyl magnesium chlorides were poor partners (Fig. 3.15) [38].

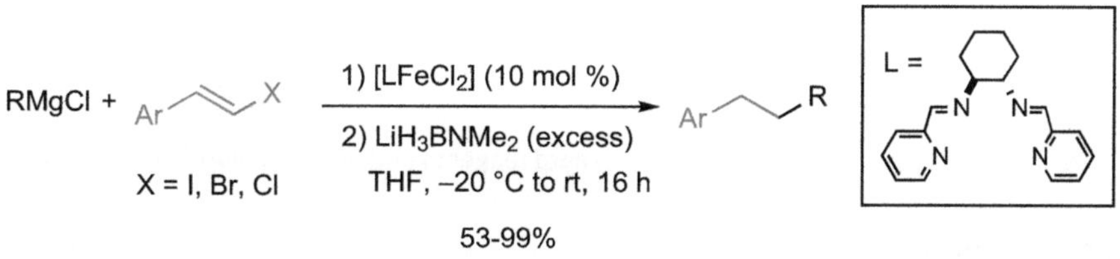

Fig. 3.15: Fe-catalyzed, reductive cross-coupling reaction between vinyl halides and Grignard reagents.

In the total synthesis of *cis*- and *trans*-solamins, two annonaceous acetogenins, the iron-catalyzed coupling reaction of an homoallylic Grignard with a (*Z*)-bromoalkene led to a key diene with the well-defined *Z,E* configuration (Fig. 3.16) [39].

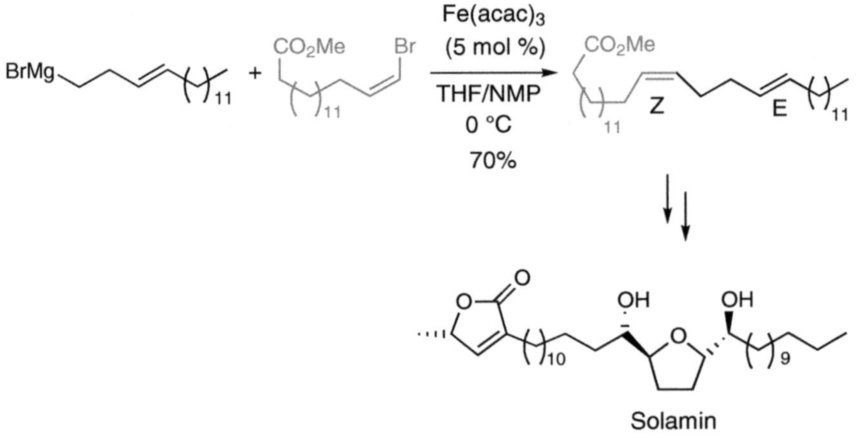

Fig. 3.16: Fe-catalyzed coupling of homoallylic magnesium bromide with a (*Z*)-bromoolefin for the synthesis of solamin.

Iron catalysts show very good compatibility with *N*-containing heterocycles, which are widely found in the pharmacopeia. Thus, Olsson *et al.* demonstrated the possibility to introduce various moieties on an imidoyl chloride through organomagnesium reagents to access benzodiazepine analogs. For example, a 4-substituted *N*-methylpiperidine group was introduced on a benzodiazepine, from the corresponding Grignard reagent, to access clozapines (Fig. 3.17) [40].

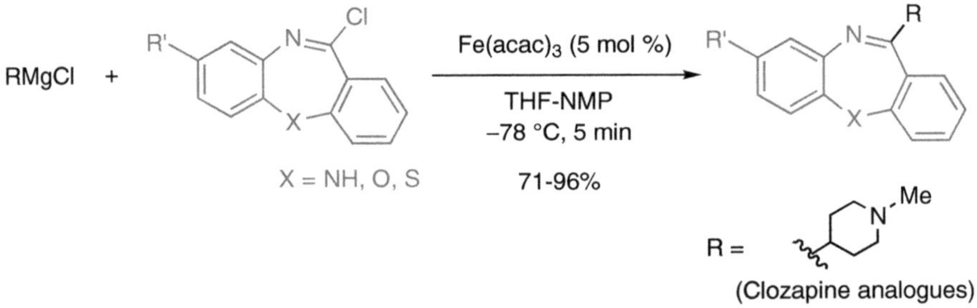

Fig. 3.17: Fe-catalyzed cross-coupling between Grignard reagents and imidoyl chlorides.

Vinyl pseudohalides

In 2009, Shi *et al.* reported that vinyl pivalates were able to couple with alkyl Grignard reagents in presence of $FeCl_2$ (1 mol %) and LiCl (6 equiv) to afford the coupling products in good to excellent yields (65–95 %). Interestingly the vinyl pivalate can contain carbonyl/carboxylate moieties that remain completely unchanged during the process, demonstrating the chemoselectivity of this process (Fig. 3.18) [41].

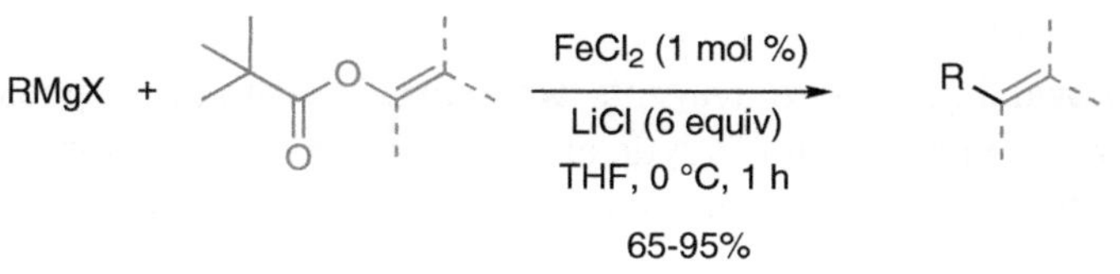

Fig. 3.18: Vinyl pivalates as partners in Fe-catalyzed cross-coupling with alkyl Grignard reagents.

Cahiez *et al.* reported that dienol phosphates can also react with Grignard reagents at low catalyst loading (1 mol %) while maintaining the double bond geometry [42, 43]. This strategy has been used to access an insect pheromone (Fig. 3.19) [43].

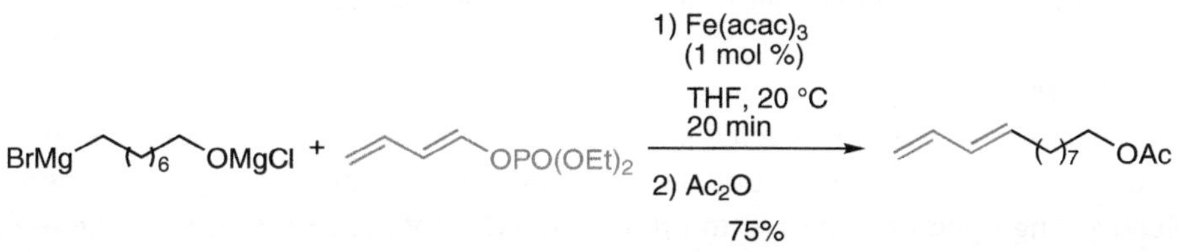

Fig. 3.19: Enol phosphates as partners in Fe-catalyzed cross-coupling with aryl Grignard reagents.

Vogel *et al.* reported that styrenyl sulfonyl chlorides reacted with alkyl Grignard reagents in the presence of $Fe(acac)_3$ in a hot THF/NMP mixture to afford the desulfinative C-C cross-coupling product in satisfactory yields (Fig. 3.20) [44]. It is worth noting that allyl or benzyl Grignard reagents did not afford the expected coupled products, while the best results were obtained with aryl magnesium halides and alkyl sulfonyl chlorides.

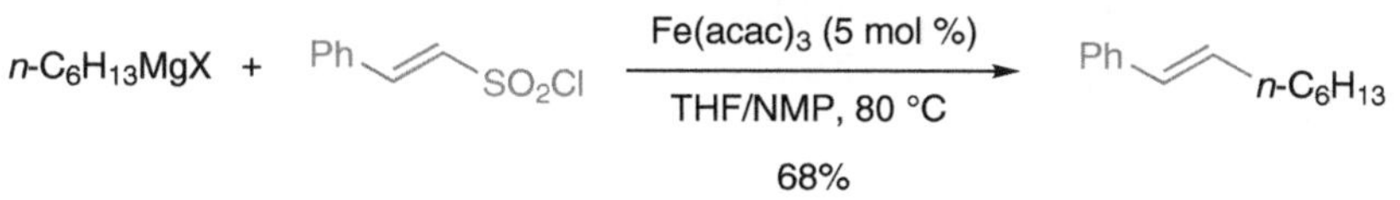

Fig. 3.20: Iron-catalyzed cross-coupling between Grignard reagents and sulfonyl chlorides.

3.2.4 With allyl and propargyl derivatives

Allyl and propargyl electrophiles behave through two different pathways, either a direct S_N2 substitution occurs at the carbon bearing the (pseudo)halogen, or the double bond participates to the reaction through a S_N2' pathway [45].

Allyl phosphonates

Yamamoto *et al.* have shown that Grignard reagents, when reacting with allylic phosphonates in THF and in the presence of catalytic amounts of iron acetylacetonate, gave almost exclusively the S_N2 coupled products in good yields (Fig. 3.21) [46, 47].

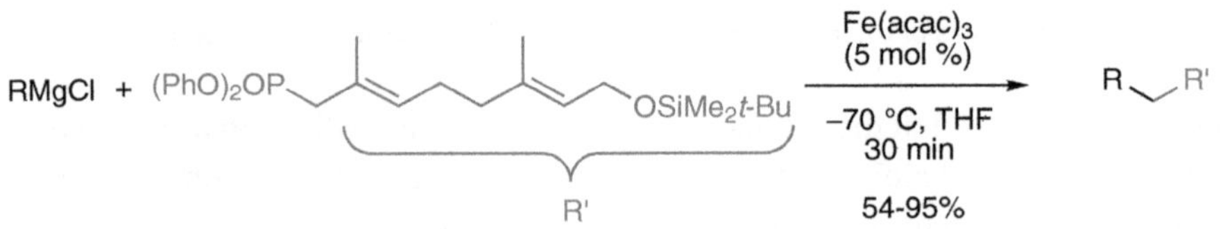

Fig. 3.21: S_N2 coupling between Grignard reagents and allylphosphates.

Allyl ethers

When oxabicyclic alkenes reacted with primary aliphatic Grignard reagents in the presence of a catalytic amount of $FeCl_3$ and TMEDA, a surprising regio- and stereoselective ring-opening occurred to afford the S_N2' product resulting from the addition of a formal alkenyl Grignard reagent on the oxabicyclic alkenes. High yields and excellent regio- and stereoselectivity were obtained for the coupling products (Fig. 3.22) [48]. The structure of the incoming group suggested that a β-hydride elimination takes place with the formation of a 2-metallo-1-alkenyl species that further underwent coupling. Moreover, when a secondary alkyl Grignard reagent (*i*PrMgBr) was reacting with the oxabicyclic alkene, a hydride reduction took place in excellent yield (92%).

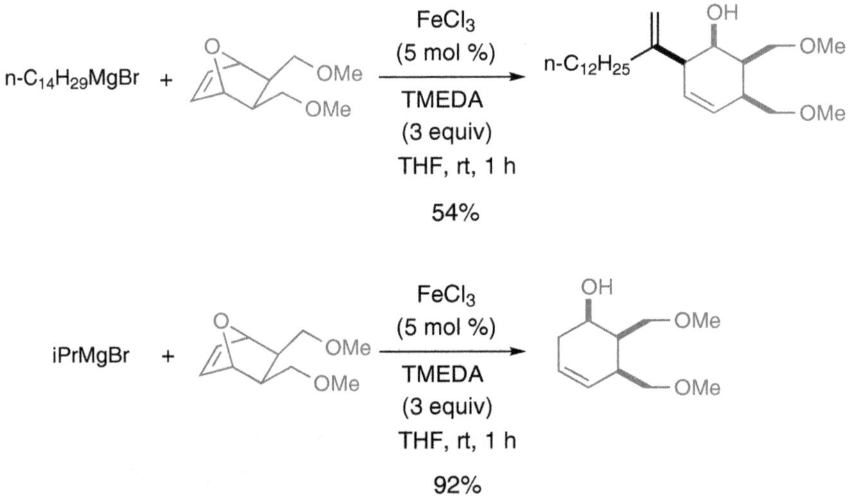

Fig. 3.22: S_N2' coupling between alkyl Grignard reagents and oxabicyclic alkenes.

Propargyl halides

Pasto *et al.* reported that propargyl chlorides react with alkyl Grignard reagents (primary, and secondary) in the presence of a catalytic amount of Fe(III) chloride or acetylacetonate to give only allene derivatives, through a S_N2' pathway, in 45–90 % yield [49, 50]. Tertiary alkyl Grignards did not react and the methyl Grignard gave lower yields and a 1:1 mixture of S_N2 and S_N2' products (Fig. 3.23).

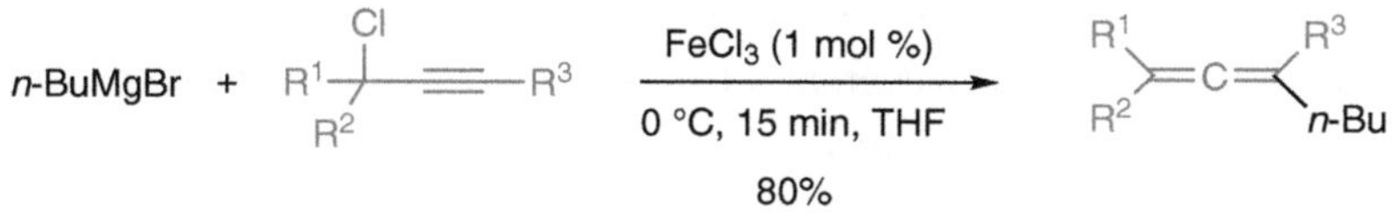

Fig. 3.23: S_N2' coupling between butylmagnesium bromide and propargyl chlorides.

Propargyl ethers

Fürstner *et al.* used a $Fe(acac)_3$-catalyzed alkyl Grignard S_N2' opening of a propargyl epoxide for the diastereoselective preparation of a *syn*-allenol (dr > 8:1) (Fig. 3.24), a key intermediate in the total synthesis of amphidinolide X [51].

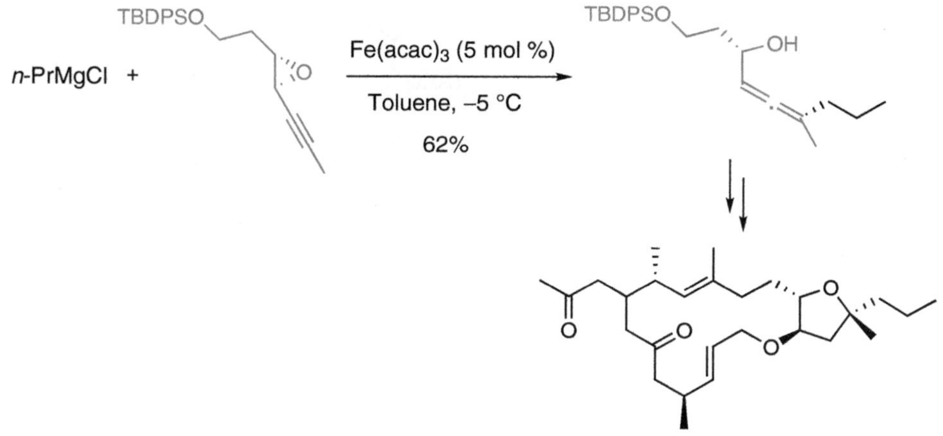

Amphidinolide X

Fig. 3.24: Fe-catalyzed propargyl opening by propylmagnesium chloride for the synthesis of amphidinolide X.

3.2.5 With carbonyl derivatives

The synthesis of ketones through "addition" of organomagnesium reagents onto carboxylic derivatives is often doomed to failure due to facile over-addition onto the carbonyl products, leading thus to tertiary alcohols. In this context, in the 1980s Marchese *et al.* demonstrated that Grignard reagents in the presence of $Fe(acac)_3$ catalyst were able to perform the formal nucleophilic substitution onto carbonyl derivatives to afford a variety of ketones [52–57].

Acyl chlorides

Acyl chlorides are very efficient partners in the iron-catalyzed cross-coupling with alkyl magnesium halides. Thus, primary, secondary and tertiary magnesium halides react selectively with acyl chlorides, even in the presence of an ester or a cyano group on the substrate (Fig. 3.25) [52, 54, 56, 57].

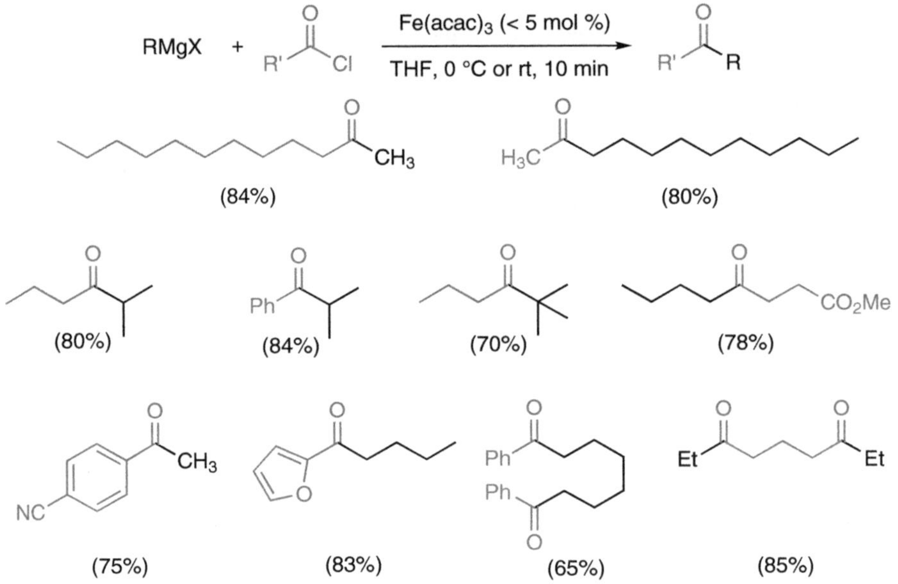

Fig. 3.25: Fe-catalyzed cross-coupling between alkyl Grignard reagents and acyl chlorides.

In 2004, Fürstner *et al.* extended the method to functionalized sophisticated molecules such as an enantiopure thiazolidinone bearing an acyl chloride moiety. Cross-coupling with MeMegBr afforded the corresponding ketone in 80 % yield (Fig. 3.26) [58].

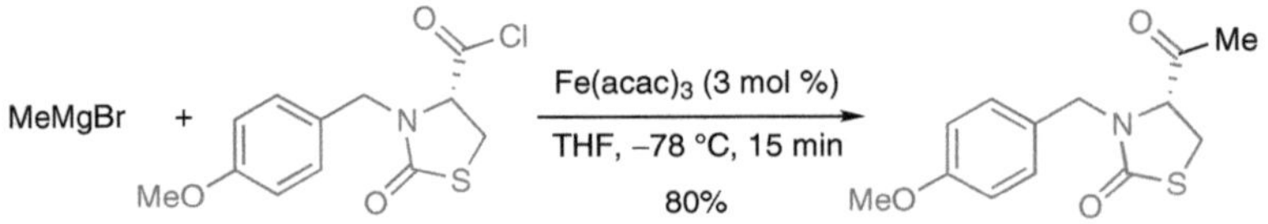

Fig. 3.26: Fe-catalyzed cross-coupling between MeMgBr and a functionalized enantiopure acyl chloride.

Thioesters

Thioesters are also able to afford ketones by reacting with Grignard reagents, albeit in a lesser extent than acyl chlorides (Fig. 3.27) [53, 55]. Indeed, much less examples have been reported in the literature and there is no mention of assessment of tertiary alkyl Grignard reagents with thioesters, neither on the compatibility with other functional groups.

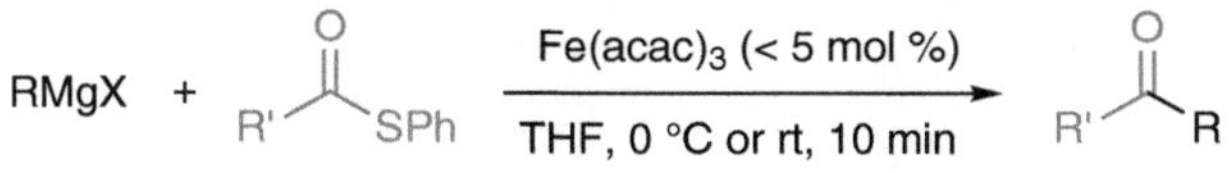

Fig. 3.27: Fe-catalyzed cross-coupling between alkyl Grignard reagents and thioesters.

3.3 Coupling reactions of aryl Grignard reagents

Aryl Grignard reagents are powerful nucleophilic partners that react with a wide variety of electrophiles through Fe-catalyzed cross-coupling reactions.

3.3.1 With aryl derivatives

Aryl halides

The first successful results have been reported by Fürstner *et al.* between phenyl/heteroaryl Grignard reagents and heteroaryl chlorides with $Fe(acac)_3$ within a few minutes (Fig. 3.28) [15].

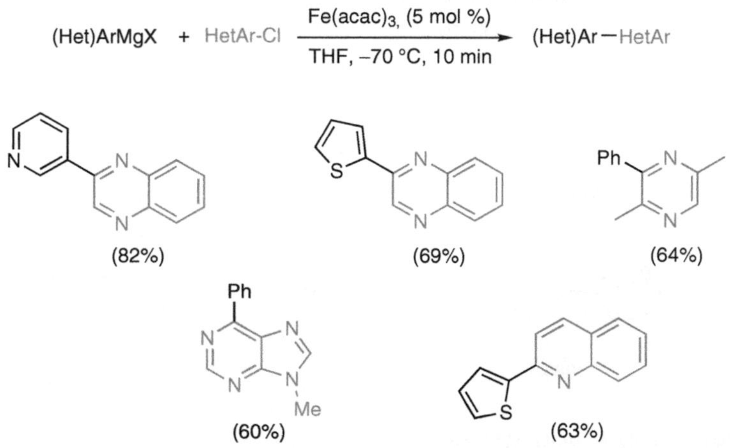

Fig. 3.28: Cross-coupling of (hetero)aryl Grignard reagents and heteroaryl chlorides.

This method has then been widely applied to halo *O*- and *N*-heteroarenes [59–63]. Notably, it can be successfully used to synthesize quinolines substituted at the C2 or C3 position (Fig. 3.29) [64, 65].

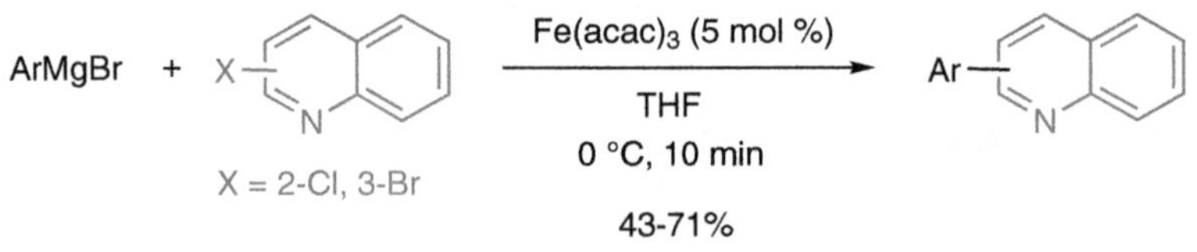

Fig. 3.29: Fe-catalyzed cross-coupling aryl Grignard reagents and halogenoquinolines.

However, the direct coupling of aryl halides and aryl Grignard reagents in the presence of ferric or ferrous precatalysts is often low yielding. In this context, Nakamura *et al.* have described an effective biaryl cross-coupling method, without homocoupling products. This coupling involves iron fluoride (FeF_3) as the precatalyst, a *N*-heterocyclic carbene ligand precursor (SIPr·HCl) and a catalytic base to activate the ligand deprotonation (EtMgBr). To explain the specific catalytic "fluoride effect" of the iron catalyst, the authors proposed a mechanism in which a strong coordination of the fluoride ion to the bimetallic adduct center suppresses the reduction of the metal *via* the usual transmetallation reduction/elimination process, promoting thus the formation of a high-valent metallate complex can act as the effective catalyst (Fig. 3.30) [25, 26].

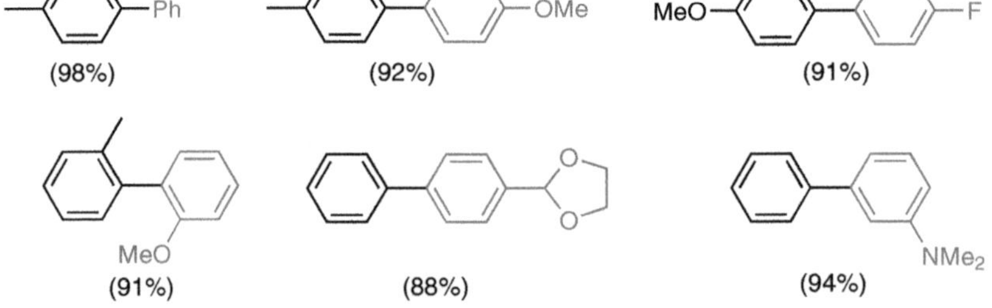

Fig. 3.30: FeF_3/NHC ligand-catalyzed biaryl synthesis.

A specific Fe-catalyzed biaryl synthesis has also been reported by von Wangelin *et al.*, in which simple $Fe(acac)_3$ in THF/NMP at 30 °C promoted the coupling of aryl Grignard reagents with chlorostyrenes. This reaction proceeded very well with *o*-chlorostyrenes but was also extended to *meta*- and *para*-isomers. Since *o*-allyl- and *o*-cyclopropyl-chlorobenzene did not undergo coupling, it was suggested that activation of aryl chlorides proceeded through a coordination of the catalyst to the vinyl substituent and subsequent migration along the conjugated π-system to the site of C-Cl bond insertion (Fig. 3.31) [66].

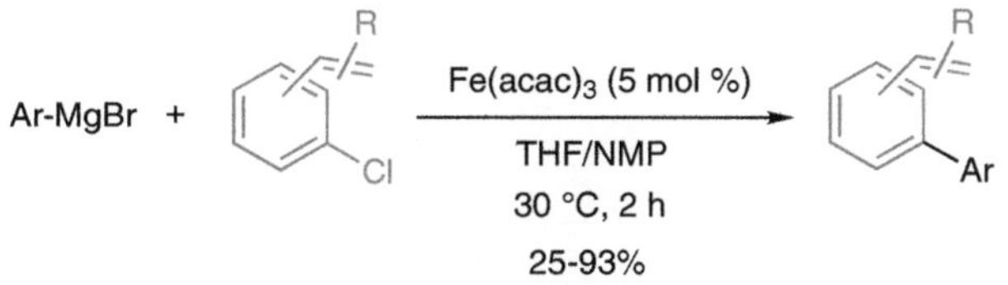

Fig. 3.31: Biaryl synthesis from chlorostyrenes.

It is worth noting that it is also possible to perform an efficient homocoupling of aryl magnesium bromides with $FeCl_3$ in the presence 1,2-dihaloethane (Cl, Br, I) as the oxidizing agent, as the corresponding biaryls were obtained in good yields (Fig. 3.32) [67].

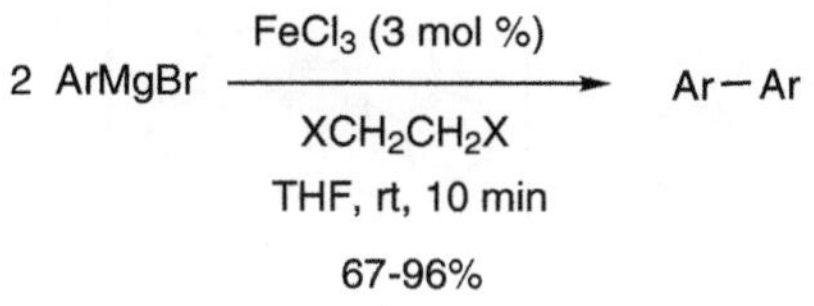

Fig. 3.32: Biaryl synthesis through Fe-catalyzed homocoupling of ArMgBr.

Cahiez *et al.* have developed a biaryl synthesis of *N*-methylcrinasiadine, a natural product extracted from *Lapiedra martinezii* (*Amaryllidaceae*), *via* a Fe-catalyzed homocoupling of aryl Grignard reagents [67]. They synthesized 2,2'-diiodo-*N*-methyl-4,5-methylenedioxybenzanilide, which was then reacted with *i*PrMgBr (2.2 equiv) to generate the di-Grignard reagent that further reacted with $FeCl_3$ to afford the target compound (Fig. 3.33).

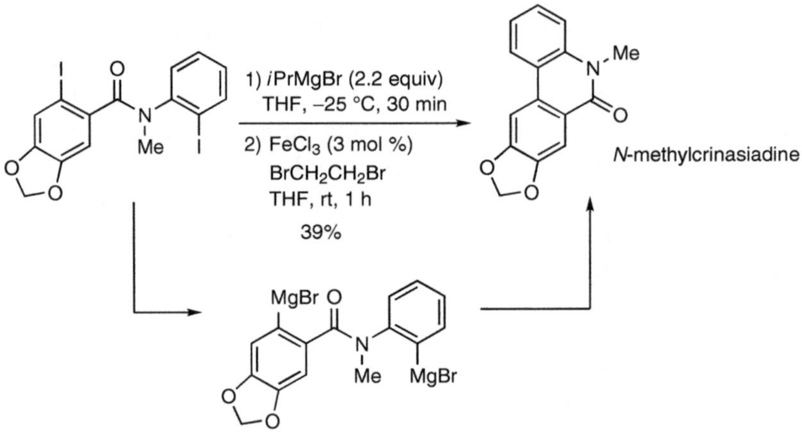

Fig. 3.33: Synthesis of *N*-methylcrinasiadine.

Aryl sulfamates and tosylates

The effectiveness of FeF_3/NHC ligand has been exploited by Agrawal and Cook to perform the coupling between aryl Grignard reagents and aryl sulfamates or tosylates (Fig. 3.34) [27, 28].

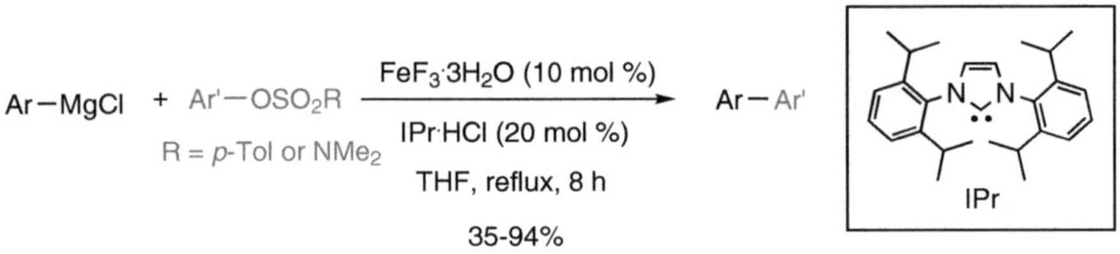

Fig. 3.34: FeF_3/NHC ligand-catalyzed biaryl synthesis from aryl sulfamates/tosylates.

3.3.2 With alkyl derivatives

Alkyl halides

Several reports appeared in the literature dealing with cross-coupling reaction of aryl Grignard reagents with alkyl halides (chlorides, bromides and iodides) in the presence of iron complexes [68–70]. One of the broadest scope has been reported by Martin and Fürstner who used a low valent iron species $[Li(TMEDA)]_2[Fe(C_2H_4)_4]$ [71]. The choice of this catalyst was based on the hypothesis that highly reduced iron–magnesium clusters generated *in situ* from $FeX_{2,3}$ may play a decisive role in the catalytic cycle. Thus, the catalyst $[Li(TMEDA)]_2[Fe(C_2H_4)_4]$ at 5 mol % promoted the coupling between PhMgBr and various alkyl iodides, including the functionalized ones (Fig. 3.35) [71].

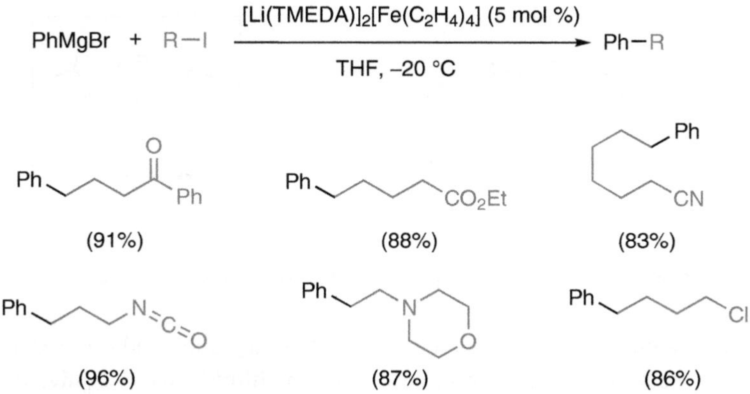

Fig. 3.35: Iron-catalysis between phenyl Grignard reagent and alkyl iodides.

In 2014, Cossy *et al.* [72] then Rueping *et al.* [32] addressed the coupling between various aryl magnesium bromides and *N*-protected 3-iodoazetidines. Interestingly, this reaction offers an easy entry to biologically active compounds bearing this small heterocycle (Fig. 3.36).

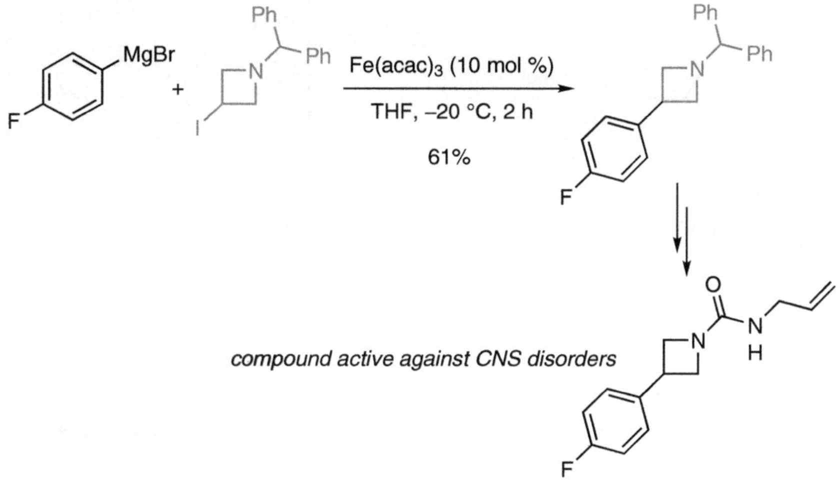

Fig. 3.36: Iron-catalysis between an aryl Grignard reagent and 3-iodoazetidine.

Moreover, primary and secondary alkyl chlorides were found to be good electrophiles for the coupling with aryl Grignard reagents, providing that the reaction was run at 40 °C in THF, in the presence of a *N*-heterocyclic carbene (IPr) and $FeCl_3$. Interestingly even the tertiary alkyl adamantyl chloride was able to react with ArMgBr (>90 % yield) whereas *t*-BuCl was much less reactive (12 % yield) (Fig. 3.37) [73].

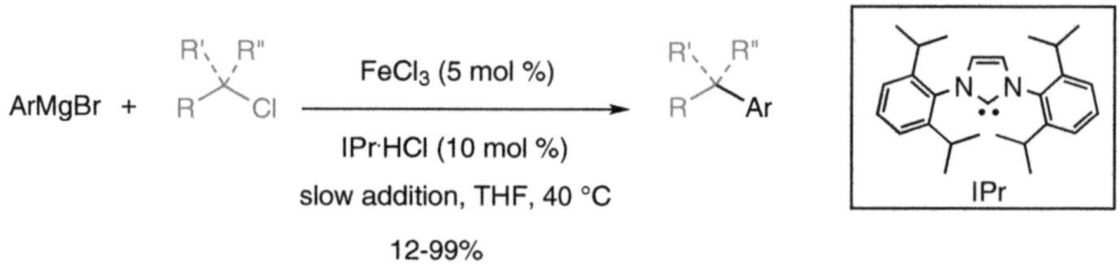

Fig. 3.37: Iron-catalyzed cross-coupling between aryl Grignard reagents and alkyl chlorides.

A few years later, Knochel *et al.* discovered that secondary alkyl iodides reacted with aryl Grignard reagents in the presence of iron dichloride (0.85 equiv) and 4-fluorobenzene in THF at –50 °C to afford the expected compound in good yields (60–90 %) [74]. Kozak *et al.* described the coupling reaction of aryl and benzyl Grignards with dichloromethane in the presence of $FeCl_3$ or an amine-bis(phenolate)iron complex, to afford the diarylmethane derivative in good yields (Fig. 3.38) [75].

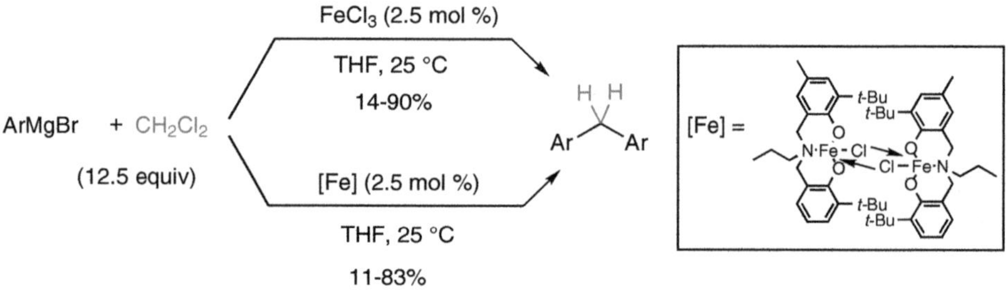

Fig. 3.38: Iron-catalyzed cross-coupling between aryl Grignard reagents and dichloromethane.

Iron-catalyzed arylation of alkyl fluorides has also been recently reported with aryl Grignard reagents in THF and in the presence of a dinuclear iron complex [76].

It has also been shown that cross-coupling of aryl Grignards with alkyl bromides could be run in Et_2O with iron-containing ionic liquid [77, 78] or in the presence of *N*-heterocyclic carbene [79, 80]. Other reports described such coupling reactions with different iron salts [81–83].

Sulfides

Denmark *et al.* found that $Fe(acac)_3$ catalyzed the cross-coupling of aryl Grignard reagents with alkyl pyridinyl thioethers to produce the coupled products in moderate to good yields (Fig. 3.39) [84].

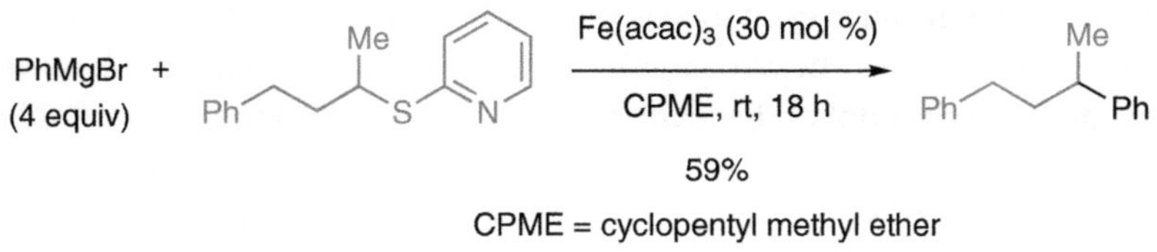

Fig. 3.39: Iron catalyzed cross-coupling between aryl Grignard reagents and alkyl pyridinyl thioethers.

Sulfones and sulfonyl chlorides

The method developed by Denmark *et al.* to couple aryl Grignard reagents with alkyl pyridinyl thioethers (*vide supra*) was extended to aryl sulfones, to produce the coupled products in modest to good yields (Fig. 3.40).[84]

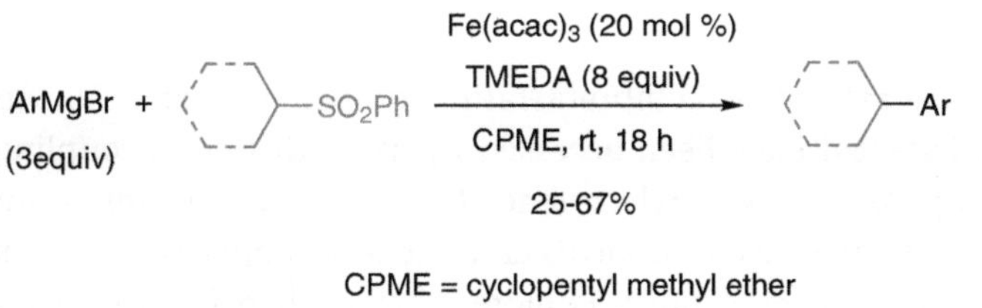

Fig. 3.40: Iron-catalyzed cross-coupling between aryl Grignard reagents and aryl sulfones.

Vogel *et al.* reported that alkyl sulfonyl chlorides reacted with aryl Grignard reagents in the presence of $Fe(acac)_3$ in a hot THF/NMP mixture to afford the desulfinative C–C cross-coupling product in moderate to good yields (Fig. 3.41) [44]. Allyl or benzyl Grignard reagents did not afford the expected coupled products, while some interesting results were obtained from alkyl magnesium halides.

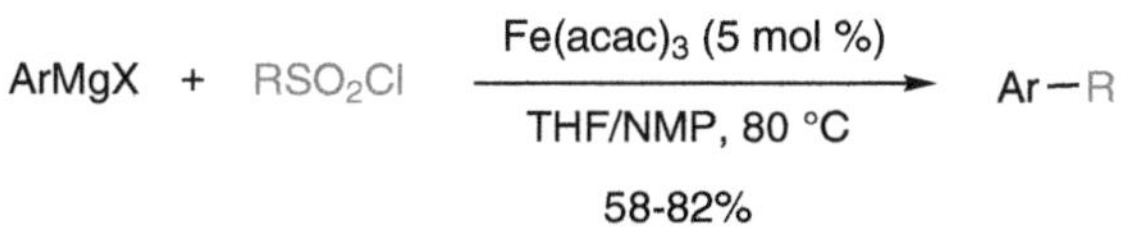

Fig. 3.41: Iron-catalyzed cross-coupling between Grignard reagents and sulfonyl chlorides.

3.3.3 With vinyl derivatives

Vinyl halides

Following the pioneer work of Kochi and Tamura in cross-coupling between Grignard reagents and vinyl halides, the field has been widely exploited and various organomagnesium reagents have been involved in the cross-coupling, even with sensitive functional groups on either the nucleophilic or the electrophilic partner. Moreover various halides, and even pseudohalides, can be used [1–3, 35, 85–95]. For example Cahiez *et al.* reported in 2001 that a functionalized aryl Grignard reagent was able to react with a vinyl iodide to afford the cross-coupling product in 69 % yield (Fig. 3.42) [92].

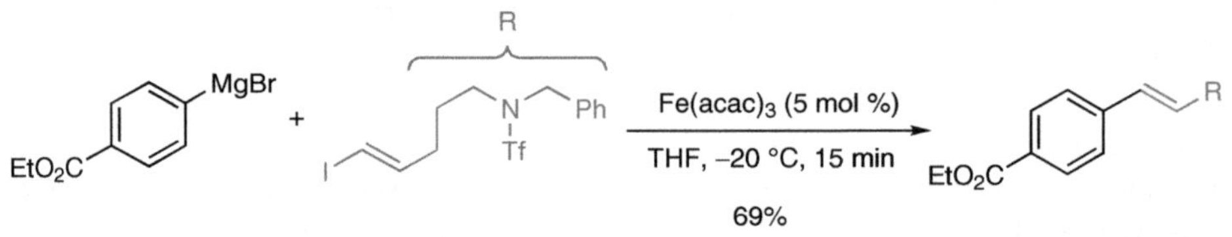

Fig. 3.42: Iron-catalyzed cross-coupling between functionalized aryl Grignard reagents and vinyl iodides.

Interestingly, the bioactive molecules combretastatins (stilbenoids) and isocombretastatins (1,1-diarylethylenes) have been accessed by using this cross-coupling. Thus, Hamze and Alami accessed 1,1-diarylethylenes from α-halogenostyrenes and aryl Grignard reagents in the presence of iron(III) chloride and copper(I) thiophene carboxylate (CuTC) catalysts, this co-catalyst is more efficient than Fe or Cu catalysts alone (55–98 % yield) (Fig. 3.43, eq 1) [34]. The prominent member of the family, combretastatin A-4 (CA4), has been accessed by coupling an aryl Grignard with a (*Z*)-bromovinyl derivative under iron catalysis, followed by a silyl deprotection to form the anti-cancer agent CA4 (Fig. 3.43, eq 2) [96].

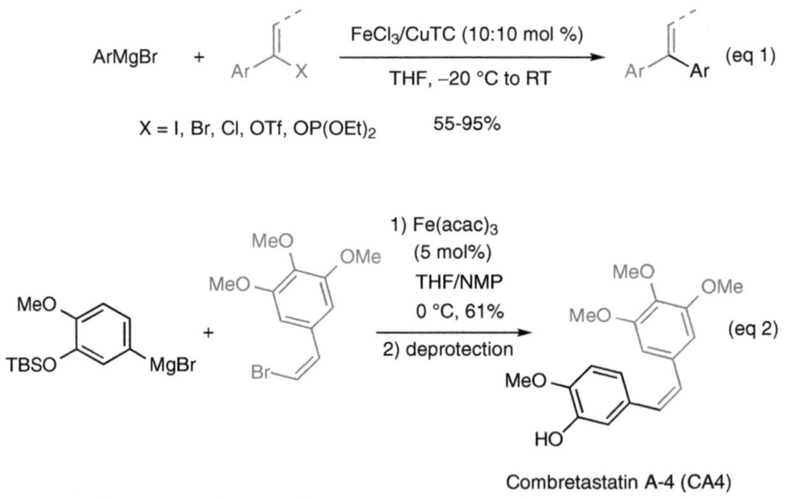

Fig. 3.43: Synthesis of isocombretastatins and combretastatins.

The coupling between aryl Grignard reagents and vinyl halides has been used as a key step in the preparation of drugs. Thus, Tewari *et al.* also reported the preparation of cinacalcet, a drug used for the treatment of secondary hyperparathyroidism, using as the key step the coupling of an aryl Grignard with a (*E*)-chloroalkene, catalyzed by $Fe(acac)_3$ in a THF/NMP mixture (Fig. 3.44) [97]. Shakhmaev *et al.* used the same key step (coupling of an aryl Grignard with a (*E*)-chloroalkene under $Fe(acac)_3$ catalysis) for the preparation of naftifine (Exoderil®), an antifungal drug (Fig. 3.45) [98].

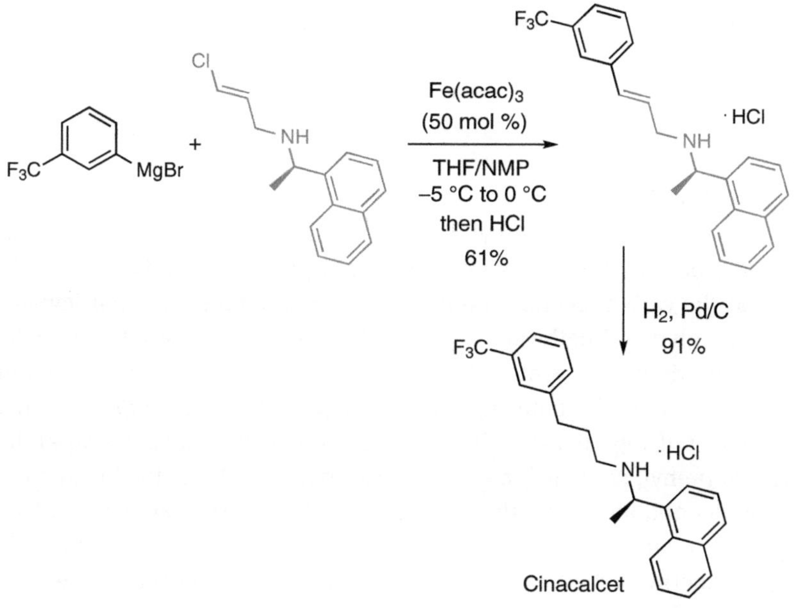

Fig. 3.44: Fe-catalyzed coupling of an aryl magnesium bromide with a chloroolefin for the synthesis of cinacalcet.

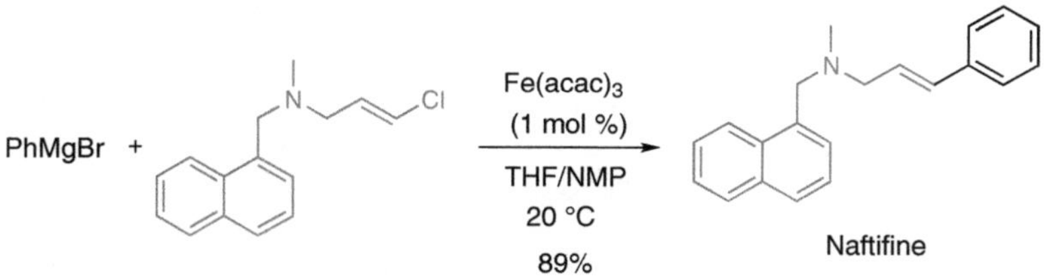

Fig. 3.45: Fe-catalyzed coupling of an aryl magnesium bromide with a chloroolefin for the synthesis of naftifine.

Vinyl sulfides

Besides halides, a variety of other groups on an alkenyl moiety can be used as efficient partners with Grignard reagents. Whereas Julia *et al.* Demonstrated earlier that sulfonyl moiety was a good leaving group [87, 90, 99], Yoshida *et al.* recently showed that phenyl vinyl sulfide was able to react with ArMgX, at room temperature in THF, to afford styrene derivatives (Fig. 3.46) [100].

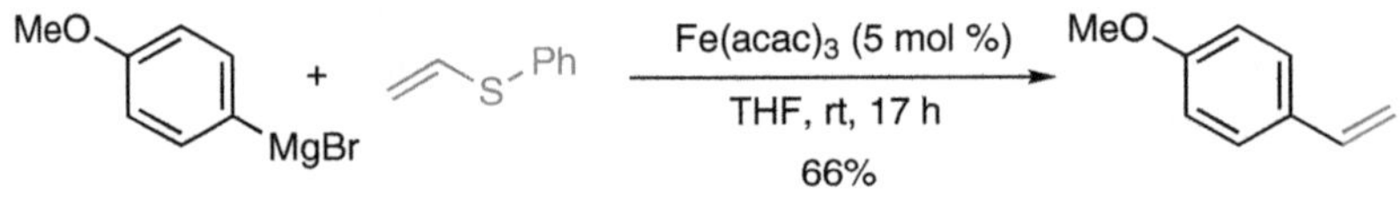

Fig. 3.46: Phenyl vinyl sulfide as partner in Fe-catalyzed cross-coupling with aryl Grignard reagents.

3.3.4 With allyl and propargyl derivatives

Allyl and propargyl halides

Among electrophiles, allyl and propargyl halides and their derivatives (phosphonates, sulfonates etc.) behave through two different pathways, either a direct S_N2 substitution occurs at the carbon bearing the halogen, or the double bond participates to the reaction through a S_N2' pathway [45]. Thus, Fürstner *et al.* showed that allylic halides reacted with Grignard reagents in the presence of a catalytic amount of an iron complex (5 mol %) through a S_N2 pathway to afford in good yield the coupled product [101]. Von Wangelin *et al.* reported that allyl acetates bearing distal substituents at the C3 position (crotyl, prenyl, cinnamyl) are good electrophiles and that the linear compounds were usually obtained through a major S_N2 pathway [102]. Moreover, allylic electrophiles bearing alternative leaving groups such as bromide, chloride, methyl carbonate, tosylate, trimethylsilyl ether and methyl thioether, showed comparable reactivity towards Grignard reagents under iron catalysis [102]. The most general method for a S_N2 process has been reported by Martin and Fürstner using the low valent iron species $[Li(TMEDA)]_2[Fe(C_2H_4)_4]$, which promoted the coupling between PhMgBr and various allyl or propargyl halides, including functionalized ones (Fig. 3.47) [71].

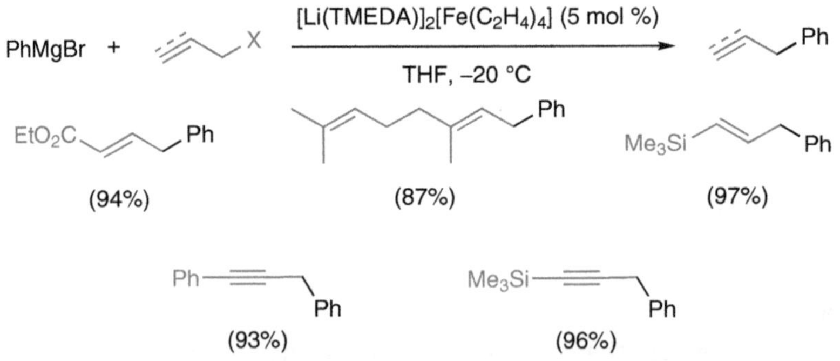

Fig. 3.47: Iron-catalysis between phenyl Grignard reagent and allyl and propargyl halides (Br, Cl).

Allyl sulfonyl chlorides

Vogel *et al.* showed that allylic sulfonyl chlorides reacted with aryl Grignard reagents in the presence of iron acetylacetonate in THF to give the desulfinylative C–C cross-coupling products as the major one, through a S_N2 pathway (Fig. 3.48) [103].

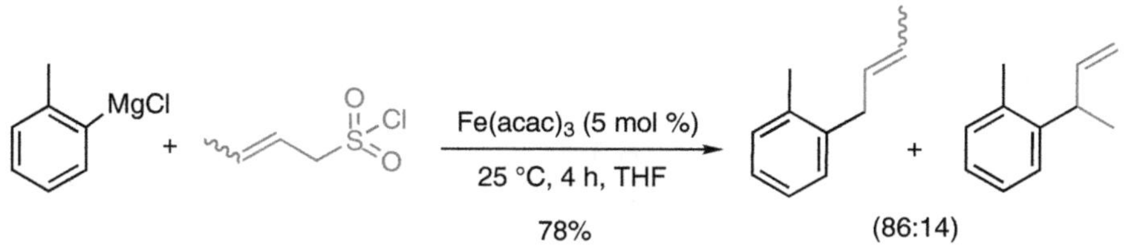

Fig. 3.48: S_N2 coupling between Grignard reagents and allylic sulfonyl chlorides.

Allyl ethers

When oxabicyclic alkenes reacted with aromatic Grignard reagents in the presence of a catalytic amount of $FeCl_3$ and TMEDA, a regio- and stereoselective ring-opening occurred to afford the S_N2' product in high yield and excellent regio- and stereoselectivity (Fig. 3.49) [48]. This reaction could also be performed with alkenyl Grignard reagents, but with lower efficiency, for example vinyl magnesium bromides reacted at reflux to afford the coupling products within *ca.* 40 % yield [48].

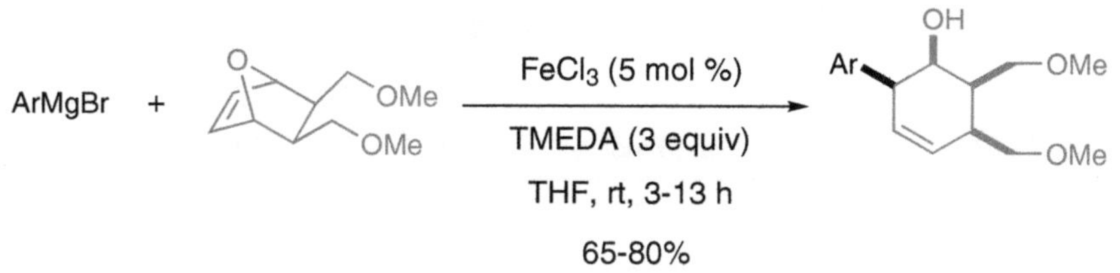

Fig. 3.49: S_N2' coupling between aryl Grignard reagents and oxabicyclic alkenes.

3.3.5 With benzyl halides

Benzylation of Grignard reagents remained a difficult task for a long time and yields of the cross-coupled products remained low [104, 105]. Thus, Meunier et al. reported in the late seventies that when phenylmagnesium bromide reacted with benzyl chloride in toluene and in the presence of an iron complex (such as CpL_2FeX, with $L = PMe_2Ph$), a mixture of homo-coupling products Ph–Ph (26 %) and $PhCH_2CH_2Ph$ (17 %) was obtained together with the heterocoupling product $PhCH_2Ph$ (36 %) [42]. To date, only aryl magnesium halides have been able to react with benzyl halides through Fe-catalyzed cross-coupling. In this context, Fürstner *et al.* finally obtained a decent yield in diphenylmethane when phenylmagnesium bromide reacted with benzyl bromide in the presence of a catalytic amount of a tetramethylethylenediamine lithium ferrate complex (62%, Fig. 3.50) [101].

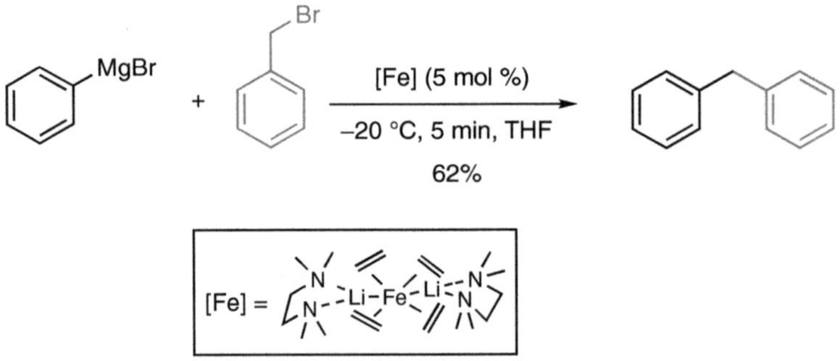

Fig. 3.50: Coupling between aryl Grignard reagents and benzylbromide.

Nakamura *et al.* showed that the reaction of aryl Grignard with benzyl chloride in the presence of an iron complex can be extended to other primary benzyl chlorides, providing the use of a diphosphine as an electron-donating ligand (Fig. 3.51) [106].

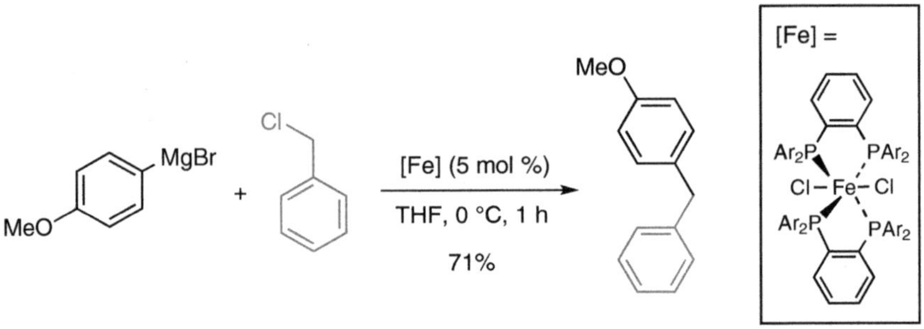

Fig. 3.51: Coupling between aryl Grignard reagents and benzylchloride.

3.3.6 With alkynyl halides

It is worth noting that there is a single known example of Fe-catalyzed coupling between a Grignard and a propargyl halide: bimetallic aryl magnesium/copper halides react with bromoarylacetylenes under iron catalysis to afford the corresponding bi(aryl)acetylene that can be reduced to yield an analogue of combretastatin A-4 (CA4, Fig. 3.52) [107].

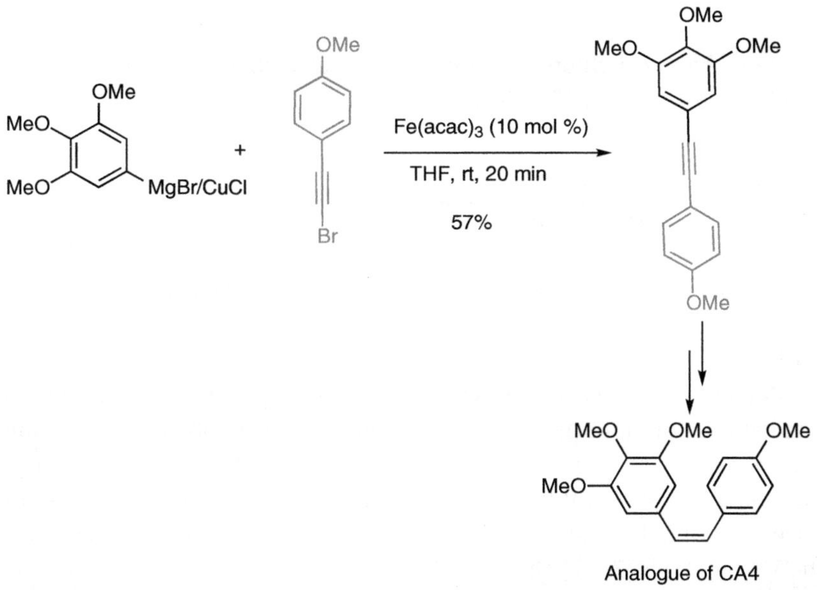

Fig. 3.52: Coupling with an akynyl bromide.

3.3.7 With carbonyl derivatives

The iron-catalyzed coupling between alkyl Grignard reagents and carbonyl derivatives (*vide supra*) was successfully extended to aryl magnesium halides [52–54, 57, 108].

Acyl chlorides

Acyl chlorides are very efficient partners in the iron-catalyzed cross-coupling with aryl magnesium halides (Fig. 3.53).

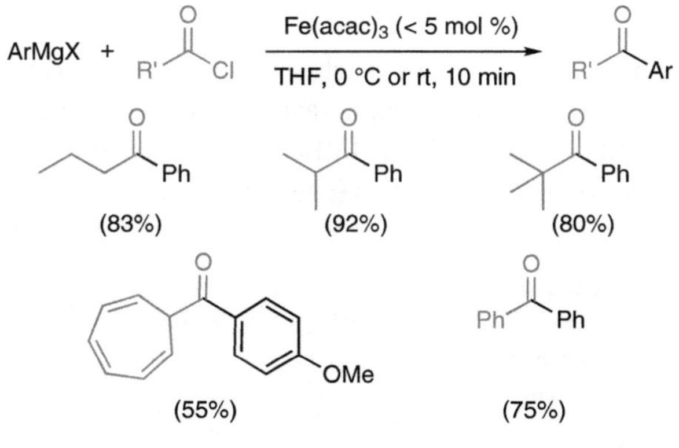

Fig. 3.53: Fe-catalyzed cross-coupling between aryl Grignard reagents and acyl chlorides.

Thioesters

In a lesser extent than acyl chlorides, thioesters are also able to afford ketones by reacting with Grignard reagents (Fig. 3.54).

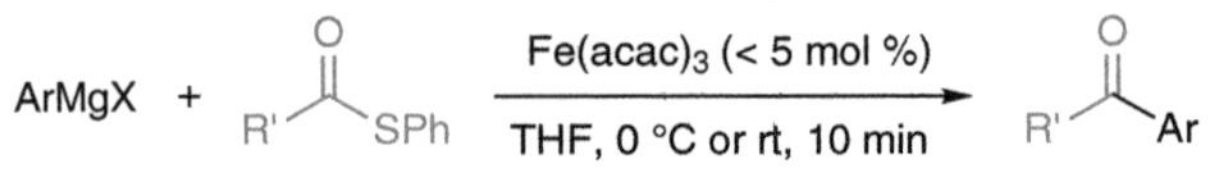

Fig. 3.54: Fe-catalyzed cross-coupling between aryl Grignard reagents and thioesters.

Aroyl cyanides

Knochel, Cahiez *et al.* showed that aroyl cyanides were also excellent partners to access diaryl ketones and that this method was very selective, tolerating esters and even halogens when the reaction was run with $Fe(acac)_3$ at –10 °C for 15 min in THF. For example, a 4-chloro substituted aroyl cyanide was coupled with an aryl Grignard reagent derived from 4-halogenated ethyl benzoate to afford the corresponding benzophenone in 75 % yield (Fig. 3.55) [109].

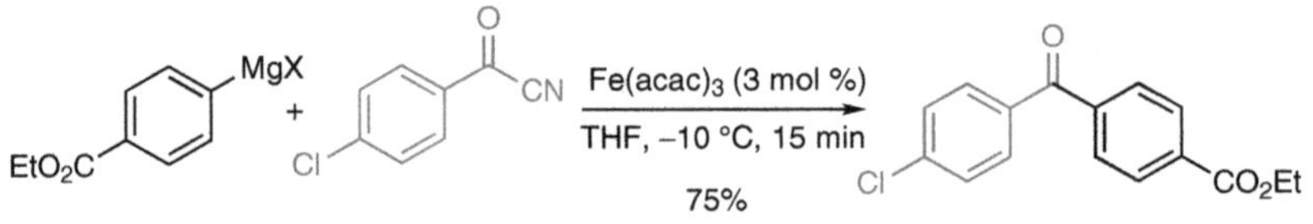

Fig. 3.55: Fe-catalyzed cross-coupling between a functionalized aryl Grignard reagent and an aroyl cyanide.

3.4 Coupling reactions of alkenyl Grignard reagents

3.4.1 With alkyl halides

Alkenyl Grignard reagents have been studied almost exclusively with alkyl electrophiles. Cossy *et al.* reported that vinyl Grignard reagents reacted with primary alkyl iodides in the presence of $FeCl_3$ (10 mol %) in THF and TMEDA (1.9 equiv), at 0 °C, to afford the expected coupled products in excellent yields (Fig. 3.56) [110].

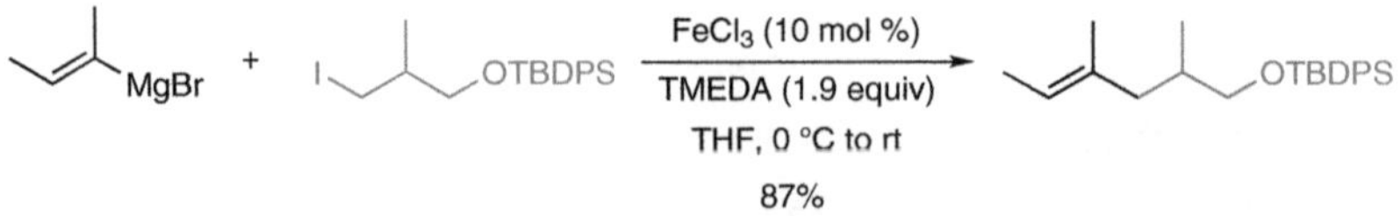

Fig. 3.56: Coupling between vinyl Grignard reagents and alkyl iodides.

Alkyl bromides are also suitable substrates and many publications appeared in the literature dealing with the cross-coupling of Grignard reagents; for instance both Cossy *et al.* and Cahiez *et al.* mentioned the same year that primary and secondary alkyl bromides were also good electrophiles for the iron-catalyzed cross-coupling with vinyl Grignard reagents (Fig. 3.57) [110, 111]. In most cases, vinyl Grignard reagents gave the cross-coupled product but an additive such as TMEDA was necessary when the reaction was run in THF [112–115] whereas in Et_2O, reflux was required [116].

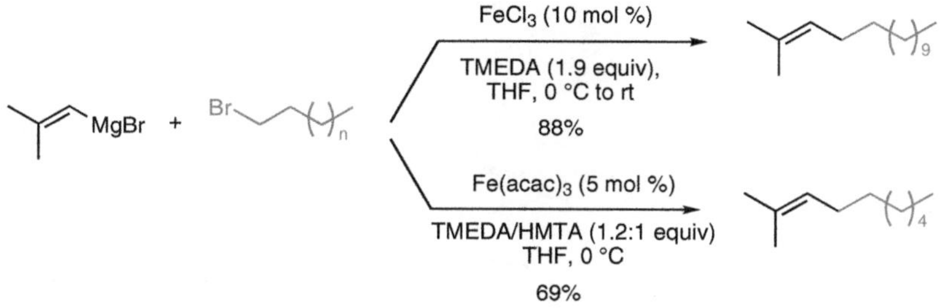

Fig. 3.57: Coupling between vinyl Grignard reagents and alkyl bromides (n = 4, ref. [111]; n = 9, ref. [110]).

Interestingly, during their synthesis of amphidinol 3, Cossy *et al.* reported the coupling of a vinyl Grignard reagent with a *C*-bromopyranoside under iron-catalyzed conditions, which led to an α-*C*-glycoside (Fig. 3.58) [117].

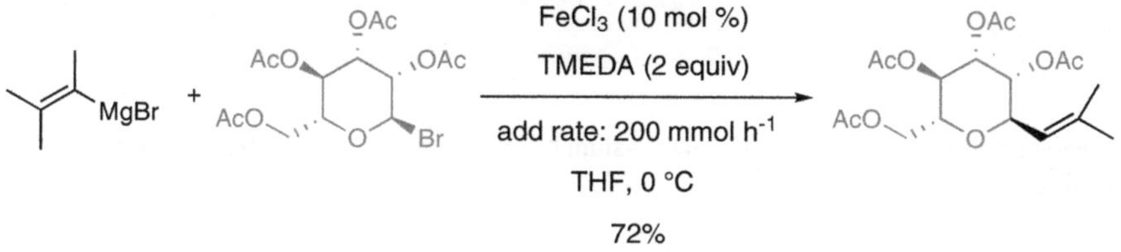

Fig. 3.58: Fe-catalyzed coupling of a vinyl magnesium bromide with a C-bromopyranoside for the synthesis of amphidinol 3.

Yamada *et al.* also reported in their formal total synthesis of (+)-allokainic acid the use of an iron-catalyzed coupling between a vinyl Grignard and a secondary bromo-alkyl derivative in modest yield (Fig. 3.59) [118].

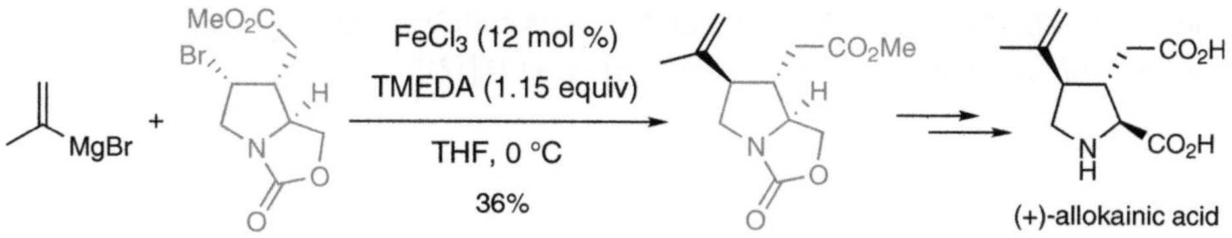

Fig. 3.59: Fe-catalyzed coupling of a vinyl magnesium bromide with a secondary bromo-alkyl derivative for the synthesis of (+)-allokainic acid.

In their total synthesis of spirangien A, both Cossy *et al.* [119] and Rizzacasa *et al.* [120] have used an iron-catalyzed coupling between a vinyl Grignard and a primary alkyl iodide as a key step, and the desired coupled products were obtained in high yields (76 % and 84 %, respectively) when $FeCl_3$ was used in the presence of TMEDA in THF (Fig. 3.60).

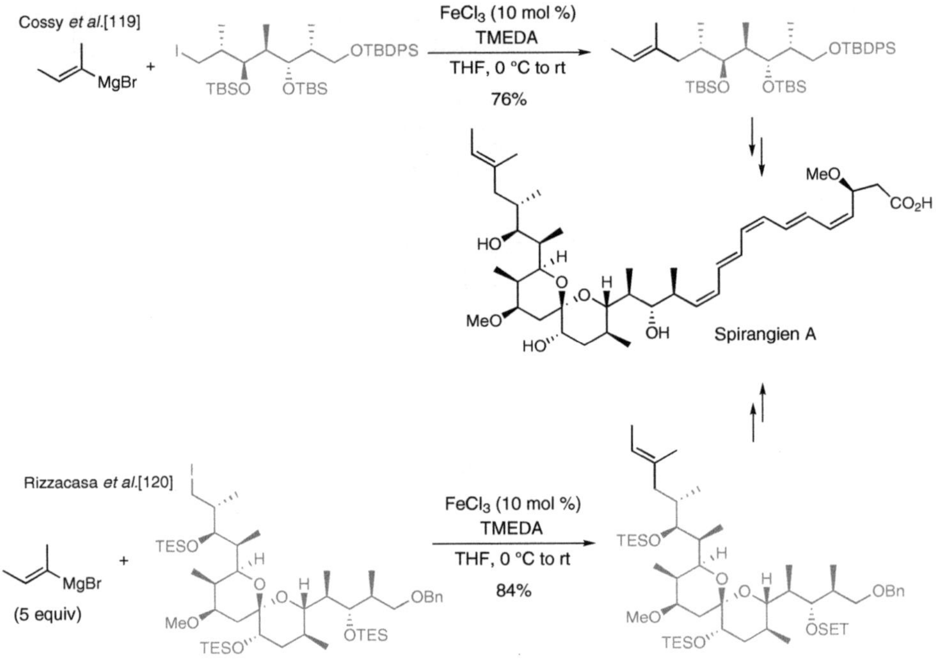

Fig. 3.60: Fe-catalyzed coupling of a vinyl magnesium bromide with a primary alkyl iodide for the synthesis of spirangien A.

3.5 Coupling reactions of alkynyl Grignard reagents

3.5.1 With vinyl halides

A method to create a C(sp^2)–C(sp) bond relies on the coupling between alkynyl magnesium halide and a vinyl halide: various enynes can be efficiently synthesized by forming *in situ* the Grignard reagent and then its coupling with a vinyl partner in the presence of low catalyst loadings of $FeCl_3$ (Fig. 3.61) [121].

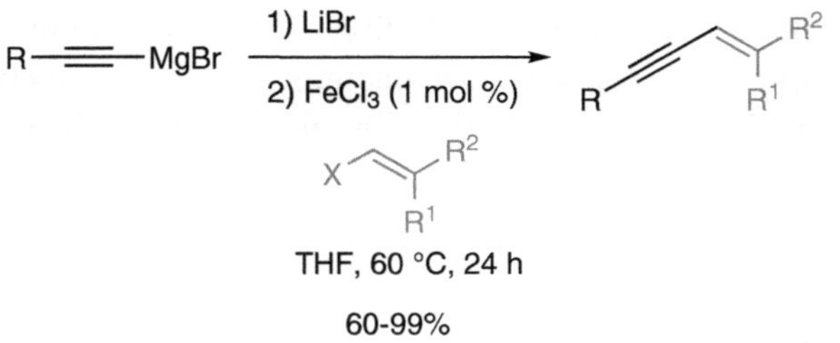

Fig. 3.61: Synthesis of enynes by Fe-catalyzed cross-coupling between alkynyl Grignard reagents and vinyl halides.

3.5.2 With alkyl halides

Independently, Hu *et al.* and Nakamura *et al.* described the cross-coupling reaction of alkynyl Grignard reagents and alkyl bromides, in the presence of iron dibromide or a bisphophine ligand iron dichloride catalyst (Fig. 3.62) [122, 123].

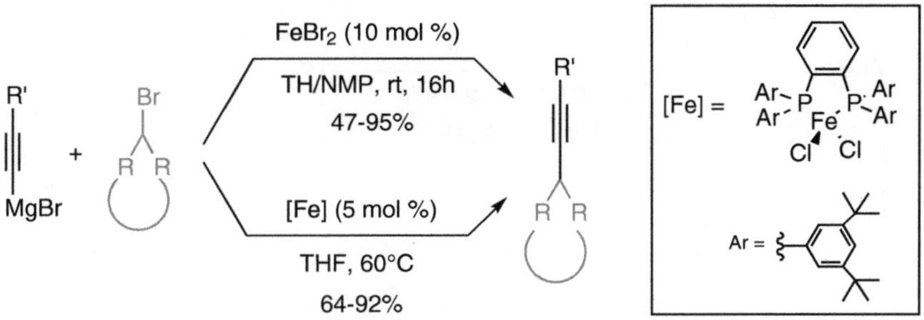

Fig. 3.62: $FeBr_2$ and $FeCl_2$-catalyzed cross-coupling between alkynyl Grignard reagents and alkyl bromides.

3.6 Miscellaneous

3.6.1 Three-component reactions

Von Wangelin *et al.* reported a very interesting hetero-coupling between aryl bromides and alkyl halides in a domino-iron catalysis process where the Grignard reagent(s) is(are) formed *in situ*. In this multicomponent reaction protocol, both halides are added at the same time to a solution containing $FeCl_3$, TMEDA and magnesium turnings in THF at 0 °C, affording alkylbenzenes, with low amounts of biaryl products. The authors stated that the formation of the Grignard species from each halide was likely to occur under the reaction conditions (Fig. 3.63) [124].

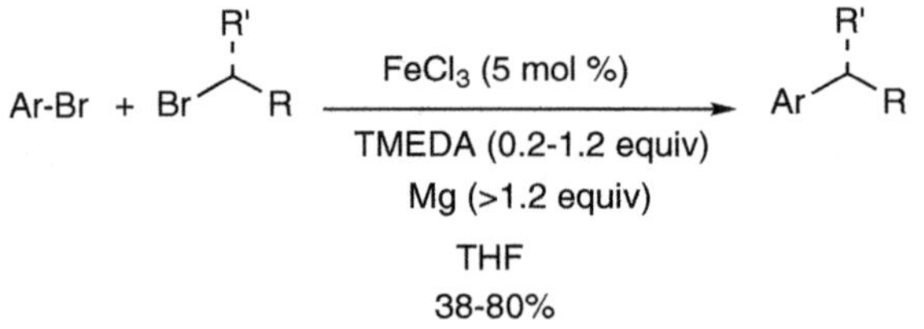

Fig. 3.63: Domino iron-catalysis with *in situ* Grignard reagent formation.

The same authors showed that this three-component reaction could also be applied to vinyl halides affording styrenes or alkylbenzenes according to the partner, with low amounts of biaryl products (Fig. 3.64) [125].

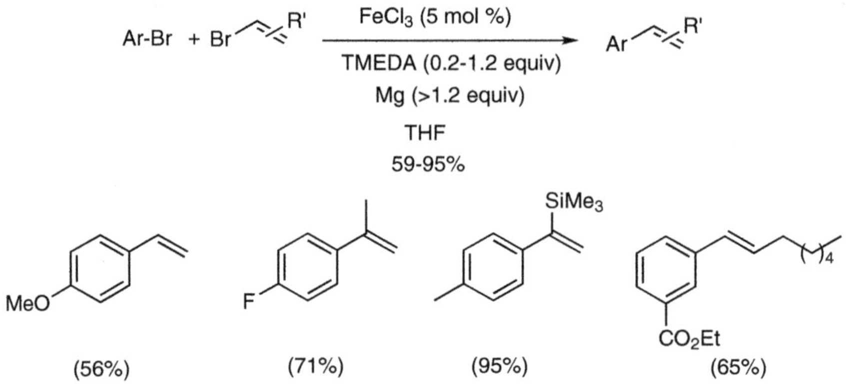

Fig. 3.64: Domino-iron catalysis with *in situ* Grignard reagent formation.

3.6.2 CH activation

In recent years, the iron-catalyzed coupling of Grignard reagents has been successfully extended to C–H functionalization. Furans [126, 127] and allylic positions [128] have been coupled with aryl Grignard reagents under iron catalysis (Fig. 3.65).

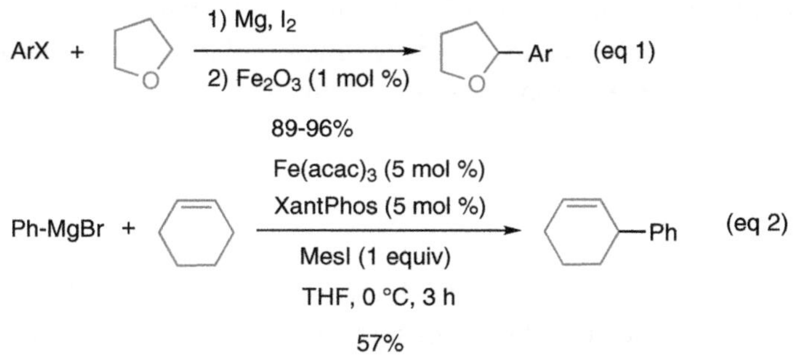

Fig. 3.65: C–C Bond forming *via* Fe-catalyzed C–H activation between aryl Grignard reagent and furan (eq 1) or cyclohexene (eq 2).

Most impressive results have been reported by Nakamura *et al.* who described several C–C bond formations *via* *N*-oriented C–H activation leading to the direct introduction of alkyl or aryl groups (Fig. 3.66) [129–132]

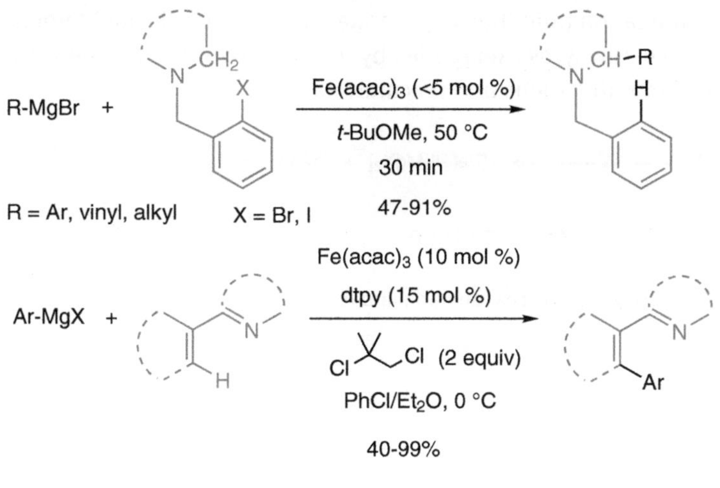

Fig. 3.66: C–C Bond forming *via* Fe-catalyzed *N*-oriented C–H activation with Grignard reagents.

3.7 Mechanistic considerations

In their pioneer report, Kochi and Tamura suggested the formation of an active "reduced form of soluble iron" generated from organomagnesium reagents and iron(II)/iron(III) chlorides [2]. Since then, several experimental [133–136] and theoretical [137–142] investigations on the mechanism of Fe-catalyzed cross-couplings have been performed and reported. As a matter of fact, the nature of the active iron species generated from Fe(III) or Fe(II) and Grignard reagents is strongly connected to the nature of the latter (Ar- or alkyl-MgBr, presence of a hydrogen atom in the β-position, use of ligand, etc.) [136]. To summarize briefly, two main pathways are generally considered involving either a double-electron transfer or a single-electron transfer mechanisms.

3.7.1 Double-electron transfer mechanism

On the basis of previous results obtained by Bogdanovic and Schwickardi [143], Fürstner *et al.* proposed the mechanism shown in Figs. 3.67 and 3.68 [10]. It was previously established that FeX_2 (*eg* X = Cl) reacts with 4 equiv of RMgX to give an "inorganic Grignard reagent" [Fe $(MgX)_2$] [143], and it is suggested that the reduction step yields a highly nucleophilic species with a formal negative charge (–II) at iron (Fig. 3.67). The nucleophilic iron species [Fe(–II)$(MgX)_2$] lacking stabilizing ligands is able to perform oxidative addition with aryl halides (Ar-X). The resulting [Ar-Fe(0)MgX] is alkylated by the excess of the Grignard reagent, RMgX, to form [Ar(R)Fe(0)MgX_2]. Subsequent reductive coupling then affords the coupling product Ar-R, and the Fe(–II) species is regenerated (Fig. 3.68). Thus, organomagnesium reagents that are not susceptible to undergo a β-hydride elimination (and thus to generate Fe(–II) species) should not be able to react through this pathway, as exemplified by the absence of reaction between MeMgBr and chloroarenes with various iron precatalysts [15].

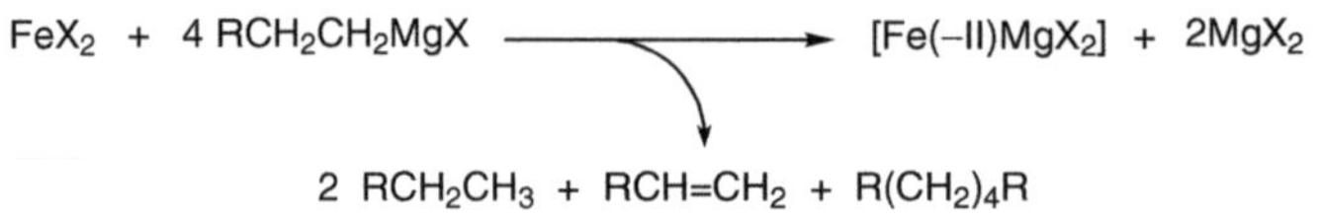

Fig. 3.67: Generation of [Fe(–II)$(MgX)_2$] from FeX_2 and Grignard reagents.

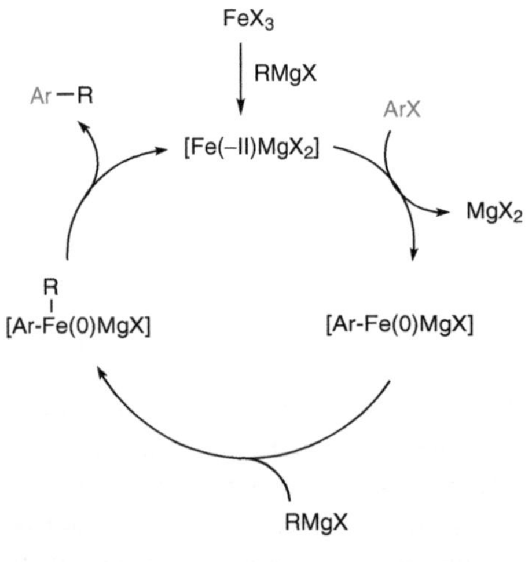

Fig. 3.68: Catalytic cycle proposed by Fürstner *et al.*

3.7.2 Single-electron transfer mechanism

In the early beginning on their work on iron catalysis with Grignard reagents, Kochi *et al.* already postulated that the catalytic process occurred *via* a redox cycle, with generation of radicals during the oxidative addition step [2, 29]. Thirty years later, during their studies with the low valent iron complex $[Li(TMEDA)]_2[Fe(C_2H_4)_4]$ as catalyst, Fürstner and Martin observed that with a 2-iodoacetal bearing an allyl chain, a 5-*exo*-trig ring-closure occurred prior to cross-coupling with PhMgBr (Fig. 3.69) [71]. This behavior could be attributed to the formation of radical intermediates.

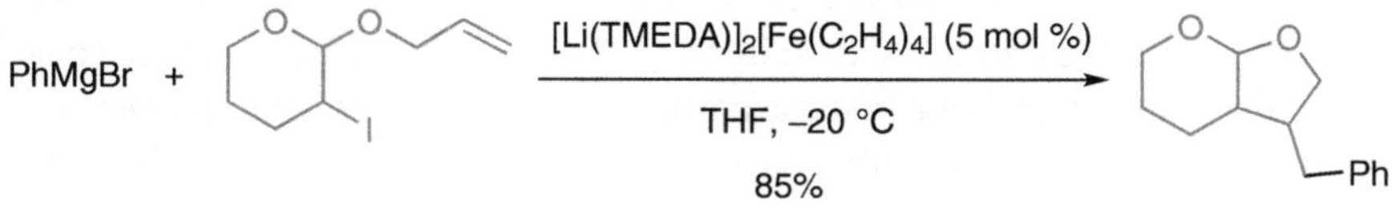

Fig. 3.69: Ring-closure followed by cross-coupling catalyzed by an iron-ate complex.

Bedford *et al.* proposed a mechanism involving SET, which begins with the reduction of the initial Fe(III) precatalyst by the organomagnesium reagent. The reduced [Fe(n)] complex then undergoes a single electron oxidation to generate a radical from the alkyl halide. Transmetalation of the arene from the Grignard reagent with the iron complex is followed by reductive elimination of the cross-coupling product Ar-R (Fig. 3.70) [144].

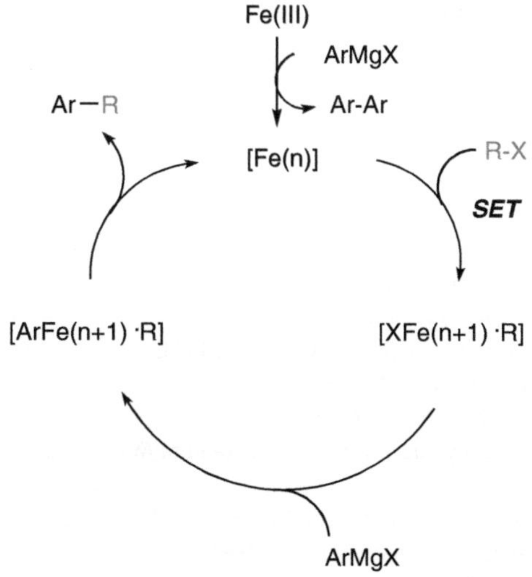

Fig. 3.70: Radical mechanism proposed by Bedford *et al.*

3.8 Conclusion

Iron-catalyzed cross-coupling reactions between Grignard reagents and electrophilic partners were initiated 40 years ago by Kochi and Tamura, and it seemed for a long time limited to un-functionalized molecules, and this cross-coupling appeared thus much less attractive than Kumada-Corriu coupling. However, decisive improvements have been brought since the late 1990s. This has led to a tremendous expansion in the field of application of iron catalysis with organomagnesium compounds. Thus, coupling between functionalized molecules is now possible, starting from a broad range of halides, tosylates, triflates, etc. as partners to create C–C bonds from sp^2, sp^3 and even sp carbon atoms. Beside the easy use of iron catalysts and their non-toxic properties, these coupling reactions are only at their infancy. Their childhood has commenced with the expansion of substrate scope in these cross-coupling reactions, including key steps in total synthesis of elaborated substances, and the recent promising results in C–H activation clearly highlight the high potential of Grignard reagent/iron catalyst coupling.

References

[1] Tamura M, Kochi J, Synthesis 1971, 1971, 303–5.
[2] Tamura M, Kochi JK, J Am Chem Soc 1971, 93, 1487–9.
[3] Cahiez G, Avedissian H, Synthesis 1998, 1199–205.
[4] Cahiez G, Gager O, Buendia J, Patinote C, Chem – Eur J 2012, 18, 5860–3.
[5] Bolm C, Legros J, Le Paih J, Zani L, Chem Rev 2004, 104, 6217–54.
[6] Bauer I, Knölker H-J, Chem Rev 2015, 115, 3170–387.
[7] Shinokubo H, Oshima K, Eur J Org Chem 2004, 2081–91.
[8] Takahashi T, Liu Y, Sci Synth 2004, 7, 597–628.
[9] Fürstner A, Martin R, Chem Lett 2005, 34, 624–9.
[10] Sherry BD, Fürstner A, Acc Chem Res 2008, 41, 1500–11.
[11] Fürstner A, Angew Chem Int Ed 2009, 48, 1364–7.
[12] Czaplik WM, Mayer M, Grupe S, von Wangelin AJ, Pure Appl Chem 2010, 82, 1545–53.
[13] Bedford RB, Acc Chem Res 2015, 48, 1485–93.
[14] Kuzmina O, Steib A, Moyeux A, Cahiez G, Knochel P, Synthesis 2015, 1696–705.
[15] Fürstner A, Leitner A, Méndez M, Krause H, J Am Chem Soc 2002, 124, 13856–63.
[16] Fürstner A, Leitner A, Angew Chem Int Ed 2002, 41, 609–12.
[17] Fürstner A, Leitner A, Seidel G. 4-nonylbenzoic acid. Org Synth 2005, 81, 33–41.
[18] Furstner A, Leitner A, Mendez M, Patent US2003220498 (A1), 2003.
[19] Rushworth PJ, Hulcoop DG, Fox DJ, J Org Chem 2013, 78, 9517–21.
[20] Agata R, Iwamoto T, Nakagawa N, Isozaki K, Hatakeyama T, Takaya H, Nakamura M Synthesis 2015, 47, 1733–40.
[21] Malhotra S, Seng PS, Koenig SG, Deese AJ, Ford KA, Org Lett 2013, 15, 3698–701.
[22] Risatti C, Natalie KJ, Shi Z, Conlon DA, Org Process Res Dev 2013, 17, 257–64.
[23] Fürstner A, Leitner A, Angew Chem Int Ed 2003, 42, 308–11.
[24] Seidel G, Laurich D, Fürstner A, J Org Chem 2004, 69, 3950–2.
[25] Hatakeyama T, Nakamura M, J Am Chem Soc 2007, 129, 9844–5.

[26] Hatakeyama T, Hashimoto S, Ishizuka K, Nakamura M, J Am Chem Soc 2009, 131, 11949–63.
[27] Agrawal T, Cook SP, Org Lett 2013, 15, 96–9.
[28] Agrawal T, Cook SP, Org Lett 2014, 16, 5080–3.
[29] Tamura M, Kochi J, J Organomet Chem 1971, 31, 289–309.
[30] Kochi JK, J Organomet Chem 2002, 653, 11–9.
[31] Dongol KG, Koh H, Sau M, Chai CLL, Adv Synth Catal 2007, 349, 1015–8.
[32] Parmar D, Henkel L, Dib J, Rueping M, Chem Commun 2015, 51, 2111–3.
[33] Luo S, Yu D-G, Zhu R-Y, Wang X, Wang L, Shi Z-J, Chem Commun 2013, 49, 7794–6.
[34] Hamze A, Brion J-D, Alami M, Org Lett 2012, 14, 2782–5.
[35] Fakhfakh MA, Franck X, Hocquemiller R, Figadère B, J Organomet Chem 2001, 624, 131–5.
[36] Dos Santos M, Franck X, Hocquemiller R, Figadère B, Peyrat J-F, Provot O, Brion JD, Alami M, Synlett 2004, 2697–700.
[37] Seck M, Franck X, Hocquemiller R, Figadère B, Peyrat J-F, Provot O, Brion JD, Alami M, Tetrahedron Lett 2004, 45, 1881–4.
[38] Bailly BAFL, Greenhalgh MD, Thomas SP, Chem Commun 2012, 48, 1580–2.
[39] Defretin J, Gleye C, Cortes D, Franck X, Hocquemiller R, Figadère B, Lett Org Chem 2004, 1, 316–22.
[40] Ottesen LK, Ek F, Olsson R, Org Lett 2006, 8, 1771–3.
[41] Li B-J, Xu L, Wu Z-H, Guan B-T, Sun C-L, Wang B-Q, Shi, ZJ, J Am Chem Soc 2009, 131, 14656–7.
[42] Cahiez G, Gager O, Habiak V, Synthesis 2008, 2636–44.
[43] Cahiez G, Habiak V, Gager O, Org Lett 2008, 10, 2389–92.
[44] Rao Volla CM, Vogel P, Angew Chem Int Ed 2008, 47, 1305–7.
[45] Figadère, B, Franck X. S_N2 versus S_N2' Reactions. In: Silverman GS, Rakita PE. Handbook of Grignard Reagents. CRC Press, 1996.
[46] Yanagisawa A, Nomura N, Yamamoto H, Synlett 1991, 1991, 513–4.
[47] Yanagisawa A, Nomura N, Yamamoto H, Tetrahedron 1994, 50, 6017–28.
[48] Nakamura M, Matsuo K, Inoue T, Nakamura E, Org Lett 2003, 5, 1373–5.
[49] Pasto DJ, Hennion GF, Shults RH, Waterhouse A, Chou S-K, J Org Chem 1976, 41, 3496–3496.
[50] Pasto DJ, Chou S-K, Waterhouse A, Shults RH, Hennion GF, J Org Chem 1978, 43, 1385–8.
[51] Lepage O, Kattnig E, Fürstner A, J Am Chem Soc 2004, 126, 15970–1.
[52] Fiandanese V, Marchese G, Martina V, Ronzini L, Tetrahedron Lett 1984, 25, 4805–8.
[53] Cardellicchio C, Fiandanese V, Marchese G, Ronzini L, Tetrahedron Lett 1985, 26, 3595–8.
[54] Cardellicchio C, Fiandanese V, Marchese G, Ronzini L, Tetrahedron Lett 1987, 28, 2053–6.
[55] Fiandanese V, Marchese G, Naso F, Tetrahedron Lett 1988, 29, 3587–90.
[56] Babudri F, D'Ettole A, Fiandanese V, Marchese G, Naso F, J Organomet Chem 1991, 405, 53–8.
[57] Dell'Anna MM, Mastrorilli P, Nobile CF, Marchese G, Taurino MR, J Mol Catal Chem 2000, 161, 239–43.
[58] Scheiper B, Bonnekessel M, Krause H, Fürstner A, J Org Chem 2004, 69, 3943–9.
[59] Hocek M, Dvořáková H, J Org Chem 2003, 68, 5773–6.
[60] Boully L, Turck A, Plé N, Darabantu M, J Heterocycl Chem 2005, 42, 1423–8.
[61] Dickschat JS, Reichenbach H, Wagner-Döbler I, Schulz S, Eur J Org Chem 2005, 2005, 4141–53.
[62] Colacino E, Benakki H, Guenoun F, Martinez J, Lamaty F, Synth Commun 2009, 39, 1583–91.
[63] Haner J, Jack K, Nagireddy J, Abdul Raheem M, Durham R, Tam W, Synthesis 2011, 731–8.
[64] Quintin J, Franck X, Hocquemiller R, Figadère B, Tetrahedron Lett 2002, 43, 3547–9.
[65] Gøgsig TM, Lindhardt AT, Skrydstrup T, Org Lett 2009, 11, 4886–8.
[66] Gülak S, von Wangelin AJ, Angew Chem Int Ed 2012, 51, 1357–61.
[67] Cahiez G, Chaboche C, Mahuteau-Betzer F, Ahr M, Org Lett 2005, 7, 1943–6.
[68] Yamaguchi Y, Ando H, Nagaya M, Hinago H, Ito T, Asami M, Chem Lett 2011, 40, 983–5.
[69] Jin M, Nakamura M, Chem Lett 2011, 40, 1012–4.
[70] Bauer G, Cheung C, Hu X, Synthesis 2015, 47, 1726–32.
[71] Martin R, Fürstner A, Angew Chem Int Ed 2004, 43, 3955–7.

[72] Barré B, Gonnard L, Campagne R, Reymond S, Marin J, Ciapetti P, Brellier M, Guérinot A, Cossy J, Org Lett 2014, 16, 6160–3.
[73] Ghorai SK, Jin M, Hatakeyama T, Nakamura M, Org Lett 2012, 14, 1066–9.
[74] Steib AK, Thaler T, Komeyama K, Mayer P, Knochel P, Angew Chem Int Ed 2011, 50, 3303–7.
[75] Qian X, Kozak CM, Synlett 2011, 852–6.
[76] Mo Z, Zhang Q, Deng L, Organometallics 2012, 31, 6518–21.
[77] Bica K, Gaertner P, Org Lett 2006, 8, 733–5.
[78] Xia C-L, Xie C-F, Wu Y-F, Sun H-M, Shen Q, Zhang Y, Org Biomol Chem 2013, 11, 8135–44.
[79] Gao H, Yan C, Tao X-P, Xia Y, Sun H-M, Shen Q, Zhang Y, Organometallics 2010, 29, 4189–92.
[80] Meyer S, Orben CM, Demeshko S, Dechert S, Meyer F, Organometallics 2011, 30, 6692–702.
[81] Bedford RB, Bruce DW, Frost RM, Goodby JW, Hird M, Chem Commun 2004, 2822–3.
[82] Bedford RB, Betham M, Bruce DW, Davis SA, Frost RM, Hird M, Chem Commun 2006, 1398–400.
[83] Chowdhury RR, Crane AK, Fowler C, Kwong P, Kozak CM. Chem Commun 2008, 94–6.
[84] Denmark SE, Cresswell AJ, J Org Chem 2013, 78, 12593–628.
[85] Neumann SM, Kochi JK, J Org Chem 1975, 40, 599–606.
[86] Smith RS, Kochi JK, J Org Chem 1976, 41, 502–9.
[87] Fabre J-L, Julia M, Verpeaux J-N, Tetrahedron Lett 1982, 23, 2469–72.
[88] Molander GA, Rahn BJ, Shubert DC, Bonde SE, Tetrahedron Lett 1983, 24, 5449–52.
[89] Fiandanese V, Miccoli G, Naso F, Ronzini L, J Organomet Chem 1986, 312, 343–8.
[90] Alvarez E, Cuvigny T, du Penhoat CH, Julia M, Tetrahedron 1988, 44, 119–26.
[91] Cahiez G, Marquais S, Tetrahedron Lett 1996, 37, 1773–6.
[92] Dohle W, Kopp F, Cahiez G, Knochel P, Synlett 2001, 2001, 1901–4.
[93] Le Marquand P, Tsui GC, Whitney JCC, Tam W, J Org Chem 2008, 73, 7829–32.
[94] Sherry BD, Fürstner A, Chem Commun 2009, 7116–8.
[95] Nishikado H, Nakatsuji H, Ueno K, Nagase R, Tanabe Y, Synlett 2010, 2087–92.
[96] Camacho-Davila AA, Synth Commun 2008, 38, 3823–33.
[97] Tewari N, Maheshwari N, Medhane R, Nizar H, Prasad M, Org Process Res Dev 2012, 16, 1566–8.
[98] Shakhmaev RN, Sunagatullina AS, Zorin VV, Russ J Org Chem 2014, 50, 322–31.
[99] Alvarez E, Cuvigny T, Hervé du Penhoat C, Julia M, Tetrahedron 1988, 44, 111–8.
[100] Itami K, Higashi S, Mineno M, Yoshida J, Org Lett 2005, 7, 1219–22.
[101] Fürstner A, Martin R, Krause H, Seidel G, Goddard R, Lehmann CW, J Am Chem Soc 2008, 130, 8773–87.
[102] Mayer M, Czaplik WM, von Wangelin AJ, Adv Synth Catal 2010, 352, 2147–52.
[103] Volla CMR, Markovic D, Dubbaka SR, Vogel P, Eur J Org Chem 2009, 6281–8.
[104] Felkin H, Meunier B, J Organomet Chem 1978, 146, 169–78.
[105] Weber K, Schnoeckelborg E-M, Wolf R, ChemCatChem 2011, 3, 1572–7.
[106] Kawamura S, Nakamura M, Chem Lett 2013, 42, 183–5.
[107] Castagnolo D, Botta M, Eur J Org Chem 2010, 2010, 3224–8.
[108] Ritter K, Hanack M, Tetrahedron Lett 1985, 26, 1285–8.
[109] Duplais C, Bures F, Sapountzis I, Korn TJ, Cahiez G, Knochel P, Angew Chem Int Ed 2004, 43, 2968–70.
[110] Guérinot A, Reymond S, Cossy J, Angew Chem Int Ed 2007, 46, 6521–4.
[111] Cahiez G, Duplais C, Moyeux A, Org Lett 2007, 9, 3253–4.
[112] Nakamura M, Matsuo K, Ito S, J Am Chem Soc 2004, 126, 3686–7.
[113] Bedford RB, Bruce DW, Frost RM, Hird M, Chem Commun 2005, 4161–3.
[114] Cahiez G, Habiak V, Duplais C, Moyeux A, Angew Chem Int Ed 2007, 46, 4364–6.
[115] Noda D, Sunada Y, Hatakeyama T, Nakamura M, Nagashima H, J Am Chem Soc 2009, 131, 6078–9.
[116] Nagano T, Hayashi T, Org Lett 2004, 6, 1297–9.
[117] Bensoussan C, Rival N, Hanquet G, Colobert F, Reymond S, Cossy J, Tetrahedron 2013, 69, 7759–70.

[118] Yamada K, Sato T, Hosoi M, Yamamoto Y, Tomioka K, Chem Pharm Bull 2010, 58, 1511–6.
[119] Guerinot A, Lepesqueux G, Sable S, Reymond S, Cossy J, J Org Chem 2010, 75, 5151–63.
[120] Gregg C, Gunawan C, Ng AWY, Wimala S, Wickremasinghe S, Rizzacasa MA, Org Lett 2013, 15, 516–9.
[121] Hatakeyama T, Yoshimoto Y, Gabriel T, Nakamura M, Org Lett 2008, 10, 5341–4.
[122] Cheung CW, Ren P, Hu X, Org Lett 2014, 16, 2566–9.
[123] Hatakeyama T, Okada Y, Yoshimoto Y, Nakamura M, Angew Chem Int Ed 2011, 50, 10973–6.
[124] Czaplik WM, Mayer M, von Wangelin AJ, Angew Chem Int Ed 2009, 48, 607–10.
[125] Czaplik WM, Mayer M, von Wangelin AJ, ChemCatChem 2011, 3, 135–8.
[126] Singh PP, Gudup S, Ambala S, Singh U, Dadhwal S, Singh B, Sawant, SD, Vishwakarma, RA, Chem Commun 2011, 47, 5852–4.
[127] Singh PP, Gudup S, Aruri H, Singh U, Ambala S, Yadav M, Sawant, SD, Vishwakarma, RA, Org Biomol Chem 2012, 10, 1587–97.
[128] Sekine M, Ilies L, Nakamura E, Org Lett 2013, 15, 714–7.
[129] Yoshikai N, Mieczkowski A, Matsumoto A, Ilies L, Nakamura E, J Am Chem Soc 2010, 132, 5568–9.
[130] Ilies L, Asako S, Nakamura E, J Am Chem Soc 2011, 133, 7672–5.
[131] Yoshikai N, Asako S, Yamakawa T, Ilies L, Nakamura E, Chem – Asian J 2011, 6, 3059–65.
[132] Ilies L, Kobayashi M, Matsumoto A, Yoshikai N, Nakamura E, Adv Synth Catal 2012, 354, 593–6.
[133] Fürstner A, Krause H, Lehmann CW, Angew Chem Int Ed 2006, 45, 440–4.
[134] Fürstner A, Martin R, Krause H, Seidel G, Goddard R, Lehmann CW, J Am Chem Soc 2008, 130, 8773–87.
[135] Bedford RB, Brenner PB, Carter E, Cogswell PM, Haddow MF, Harvey JN, Murphy, DM, Nunn, J, Woodall C, Angew Chem Int Ed 2014, 53, 1804–8.
[136] Lefèvre G, Jutand A, Chem – Eur J 2014, 20, 4796–805.
[137] Kleimark J, Hedström A, Larsson P-F, Johansson C, Norrby P-O, ChemCatChem 2009, 1, 152–61.
[138] Ren Q, Guan S, Jiang F, Fang J, J Phys Chem A 2013, 117, 756–64.
[139] Hedström A, Lindstedt E, Norrby P-O, J Organomet Chem 2013, 748, 51–5.
[140] Hedström A, Bollmann U, Bravidor J, Norrby P-O, Chem - Eur J 2011, 17, 11991–3.
[141] Kleimark J, Larsson P-F, Emamy P, Hedström A, Norrby P-O, Adv Synth Catal 2012, 354, 448–56.
[142] Hedström A, Izakian Z, Vreto I, Wallentin C-J, Norrby P-O, Chem – Eur J 2015, 21, 5946–53.
[143] Bogdanović B, Schwickardi M, Angew Chem Int Ed 2000, 39, 4610–2.
[144] Bedford RB, Betham M, Bruce DW, Danopoulos AA, Frost RM, Hird M, J Org Chem 2006, 71, 1104–10.

Alice Rérat, Corinne Gosmini

4 Grignard Reagents and Cobalt

4.1 Introduction

Transition metal-catalyzed C–C cross-couplings are among the most powerful methods that have received considerable attention from organic chemists over the last few years. These approaches involve an organometallic species as the nucleophilic reagent, and an electrophilic compound such as an organic or pseudo-halide. Classically, various organometallic reagents such as organomagnesium, organoboron, organostannane or organozinc reagents can be employed. However, Grignard reagents are among the most reactive and most frequently employed substrates to form carbon–carbon bonds with various catalysts [1]. Although Kharasch's first seminal works on cobalt-catalyzed C–C bond formation with organomagnesium reagents were reported in 1943 [2], Kumada and Corriu independently developed the cross-coupling of Grignard reagents with a nickel catalyst only 30 years later. Now, the most common catalysts are palladium and nickel. Moreover, the development of broadly applicable, more eco-compatible and economical catalysts are an important goal in modern chemistry. In fact, these palladium and nickel catalysts have some drawbacks even if they are efficient, such as the high cost of palladium and the toxicity of nickel catalysts. Fortunately, inexpensive alternative catalyst systems are available based upon iron and cobalt salts. However, cobalt catalysts sometimes present a higher reactivity than iron catalysts for various carbon–carbon bond-forming reactions although elegant works have been developed with iron-based catalysis. The cobalt-catalyzed cross-coupling of Grignard reagents has attracted considerable attention since Kharasch's pioneering work and several groups have reported elegant works. Cobalt and organomagnesium compounds are easily available and fairly inexpensive. In addition, the more recent discovery of mild protocols for the formation of organomagnesium compounds allows the presence of reactive functions such as ester, nitrile, nitro and other functions [3]. This chapter describes different methods using various Grignard reagents and cobalt catalysis in order to form C–C bonds.

4.2 Coupling reactions of alkylmagnesium reagents

4.2.1 With aryl halides

Since the preliminary work of Kharasch in the middle of the 20th century concerning the cobalt-catalyzed reaction of aliphatic Grignard reagents with different electrophiles [4–5], Davies *et al.* reported the first cobalt-catalyzed arylation of alkyl Grignard reagents. They showed that the cross-coupling of methylmagnesium iodide with bromopyridines was improved in the presence of cobalt chloride to form substituted picolines (Fig. 4.1) [6].

The formation of the methylpyridine is interpreted as a nucleophilic substitution of the bromide induced by cobalt chloride by coordination with the nitrogen of the pyridine. It is worth pointing out that the nucleophilic substitution occurs more slowly with 3-bromopyridines leading to lower yields. Moreover, methyl magnesium halides, which are the most reactive Grignard reagents for nucleophilic substitution of bromopyridines, lead to moderate yields (26–60 %).

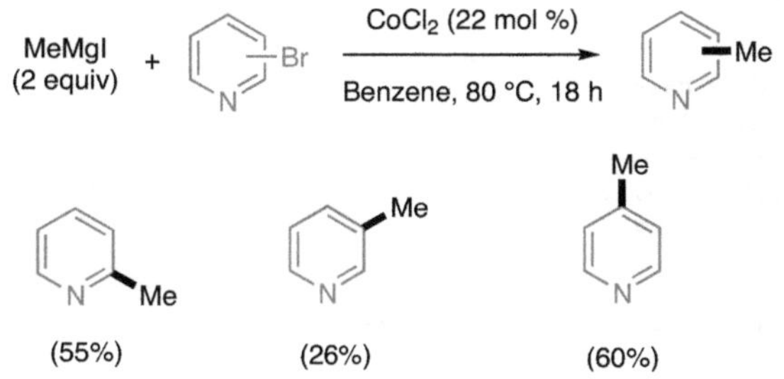

Fig. 4.1: Cobalt-catalyzed cross-coupling of bromopyridines with methylmagnesium iodide.

More recently in 2004, Oshima *et al.* reported the cross-coupling of 2-chloropyridines with benzylmagnesium chloride or trimethylsilylmethyl Grignard reagent in the presence of a catalytic amount of cobalt(II) acetylacetonate in dioxane. Selected examples are shown in Fig. 4.2. In this case, the cross-coupling products were obtained in good to excellent yields [7]. This method was extended to other chloropyridine derivatives. From a mechanistic point of view, there is probably no 2-pyridylcobalt or 2-pyridyl radical intermediate as the cyclization product with a pyridine, bearing an alkene, is not observed. A cobalt-mediated nucleophilic aromatic substitution cannot be excluded.

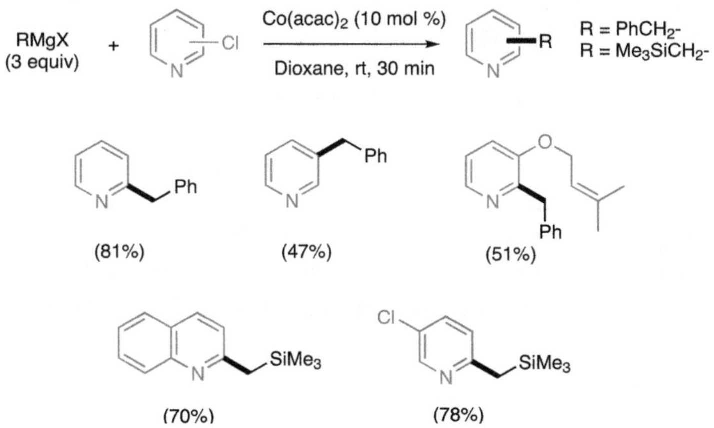

Fig. 4.2: Cobalt-catalyzed benzylation and trimethylsilylmethylation of chloropyridine derivatives.

To extend the scope, the same group reported later the cobalt-catalyzed cross-coupling of various aryl bromides with primary alkyl Grignard reagents in the presence of a stoichiometric amount of *N,N,N',N'*-TetraMethyl-1,3-PropaneDiAmine (TMPDA), a catalytic amount of cobalt chloride (5 mol %) and an *N*-heterocyclic carbene (IMes·HCl) (6 mol %) [8]. Aryl bromides bearing various substituents such as Me, MeO, CF_3, Cl, Me_2N, OTHP can be involved in the cross-coupling with different primary alkylmagnesium chlorides to produce the coupling products in good to high yields (Fig. 4.3). However, the coupling with the methyl Grignard reagent is not efficient and an excess of the Grignard reagent is required.

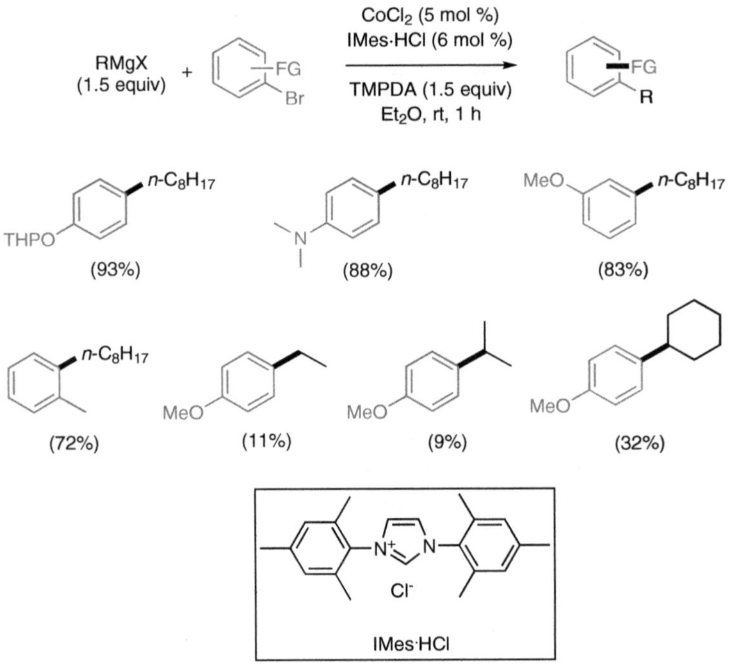

Fig. 4.3: Cobalt-catalyzed cross-coupling of aryl bromides with alkyl Grignard reagents.

More recently, different cobalt complexes containing aryl-substituted NHC ligands have been prepared and characterized employing $CoCl_2(IMes)_2$ as the precatalyst. Their catalytic efficacy has been examined in the Kumada coupling with alkyl Grignard reagents. However, substantially lower conversion was observed for coupling between alkyl Grignard reagents and aryl electrophiles than with aryl Grignard reagents [9].

4.2.2 With alkyl halides

One major limitation of the cross-coupling with alkyl halides is the presence of β-hydrogen atoms. These reagents are subject to slow oxidative addition and subsequently to rapid bhydride elimination hampering the cross-coupling process. However, the β-elimination is reduced in the case of cobalt catalytic complexes due to their easy oxidative addition into alkyl halides. Therefore, cobalt is a good metal to couple alkyl halides.

In 2004, Oshima *et al.* found that the cobalt complexes [$CoCl_2$(dppe)] or [$CoCl_2$-(dppp)] catalyzed cross-coupling between alkyl halides and benzyl Grignard reagents [10]. Selected examples are reported in Fig. 4.4. The number of methylene units in the phosphine ligand between the two phosphorous atoms is important. From a mechanistic point of view, an electron transfer was demonstrated to take place.

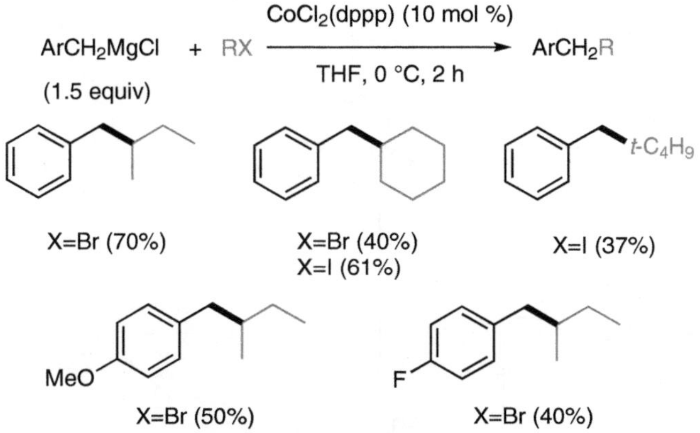

Fig. 4.4: Cobalt-catalyzed cross-coupling of benzylic Grignard reagents with alkyl halides.

One year later, the same group reported the cobalt-catalyzed regio- and stereoselective synthesis of 1,2-di- and 1,1,2-tri-silyethenes from silylmagnesium reagents and silyl-substituted dibromomethanes (Fig. 4.5) [11]. These products are used as precursors to form various alkenes bearing a silyl group.

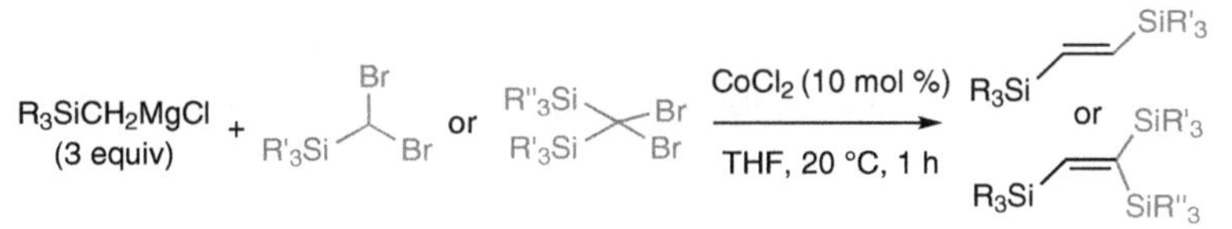

Fig. 4.5: Cobalt-mediated reaction of silylmethylmagnesium reagents with silyl-substituted dibromomethanes.

The catalytic version is only efficient for dibromosilylmethanes. In the case of dibromodisilylmethanes, a stoichiometric amount of cobaltate reagent prepared from $CoCl_2$ and 4 equivalents of Grignard reagents are necessary to access to 1,1,2-trisilyethenes. However, the attempted synthesis of tetrasilylethene from $(R_3Si)_2CBr_2$and $(R_3Si)_2CHMgCl$ was not successful. A mechanism was proposed for the stoichiometric reaction and can explained the formation of the major *E*-stereoisomer. This stereoisomer is coming from the more stable eclipsed conformer upon β-hydride elimination (Fig. 4.6).

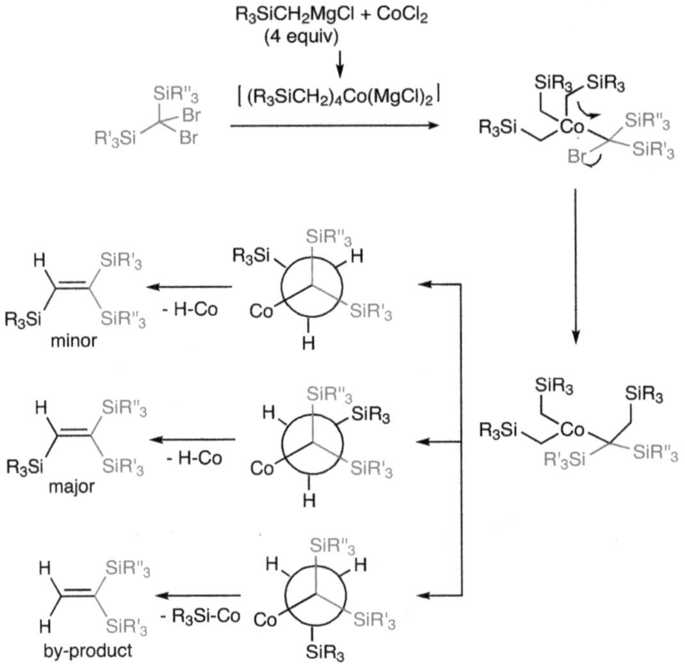

Fig. 4.6: Proposed mechanism for the reaction of silylmethylmagnesium reagents with silyl-substituted dibromomethanes.

The coupling of unactivated secondary alkyl halides remains a challenge although few couplings were described between alkyl Grignard reagents and secondary halides under nickel and palladium catalysis. The problem is that these secondary alkyl halides undergo very slow oxidative addition to low valent transition metal whereas the corresponding secondary alkyl metal intermediates undergo rapid β-elimination.

In 2008, Cahiez's group reported the first example of a cobalt-catalyzed alkyl–alkyl cross-coupling of unactivated secondary halides with Grignard reagents [12]. As described previously for the coupling of aryl Grignard reagents with secondary alkyl halides, the presence of an amine is essential to increase the yield of the coupling product. In this case, a catalytic amount of *N,N*-TetraMethylEthyleneDiAmine (TMEDA) with an ate complex $CoCl_2 \cdot 2MI$ (with $M = Li$ or Mg) leads to high yields (Fig. 4.7). The iodide anion has a beneficial influence on the cross-coupling. However, the yields depend on the steric effects, the length of the alkyl chain and the nature of the halide (no reaction with alkyl chlorides). The reaction is also possible with primary alkyl bromides but not with tertiary alkyl bromides. Only primary alkyl Grignard reagents can be used. This reaction is chemoselective as various functional groups were tolerated on the alkyl halide.

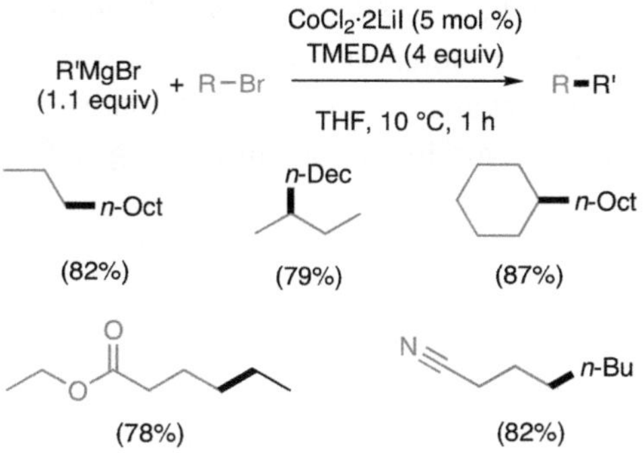

Fig. 4.7: Cobalt-catalyzed cross-coupling between alkylmagnesium reagents and primary and secondary alkyl halides.

Generally, cobalt complexes are inert towards the more readily available alkyl chlorides. However, unusual heterobimetallic complexes associating cobalt and zirconium, linked by three phosphanylamide ligands, allow the cross-coupling with alkyl chlorides [13]. Moreover, these complexes give slightly better yields in the cross-coupling products with primary and secondary alkyl bromides than with the monometallic cobalt complexes. A few examples are reported in Fig. 4.8.

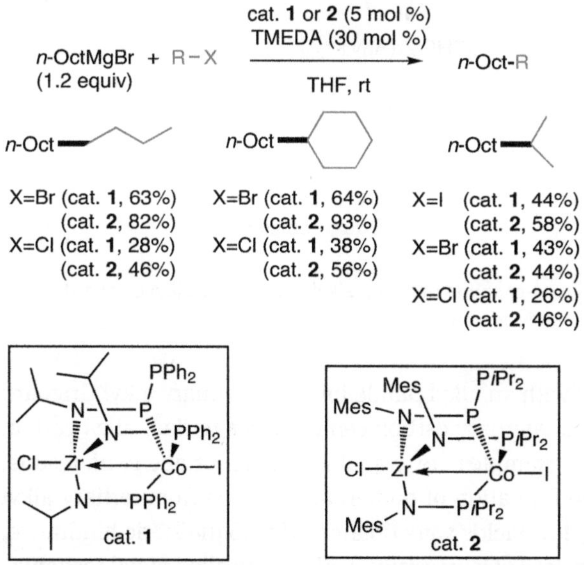

Fig. 4.8: Cross-coupling between alkyl halides and alkyl Grignard reagents catalyzed by Co/Zr.

The unusually efficient catalytic performance of these heterobimetallic complexes is certainly relied on the zirconium and cobalt interactions. In this reaction, TMEDA is also necessary to increase the yield by chelation to the Grignard reagent contrary to previous cases where TMEDA is coordinating to the transition metal salt. The slight differences between the alkyl chlorides and bromides reactivity can be explained by the mechanism. A radical mechanism in the case of alkyl chlorides is proposed compared to an oxidative addition in the case of alkyl bromides. Moreover, the low yields obtained with aryl halides can be explained by a radical mechanism due to the instability of the phenyl radical. The authors proposed a mechanism for this cross-coupling (Fig. 4.9).

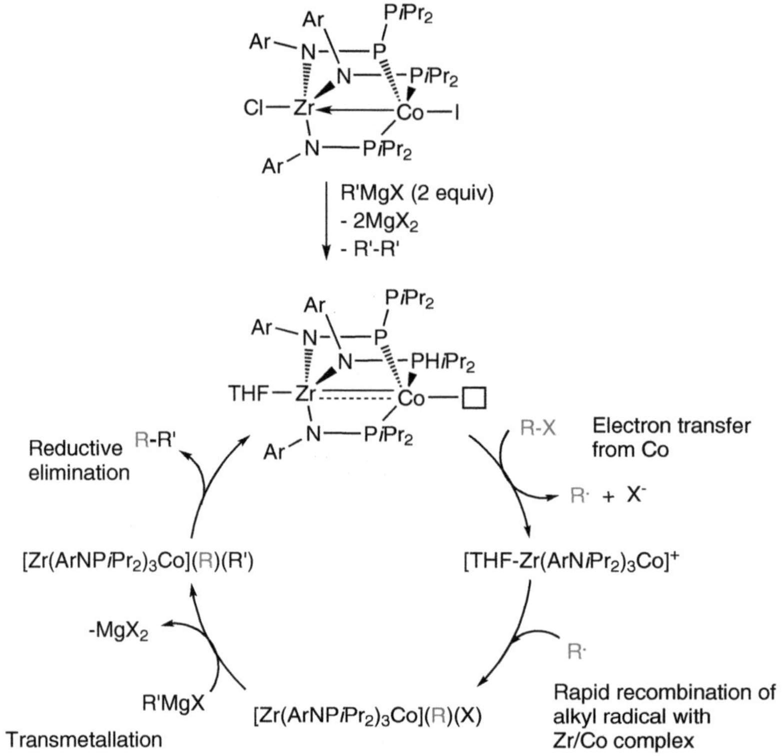

Fig. 4.9: Proposed mechanism for the cross-coupling between alkyl halides and alkyl Grignard reagents catalyzed by Co/Zr heterobimetallic complexes.

All these previous cross-coupling with an alkyl halide involve a primary alkyl Grignard reagent. Although formation of a quaternary carbon center from a tertiary alkyl carbon is a promising route, only limited examples are available. Before Kambe *et al.* results, only copper catalysts allowed the formation of quaternary carbons from tertiary alkyl Grignard [14]. However, generally low yields were obtained due to the steric hindrance, facile β-hydrogen elimination and ease of isomerization of tertiary alkyl metal reagents.

Following on their previous work on cross-coupling involving alkyl halides with alkyl Grignard reagents, Kambe's group reported a Co/LiI/1,3-butadiene catalytic system as a new access to alkyl–alkyl compounds (Fig. 4.10) [15]. This new route affords branched alkanes with quaternary carbon centers by cross-coupling of alkyl halides with tertiary alkyl Grignard reagents. Using various catalysts such as Ni, Pd, Cu or Co, the reaction with tertiary Grignard reagents was only successful with cobalt. This catalytic system is also efficient for the coupling of primary and secondary alkyl Grignard reagents. Various alkyl electrophiles are used such as fluoro-, bromo-, iodo- and tosyl- alkyls bearing various functionalized groups with a relative reactivities of I>OTs>Br>F>>Cl. Good to excellent yields were obtained with tertiary alkyl Grignard reagents even if yields are slightly better than with secondary Grignard reagents. This reaction is chemoselective and the reaction occurs only at the Csp_3 –Br when a Csp_2 –Br bond is present. After a mechanistic study, an ionic mechanism without formation of radicals as intermediates from alkyl bromides has been proposed. In fact, there was no proof of a radical when 5-hexenyl and cyclopropylmethyl groups were present. This result suggested that a S_N2 and a C–C bond formation proceeded with stereoinversion.

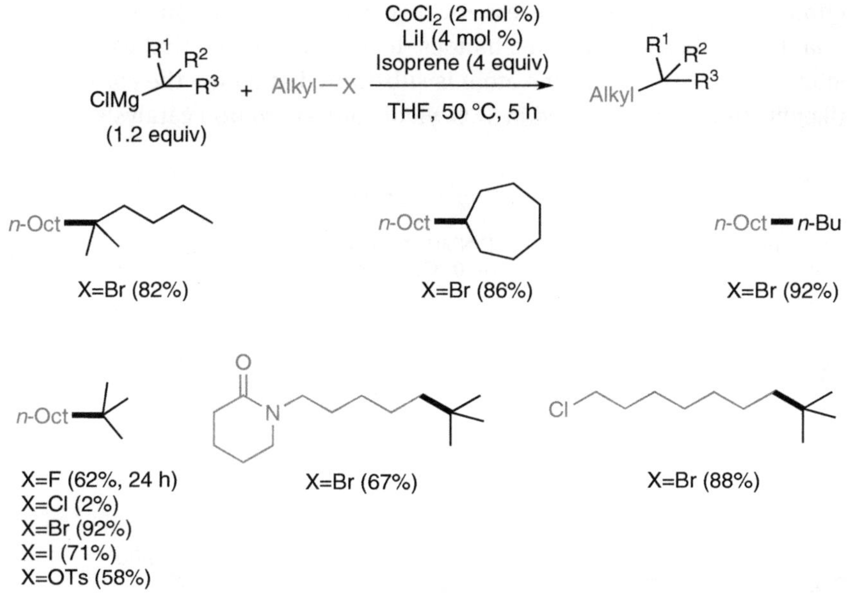

Fig. 4.10: Cobalt-catalyzed cross-coupling of alkyl halides with tertiary alkylmagnesium chlorides.

After optimization of the conditions, the catalyst loading could be reduced to 0.1 mol % with the same yield and the reaction could be easily scale up [16].

4.2.3 With alkenyl species

4.2.3.1 Alkenyl halides

In 1943, Kharasch showed the first example of a cross-coupling between an alkyl Grignard reagent and a vinyl halide in the presence of cobalt chloride [4]. The cross-coupling occurs only when the unsaturated carbon atom bearing the halogen atom also bears a hydrogen atom.

In 1998, Cahiez *et al.* reported an improvement of this cobalt-catalyzed alkenylation of Grignard reagents. He developed a procedure for the cross-coupling of alkenyl iodides, bromides and chlorides with alkyl-Grignard reagent in the presence of $Co(acac)_2$ and *N*-methylpyrrolidone (NMP) (Fig. 4.11) [17]. Previously, the authors had already shown that the use of NMP as a cosolvent had a remarkable beneficial influence on the Fe-catalyzed alkenylation of RMgX [18]. In the case of cobalt-catalyzed reactions, good yields were obtained with four to nine equivalents of NMP. Moreover, this reaction is chemo- and stereoselective. This coupling is efficient when primary and secondary alkylmagnesium reagents are involved in the reaction but not with tertiary ones since β-elimination is competitive. Moreover, this reaction is less efficient with less reactive alkenyl halides such as alkenyl chlorides and β-disubstituted alkenyl halides. It should be noted that with (*E*)-1,2-dichloroethylene, only one chlorine atom is substituted and only the corresponding (*E*)-alkenyl chloride is observed in 57 % yield contrary to iron catalysis.

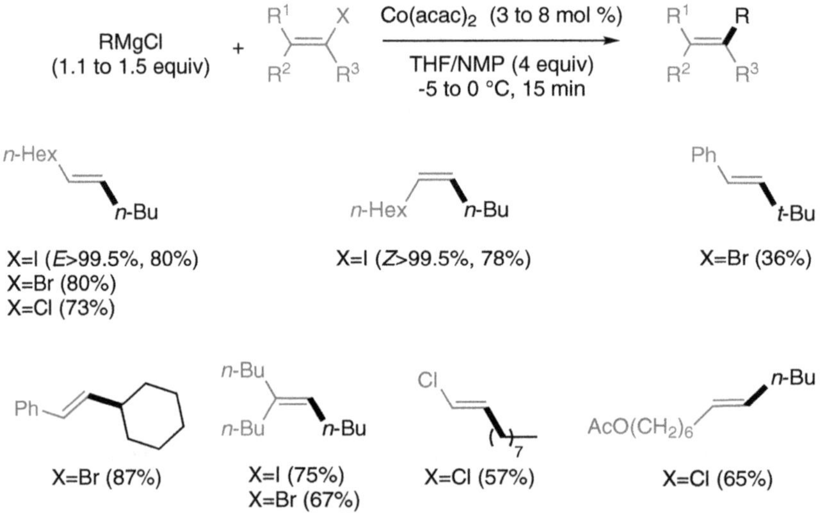

Fig. 4.11: Cobalt-catalyzed alkenylation of alkyl Grignard reagents.

The cobalt-catalyzed mono-coupling of silylated Grignard reagent (R_3SiCH_2MgCl) with 1,2dihalogenoethylene was also reported by Okamoto *et al.* in order to form allylsilanes [19]. The reaction proceeds exclusively in a mono-coupling pathway to provide 3-trimethylsilyl-1-halogeno-1-propene in the presence of 1 mol % of Co(II) or Co(III) acetylacetonate in THF or THF/NMP (Fig. 4.12). Moreover, a stereodifferenciation between (*E*)- and (*Z*)-1,2-dibromoethylenes is observed. A mixture of the two isomers gave >99 % of the (*E*) geometry in high yield.

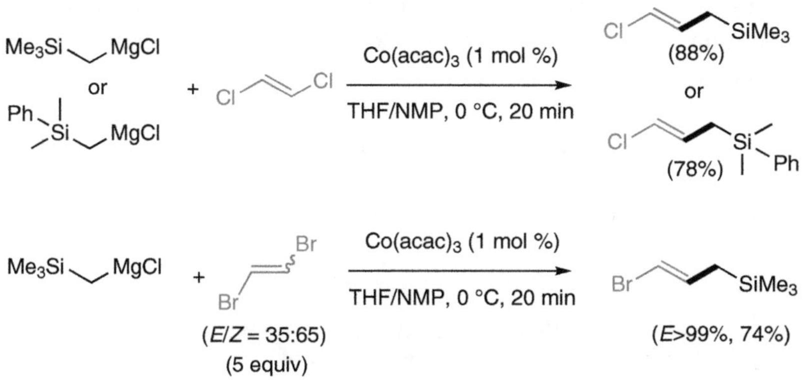

Fig. 4.12: Cobalt-catalyzed cross-coupling of silylated Grignard reagents with dihalogenoethylene.

These γ-halo (*E*)-allylsilanes were converted to various γ-substituted (*E*)-allylsilanes by Ni- or Pd-catalyzed cross-coupling with different organometallic compounds, mainly Grignard reagents. Later on, Yorimitsu *et al.* and Oshima *et al.* reported the cross-coupling of alkenyl halides with silylmethyl Grignard reagents in presence of cobalt(II) chloride. In this case, the presence of NMP is not necessary. Good to excellent yields of various substituted allylsilanes are obtained (Fig. 4.13). It should be noted that the reaction is highly stereoselective [20].

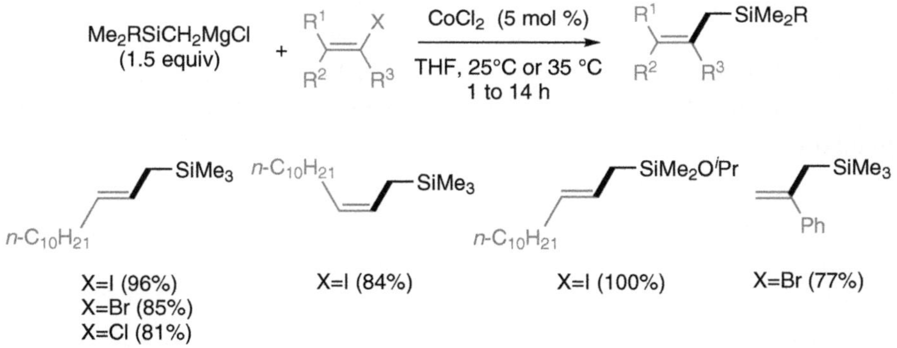

Fig. 4.13: Cobalt-catalyzed cross-coupling of alkenyl halides with silylmethyl Grignard reagents.

4.2.3.2 Alkenyl tellurides

The cross-coupling of alkenyl tellurides with alkyl Grignard reagents in the presence of $CoCl_2(PPh_3)_2$ was also reported by Uemura *et al.* under mild conditions [21–22]. This reaction requires the formation of the alkenyl telluride. The alkenyl telluride was obtained by addition of phenyltellurol, formed *in situ* by reduction of biphenyl ditelluride with sodium borohydride, to the corresponding acetylene. Nevertheless, low yields were obtained in cross-coupling products when the reaction was performed with alkyl Grignard reagents (Fig. 4.14). The results show that a formal substitution of PhTe moiety by the alkyl group occurred mainly under retention of the configuration.

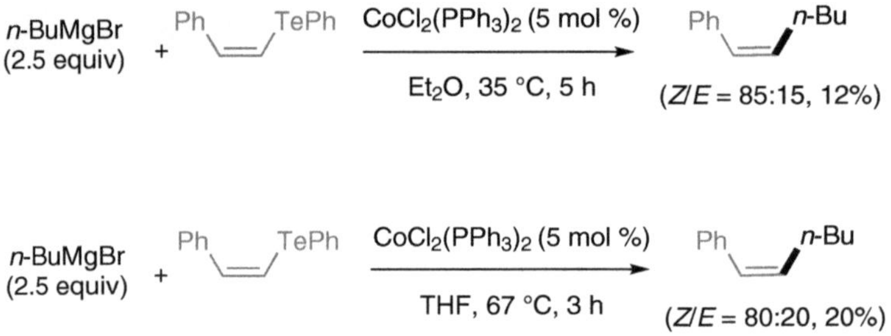

Fig. 4.14: Cobalt-catalyzed cross-coupling between alkenyl tellurides and alkyl Grignard reagents.

4.2.4 With allylic ethers

In 2004, Oshima *et al.* described the allylic substitution between allylic ethers and alkyl Grignard reagents such as Me_3SiCH_2MgCl (Fig. 4.15) starting from linear or branched allylic ethers bearing a phenyl group [23]. The linear product was obtained in excellent yield and the choice of the ligand (none, dpph or dppb) was not important. On the other hand, alkyl-substituted allylic ethers afforded a mixture of linear and branched isomers and required higher temperature. In addition, a branched substrate gives a better yield than the linear substrates. With allylic acetals, the formation of mono- or bis-trimethylsilylmethylation can be controlled by the amount of the Grignard reagents.

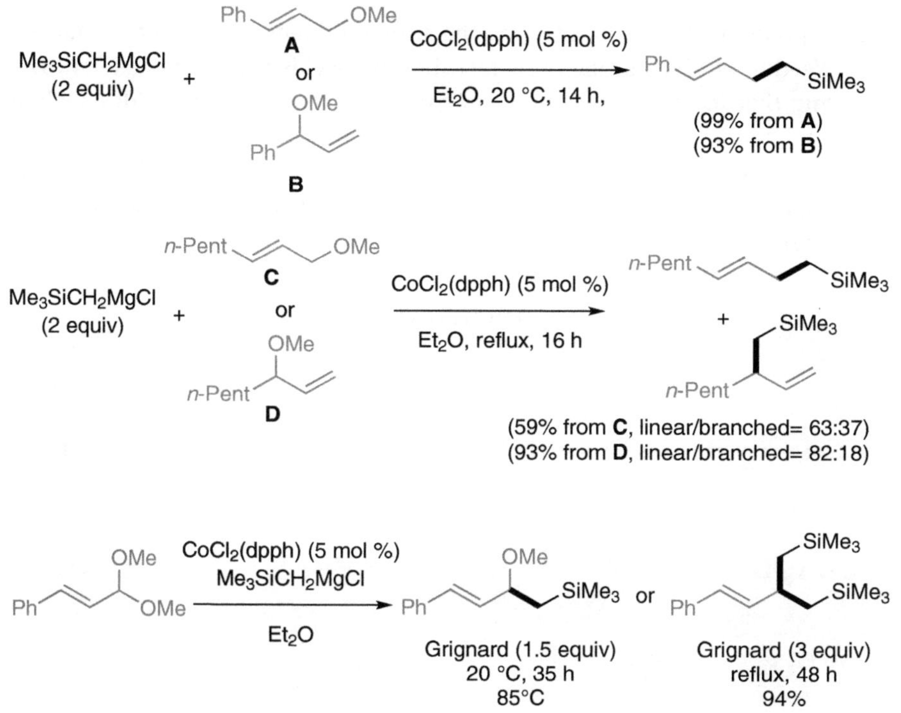

Fig. 4.15: Cobalt-catalyzed cross-coupling between allylic methyl ether and silylmethylmagnesium chlorides.

4.2.5 With alkynyl halides

Only few, not very recent, examples of cross-coupling involving alkynyl halides and Grignard reagents in the presence of a cobalt salt as the catalyst, affording disubstituted acetylenes, are reported. The first example of a cross-coupling of methylmagnesium bromide with phenylethynyl bromide in the presence of a cobalt catalyst was reported by Kharasch *et al.* [5]. They have shown that a catalytic amount of cobalt avoided the bromine–Mg–exchange on the alkyne and 1-phenyl-2-methylacetylene was obtained 62 % yield with methylmagnesium bromide (Fig. 4.16).

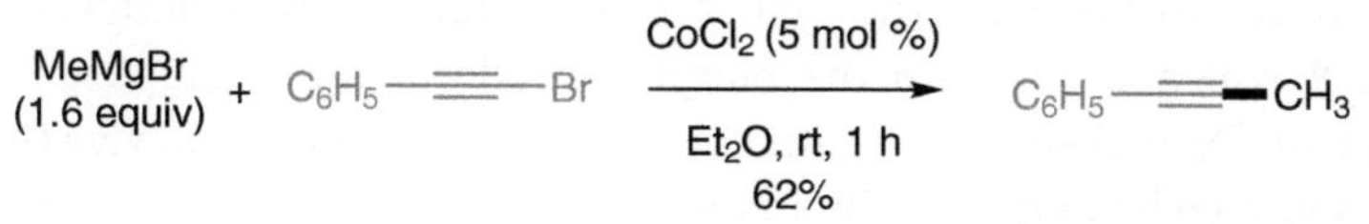

Fig. 4.16: Cobalt-catalyzed cross-coupling of alkynyl bromide with methylmagnesium bromide.

After these preliminary results, showing the strong influence of the cobalt chloride catalyst, Weedon *et al.* confirmed this fact, but low yields were obtained [24]. It is worth noting that in the absence of a catalytic amount of metal chlorides such as $CoCl_2$, alkynyl halides and alkyl Grignard reagents underwent metathesis.

Similar to the cross-coupling of alkenyl halides with silylmethyl Grignard reagents, a propargylic silane was also prepared from an iodo alkyne in moderate yield (Fig. 4.17) [20].

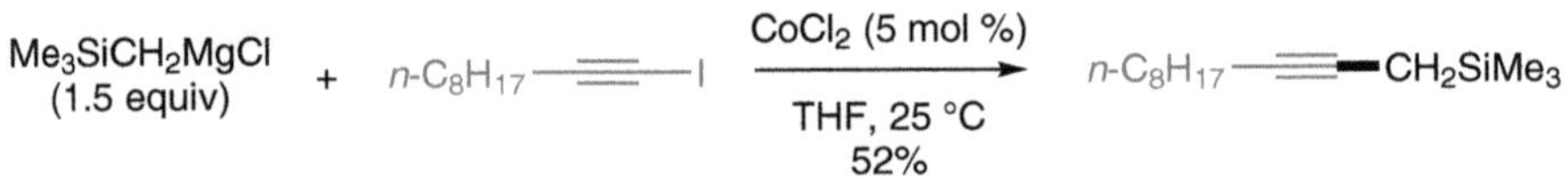

Fig. 4.17: Cobalt-catalyzed cross-coupling of alkynyl halides with silylmethyl Grignard reagents.

4.3 Coupling reactions of arylmagnesium reagents

4.3.1 With aryl halides

The first example of cobalt-catalyzed cross-coupling involving aromatic Grignard reagents was described by Uemura *et al.* in 1982 and 1983 [21–22]. They reported a coupling with biaryl tellurides in the presence of bis(triphenylphosphine) cobalt(II) chloride (Fig. 4.18). The coupling products were obtained in low yields since the homo-coupling products of both the Grignard reagents and the biaryl tellurides were produced in significant amounts, which makes the purification problematic.

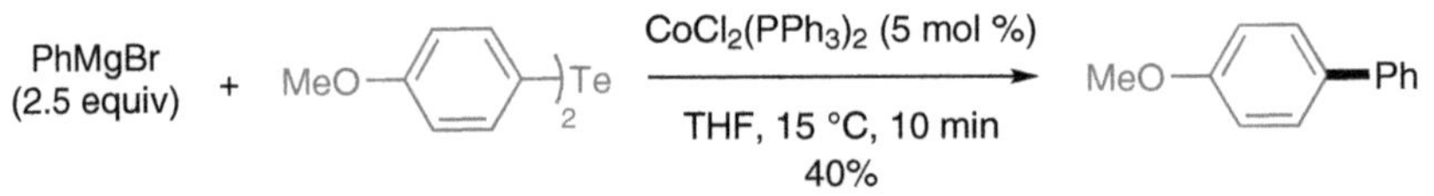

Fig. 4.18: Cobalt-catalyzed cross-coupling between aromatic Grignard reagents and biaryl tellurides.

A mechanism is proposed in Fig. 4.19. The first step is the formation of biphenylcobalt intermediate **E** which reacts with the biaryl telluride to form complex **F**. Complex **F** can then react with phenylmagnesium bromide or free tellurium to give **G**. The product is then obtained by reductive elimination.

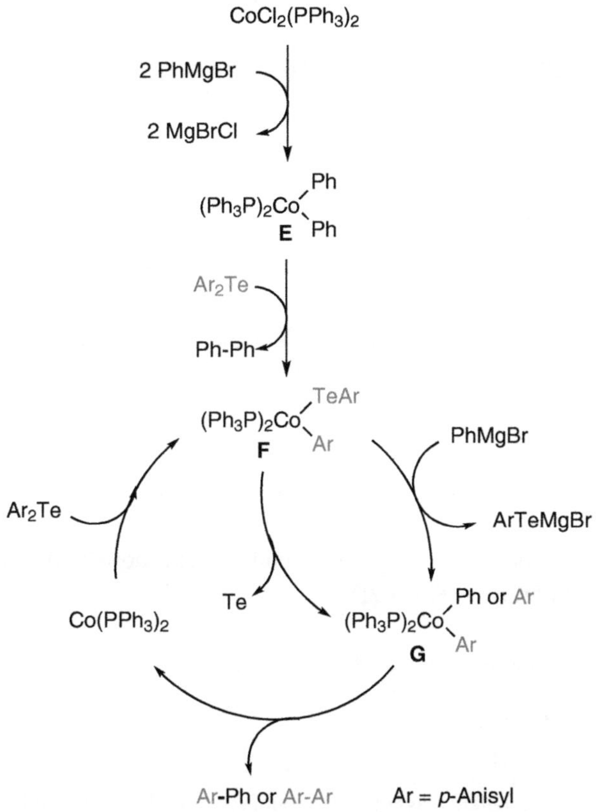

Fig. 4.19: Proposed mechanism for the cobalt-catalyzed cross-coupling of aryl tellurides with aromatic Grignard reagents.

Later, Knochel *et al.* reported the cobalt-catalyzed cross-coupling between aromatic Grignard reagents and heterocyclic chlorides using the simple catalytic system cobalt(II) chloride [25]. Selected examples are reported in Fig. 4.20. Two equivalents of the Grignard reagents were necessary to produce the products in moderate to good yields. Sterically hindered Grignard reagents decrease the rate of the reaction thus, the complete conversion of the starting materials was observed after a long reaction time. Noteworthy, the reaction tolerates the presence of functional groups such as esters as well as heteroaromatic magnesium compounds and that 2-chloropyridines are less reactive than quinoline and isoquinoline derivatives.

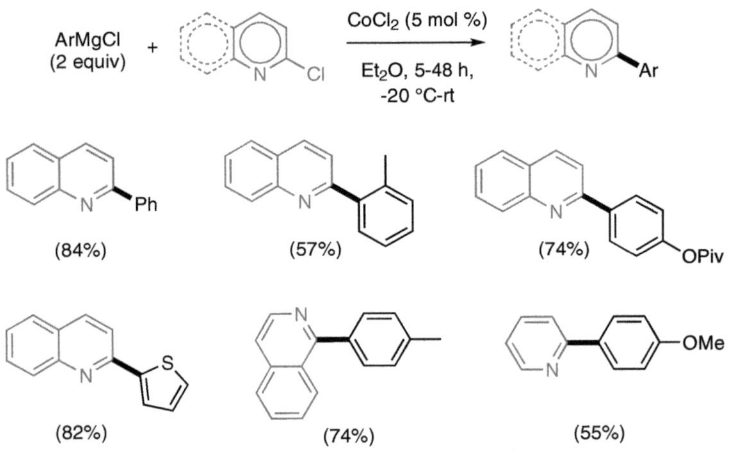

Fig. 4.20: Cobalt-catalyzed cross-coupling between aromatic Grignard reagents and heterocyclic chlorides.

Interestingly, cobalt(II) chloride can be successfully replaced by cobalt powder if the reaction is carried out at room temperature (Fig. 4.21).

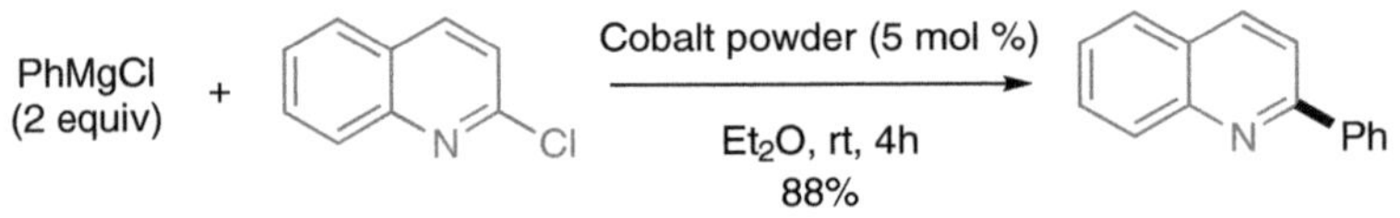

Fig. 4.21: Cross-coupling between aromatic Grignard reagents and heterocyclic chlorides catalyzed by cobalt powder.

One year later, in 2004, Oshima *et al.* reported the same cross-coupling using cobalt(II) acetylacetonate as the catalyst and dioxane as the solvent (Fig. 4.22) [7]. Although the yields are similar, the reaction is much faster in this case, even though, a larger amount of the catalyst and the Grignard reagent are used compared to the previous method developed by Knochel *et al.* [25].

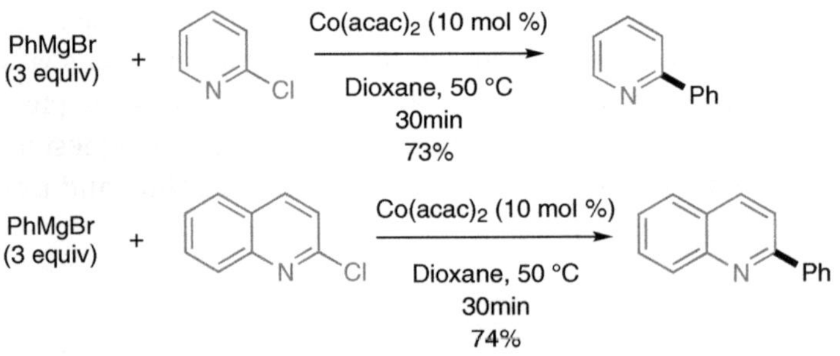

Fig. 4.22: Cobalt-catalyzed cross-coupling between phenylmagnesium bromide and heterocyclic chlorides.

Although Oshima *et al.* were the first to use *N*-heterocyclic carbene (NHC) ligands in cobalt-catalyzed cross-coupling between aliphatic Grignard reagents and aryl bromides, their conditions cannot be successfully applied to the cross-coupling between aromatic Grignard reagents and aryl bromides [8].

In 2005, Chen *et al.* reported the use of several cobalt(III) complexes containing pyrimidine NHC ligands for the cross-coupling of aryl halides with aromatic Grignard reagents [26]. It is noteworthy that only a small catalytic amount (2 mol %) of the cobalt complexes and a slight excess of the Grignard reagent (1.2 equiv) were necessary to afford the cross-coupling products in good to excellent yields. Selected examples are presented in Fig. 4.23. Interestingly, aromatic and heteroaromatic bromides and chlorides were successfully coupled. Unfortunately, with aryl bromides or Grignard reagents bearing *ortho* substituents, the corresponding biaryls were only obtained in moderate yields. Furthermore, the presence of electron-donating groups on the aryl bromides also decreased the yield.

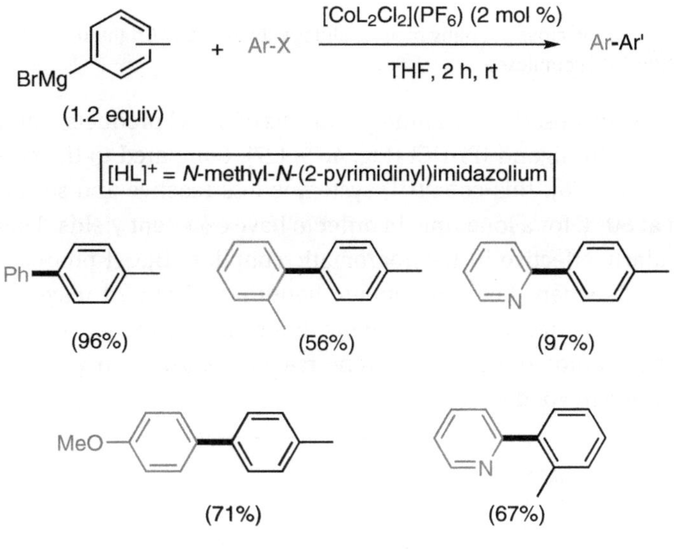

Fig. 4.23: Cross-coupling of aryl halides with aromatic Grignard reagents catalyzed by cobalt(III)-NHC complexes.

The following mechanism is proposed (Fig. 4.24). At first, the cobalt(III) complex is reduced to the active cobalt(I) by the Grignard reagent. Then, the oxidative addition of the aryl halide on the cobalt(I) center forms the intermediate ArCo(III). The strong electron-donating properties of the NHC ligand are supposed to play an important role in this step by stabilizing the cobalt(I) center and facilitating the insertion of the aryl halide. The nucleophilic substitution of the halide by the Grignard reagent, followed by reductive elimination, gives the biaryl product and regenerates the cobalt(I) complex.

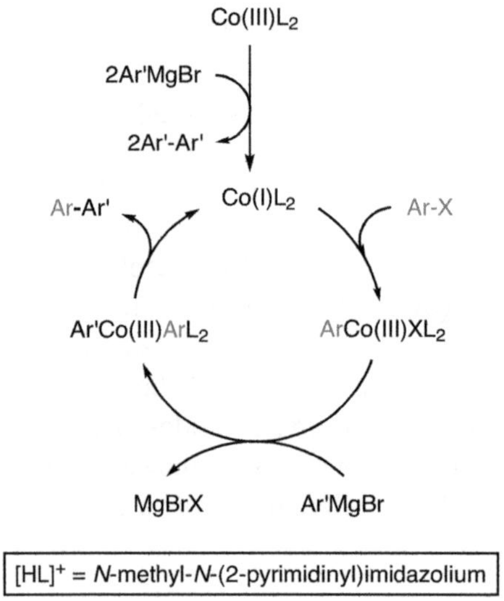

Fig. 4.24: Proposed mechanism for the cross-coupling of aryl halides with aromatic Grignard reagents catalyzed by cobalt(III)-NHC complexes.

The same year, Nakamura *et al.* described a similar synthesis of biaryl products using cobalt(II) fluoride with the NHC ligand IPr·HCl (Fig. 4.25) [27]. Compared to the catalytic system previously used [26], this cobalt(II) system is less reactive and so the reaction must be heated at 60 °C for a long time in order to have excellent yields. This cobalt catalyst is particularly effective in heteroaromatic coupling. Biaryl products are obtained in better yields when (hetero)-aromatic halides or Grignard reagents are bearing electron-poor or electron-rich substituents. Interestingly, the reaction is highly chemoselective thus, 1-bromo-4-chlorobenzene reacts selectively to give the corresponding biaryl chloride in good yield.

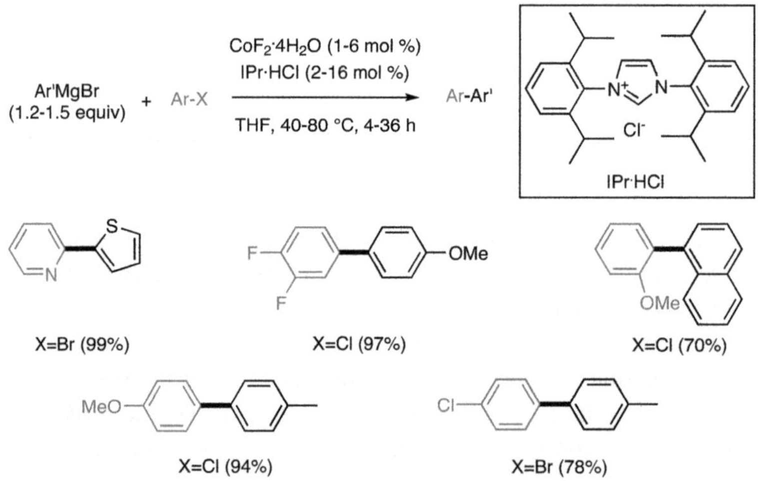

Fig. 4.25: Cross-coupling of aryl halides with aromatic Grignard reagents catalyzed by cobalt(II)-NHC complexes.

With this catalytic system, the mechanism proposed is similar to the one described in Fig. 4.24. However, a metal-fluoride-ate complex as a reactive intermediate is proposed which catalyzes the desired biaryl cross-coupling with high selectivity instead of the undesired homocoupling (Fig. 4.26).

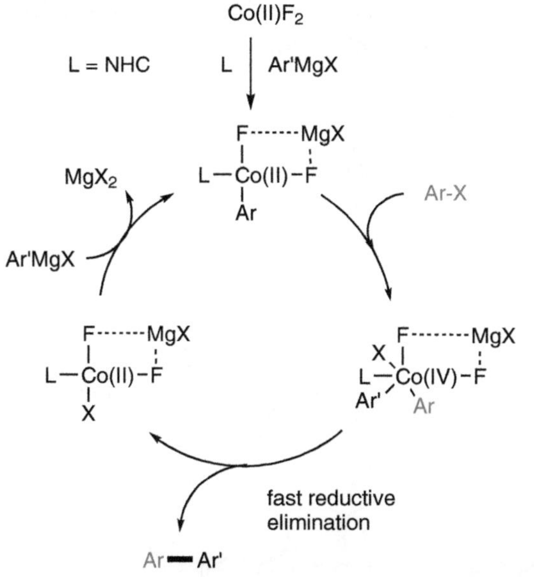

Fig. 4.26: Proposed mechanism of the cross-coupling of aryl halides with aromatic Grignard reagents catalyzed by cobalt(II)-NHC complexes.

Matsubara *et al.* studied a series of new cobalt(II) complexes bearing bulky IPr ligands. They investigated two of these IPr-cobalt(II) complexes in the coupling reaction of aromatic halides with arylmagnesium chlorides ($[Co(IPr)X]_2$-$(\mu\text{-}X)_2$ and Co(IPr) $X_2(C_5H_5N)$) [28]. These complexes gave similar results to the ones obtained by Nakamura *et al.* [27]. Tonzetich *et al.* also reported the use of IMes-cobalt(II) complexes [9], and all these different complexes led to comparable results.

Von Wangelin *et al.* reported the coupling reaction of deactivated chlorostyrenes with aromatic Grignard reagents [29]. Cheap cobalt(III) acetylacetonate was used under mild conditions to afford biaryl products in good to excellent yields. Several examples are shown in Fig. 4.27. Different mono- or di-substituted *ortho-*, *meta-* and *para*-chlorostyrenes were successfully coupled. Under these conditions, no polymerization of the chlorostyrene was observed. In the case of 2,4-dichlorostyrene, the reaction is highly regioselective and only takes place with the chloride atom that is in an *ortho* position to the vinyl group. Indeed, compared to *meta-* or *para*-chlorostyrenes, the *ortho*-chlorostyrene is much more reactive due to the proximity between the vinyl group and the C–Cl site. Moreover, it is noteworthy that the reaction is also highly chemoselective. Intramolecular competition between an aryl bromide and a chlorostyrene only leads to the substitution of the chloride although, in general, aryl bromides are more reactive than aryl chlorides in cross-couplings. Under the same conditions, cross-coupling of chloroarenes without any vinyl substituent affords biaryls in low yields. This result highlights the strong effect of the olefinic substituent on the activation of the C–Cl bond by coordination to the catalyst. Sensitive functional groups such as acetals can be tolerated.

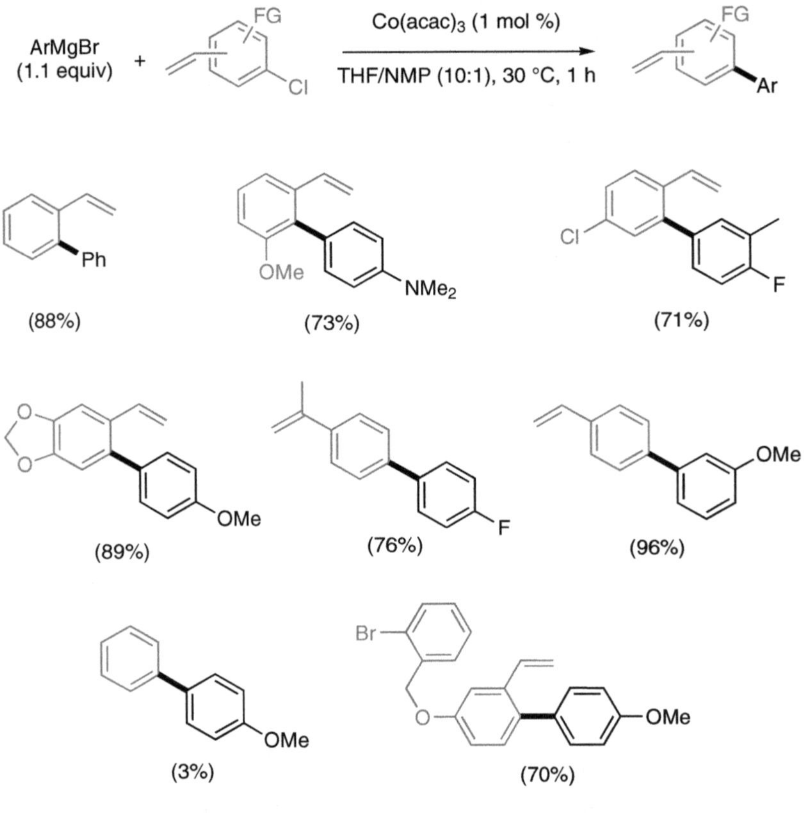

Fig. 4.27: Cobalt-catalyzed cross-coupling of chlorostyrene derivatives with aromatic Grignard reagents.

Ten years after their investigation on the arylation of *N*-heterocyclic chlorides by Grignard reagents [25], Knochel *et al.* have developed a new catalytic system for the cross-coupling between *N*-heteroaryl halides and arylmagnesium reagents (Fig. 4.28) [30]. To the cobalt(II) chloride catalyst previously used was added isoquinoline as a ligand, which prevents side reactions. As for the solvent, the use of methyl *tert*-butyl ether as a co-solvent reduces the formation of homocoupling products. These modifications enable this cobalt-catalyzed reaction to be performed at room temperature with an improved reaction rate and yield. As a rule, moderate to good yields were obtained from 2-chloro- or 2-bromo-pyridines, pyrimidines and triazines. Electron-poor or rich *N*-heterocycles and Grignard reagents can successfully be coupled. Sterically hindered heterocycles are also well tolerated. Interestingly, the reaction is chemoselective and polyfunctional pyridines react selectively.

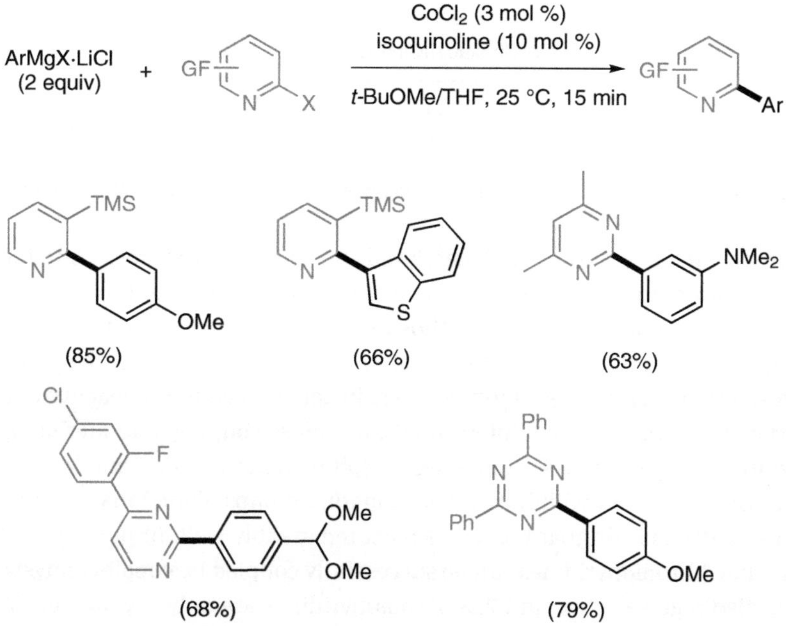

Fig. 4.28: Cobalt-catalyzed cross-coupling of *N*-heterocyclic derivatives with aromatic Grignard reagents using isoquinoline as a ligand.

Sun *et al*. have developed a new kind of cobalt complex bearing phosphorus and nitrogen ligands in order to activate C–Cl bonds for the reaction with Grignard reagents [31]. Unfortunately, this system is only working with chlorobenzene and 4-anisylmagnesium bromide (Fig. 4.29) in moderate yields, since the cobalt-complex is unstable in the presence of the Grignard reagent.

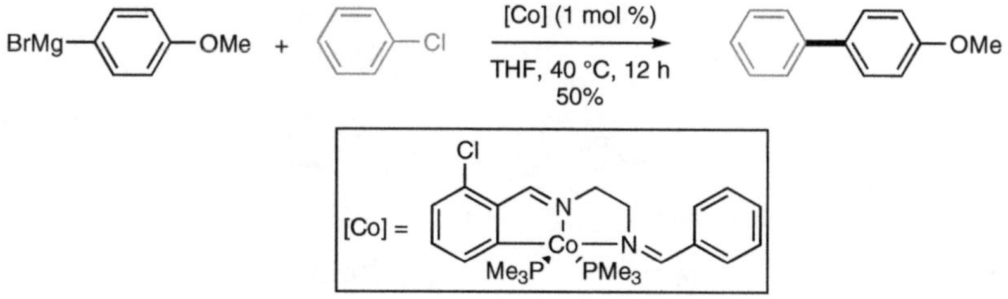

Fig. 4.29: Cross-coupling of aryl chlorides with aromatic Grignard reagents using a new cobalt-complex.

On the other hand, Elango *et al.* used different drugs as ligands such as trimethoprim, cimetidine, niacinamide and ofloxacin to synthesize and characterize various Co(II) complexes [32]. Among all the drugs tested, the ofloxacin ligand showed excellent activity for the cross-coupling between bromobenzene and phenylmagnesium bromide (Fig. 4.30). However, this is the only example reported for this cross-coupling.

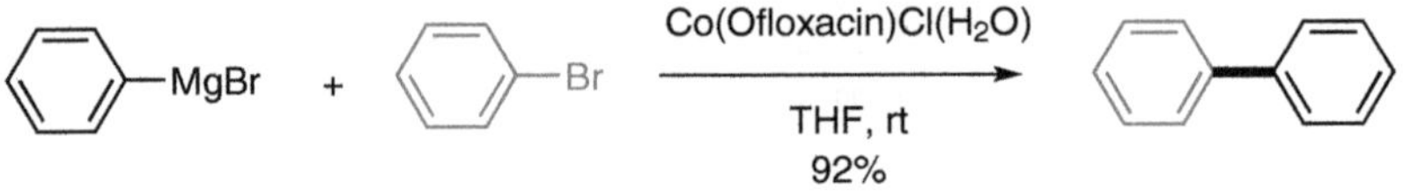

Fig. 4.30: Cross-coupling of aryl bromide with aromatic Grignard reagents using cobalt-drugs complexes.

Duan *et al.* were the first to report a cobalt-catalyzed substoichiometric cross-coupling between aromatic Grignard reagents and aryl halides (Fig. 4.31) [33]. Titanium ethoxide together with cobalt(II) chloride and tributylphosphine constitute the catalytic system. It is noteworthy that only aryl bromides give good to excellent yields. The corresponding aryl triflates, tosylates, fluorides or iodides were in general less or not reactive and aryl chlorides needed to be activated. Interestingly, this cross-coupling tolerates many functional groups such as ketones, sulfonates, sulphonamides, esters, amides and acids as well as other electron-withdrawing or -donating groups. Functionalized aryl or heteroaryl bromides and Grignard reagents react remarkably well. In particular, it is the first time that 3-bromopyridines can be successfully coupled by cobalt-catalysis. The reaction is also regioselective and 2,5-dibromopyridines are only arylated at C2, which pave the way for high regioselective diarylation.

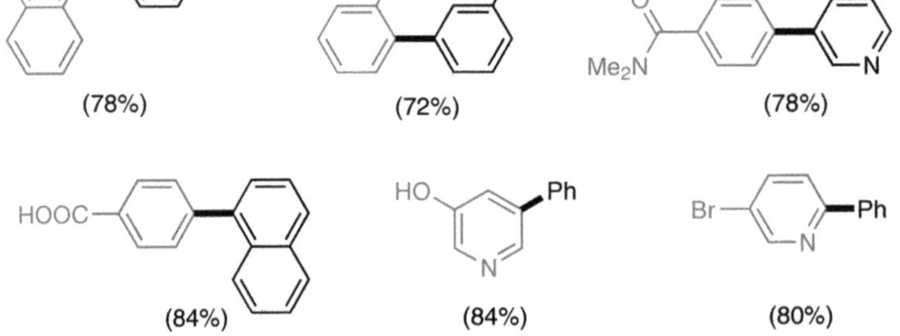

Fig. 4.31: Cobalt-titanium co-catalyzed cross-coupling of aryl bromides with aromatic Grignard reagents.

4.3.2 With alkyl halides

In the first part, the cross-coupling of alkyl Grignard reagent with aryl halides was described. However, the reverse reaction involving an aryl Grignard reagent is also possible.

In 2006, Oshima *et al.* reported the first cross-coupling between primary alkyl halides and aryl Grignard reagents in the presence of $CoCl_2$ with the use of *N,N,N',N'*-tetramethylcyclohexane-1,2-diamine as the ligand [34]. Selected examples are presented in Fig. 4.32. It must be underlined that in this case only a slight excess (1.2 equiv) of the Grignard reagent is used. The yields are higher with alkyl iodides than with alkyl bromides or alkyl chlorides. The reaction is not sensitive to steric hindrance thus, *ortho*-subsituted aromatic Grignard reagents react with alkyl iodides in excellent yields. In addition, the reaction of secondary acyclic and cyclic alkyl halides also proceeds readily. The reaction is highly chemoselective thus, ester and acetal groups are well tolerated. In addition, the reaction is also stereoselective. Phenylation of (2*R*,3*S*)-tetrahydrofurans is in favour of the 2,3-*trans* isomer.

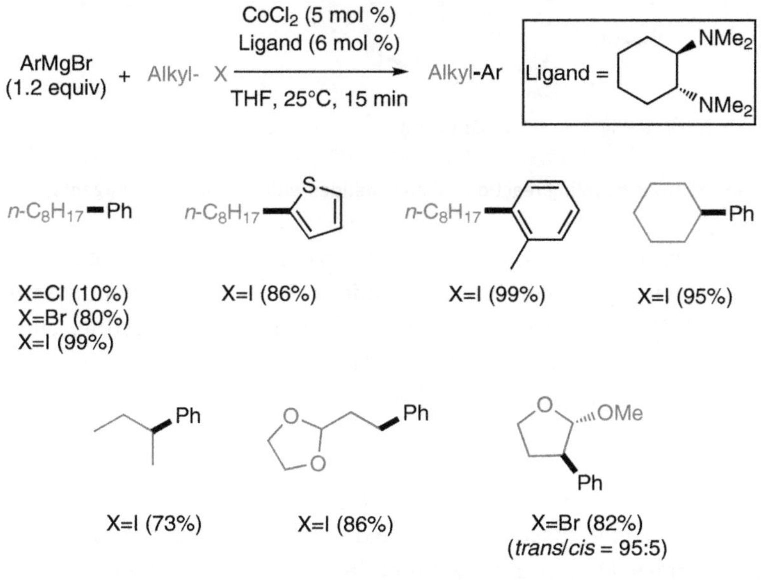

Fig. 4.32: Cobalt-diamine-catalyzed cross-coupling reactions of alkyl halides with aromatic Grignard reagents.

The same year, Oshima *et al.* also reported the cobalt-catalyzed cross-coupling between primary alkyl halides and aromatic Grignard reagents in the presence 1,3bis(diphenylphosphino)propane (dppp) instead of *N,N,N',N'*-tetramethylcyclohexane-1,2-diamine [35]. Selected examples are described in Fig. 4.33. It should be pointed out that a large excess of the Grignard reagent is necessary (3 equiv). Primary alkyl bromides afford the coupling products in moderate to good yields, even with heteroaromatic Grignard reagents. Due to steric hindrance, *ortho*-substituted arylmagnesium reagents do not react. Compared to alkyl bromides, the corresponding alkyl iodides give the cross-coupling product in a lower yield. Moreover, secondary alkyl bromides afford the product in poor yields. It is noteworthy that the presence of an ester or an acetal is tolerated.

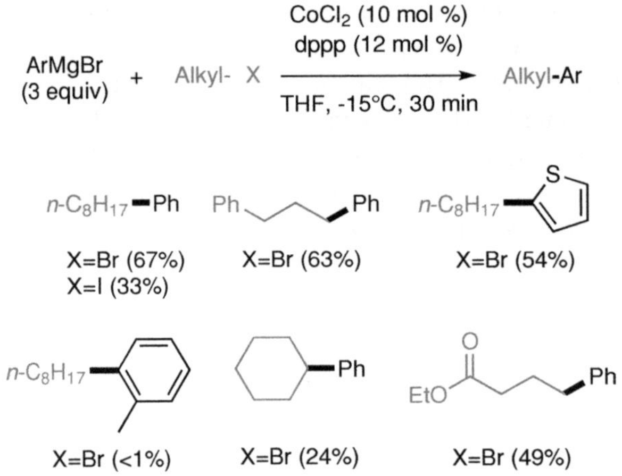

Fig. 4.33: Cobalt-catalyzed cross-coupling reactions of alkyl halides with aryl Grignard reagents.

In 2009, Cahiez *et al.* described the cobalt-catalyzed cross-coupling of aromatic Grignard reagents with primary and secondary alkyl halides [36]. A new catalytic system was developed, cobalt(III) acetylacetonate/TMEDA (1:1). This is interesting, since TMEDA is a more simple and less expensive than *N,N,N',N'*-tetramethylcyclohexane-1,2-diamine which was used by Oshima *et al.* [34]. With this catalytic system, only a slight excess (1.1 equiv) of the Grignard reagent was necessary. As shown in Fig. 4.34, various primary and secondary alkyl halides were successfully coupled. Alkyl iodides and bromides give excellent yields. Compared to the cobalt-catalyzed cross-coupling developed by Oshima *et al.* [34], these conditions led to comparable results both with alkyl bromides and iodides. Moreover, alkyl bromides tend to be less expensive and more readily available than the corresponding alkyl iodides. However, the cross-coupling affords poor yields with alkyl chlorides. Excellent yields were obtained with primary and secondary acyclic or cyclic alkyl bromides, whereas tertiary alkyl bromides do not react. The reaction is slightly sensitive to steric hindrance as it is possible to couple a secondary alkyl bromide with an *ortho*-substituted aromatic Grignard reagent in high yields. However, highly sterically hindered aromatic Grignard reagents afford poor yields.

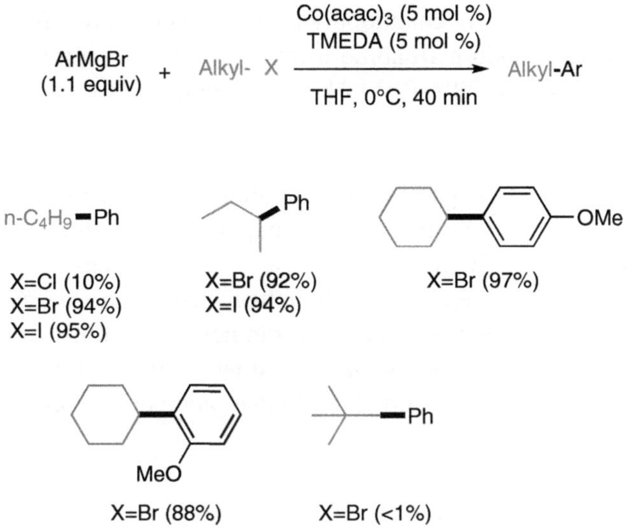

Fig. 4.34: Cobalt-catalyzed cross-coupling of simple primary and secondary alkyl halides with aryl Grignard reagents.

The reaction is also highly chemoselective (Fig. 4.35). 1-Bromo-5-chloropentane reacts selectively to give the corresponding alkyl chloride in high yield. The reaction also tolerates the presence of sensitive groups such as acetates, esters, amides or ketones.

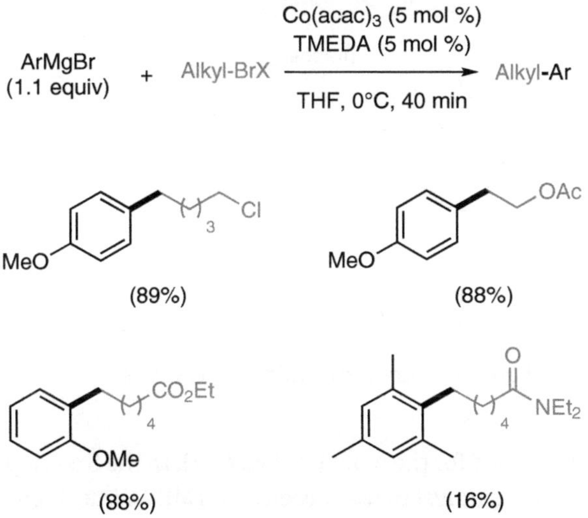

Fig. 4.35: Cobalt-catalyzed cross-coupling of functionalized alkyl halides with aryl Grignard reagents.

Wu *et al.* reported, in 2011, the synthesis of enantiopure pyrrolidine derivatives from aromatic Grignard reagents and an iodo-pyrrolidine in the presence of bis(triphenylphosphine)cobalt(II) chloride and TMEDA (1:1) [37]. Yields are good to excellent (Fig. 4.36) however, this reaction can only be applied to alkyl iodides. Different electron-rich or electron-poor phenyl groups were attached to the pyrrolidine derivative in excellent yields. Noteworthy, the influence of steric hindrance is not important. Thus, different sterically congested aryl groups such as 1-naphtyl, 9-phenanthrenyl and *o*-tolyl led to the coupling product in high yields. Heteroaryl-Grignard reagents are less reactive compared to aryl Grignard reagents: the 2-thiophenylmagnesium reagent reacted in moderate yield whereas the 2-pyridinyl reagent did not react. In this coupling, the stereochemistry of the pyrrolidines was unaffected by the reaction conditions and the coupling proceeded under retention of the configuration of the starting material.

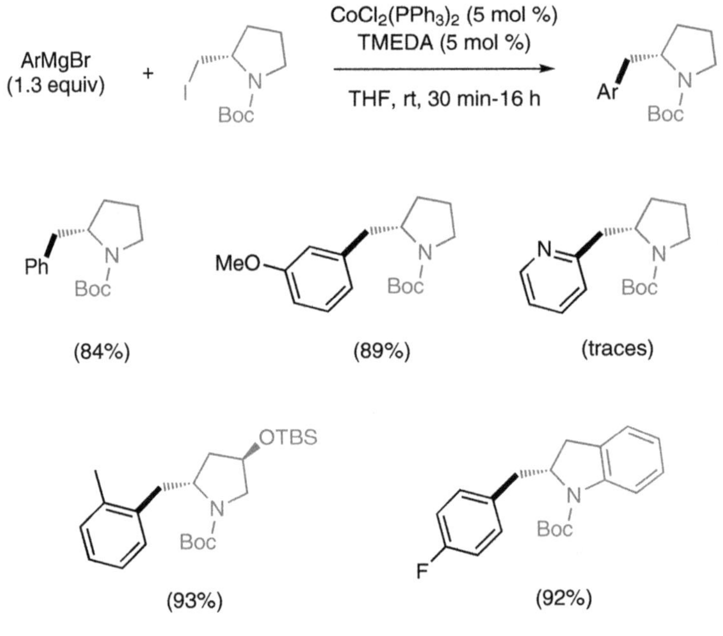

Fig. 4.36: Cobalt-catalyzed synthesis of enantiopure pyrrolidine derivatives.

The following catalytic cycle was proposed for the reaction (Fig. 4.37). In the first step, the Grignard reagent reduces the cobalt catalyst in the presence of TMEDA into a reactive CoL_2 species **H.** This one reacts with pyrrolidine **I** *via* a single electron transfer to give a mixture of complex **K** and radical **J.** These latter two species combine to afford intermediate **L**, which is transformed into cobalt complex **N** through phenylation with Grignard reagent **M**. Finally, reductive elimination of **N** leads to the desired product **O** and the cobalt complex **H**.

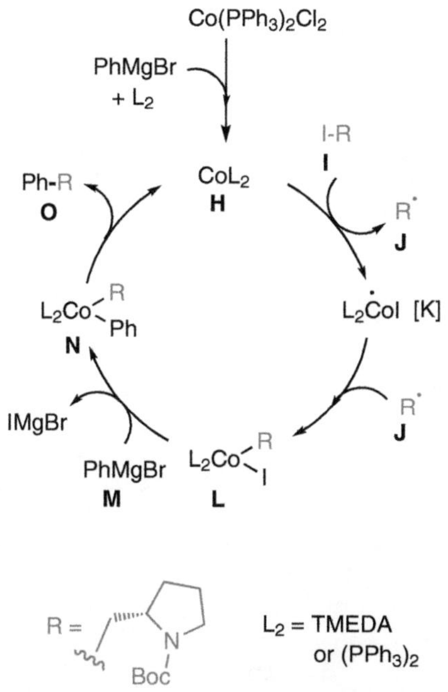

Fig. 4.37: Proposed mechanism for the formation of pyrrolidine derivatives.

In 2012, Cossy *et al.* described the diastereoselective cobalt-catalyzed cross-coupling of aryl Grignard reagents with *C*-bromo-glycosides in the presence of cobalt(III) acetylacetonate and TMEDA (1:1) [38]. Selected results are presented in Fig. 4.38. With 1-bromo mannose derivative **P** and *O*-acetyl-α-bromo-D-galactose **R** the reaction was diastereoselective; thus, only the α-isomer was observed. Electron-rich or electron-poor Grignard reagents led to good yields, however more substituted or sterically hindered aryl Grignard reagents only afford poor yields in the coupling products. Moreover, heteroaryl Grignard reagents were not used successfully. With 1-bromo glycoside **Q**, the coupling was still diastereoselective in favour of the α-isomer and the products were obtained in good yields, but in a lower α/β ratio.

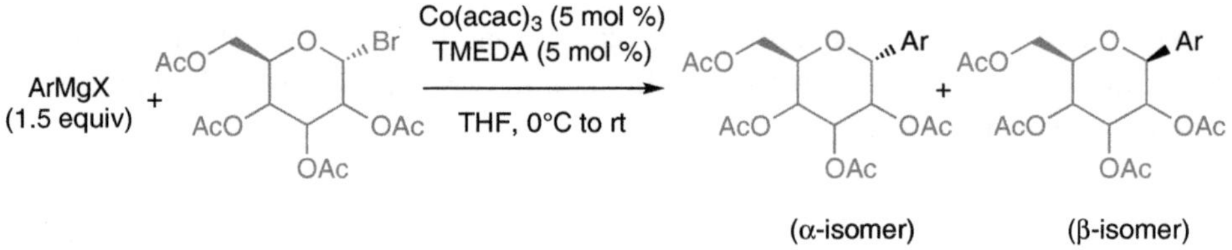

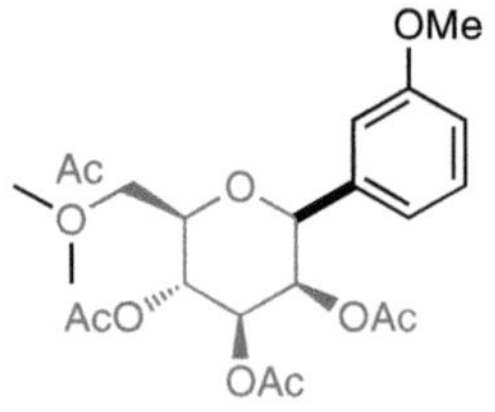

P1

(82%, α/β = >9:1)

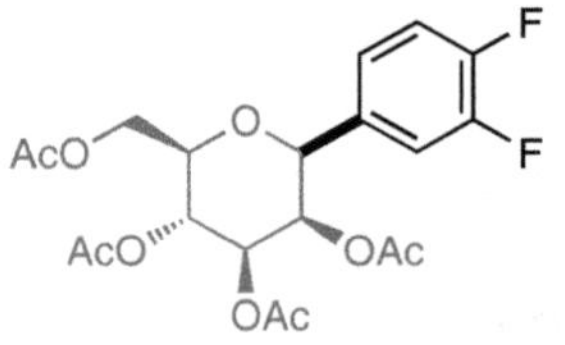

P2

(72%, α/β = >9:1)

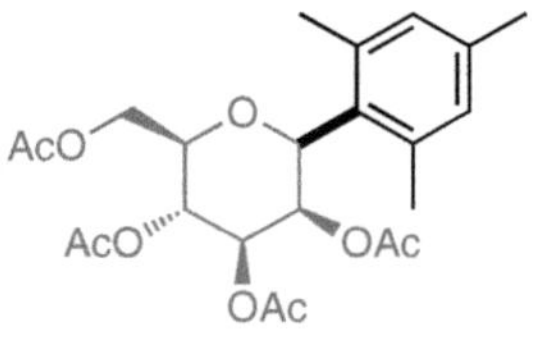

P3

(37%, α/β = >9:1)

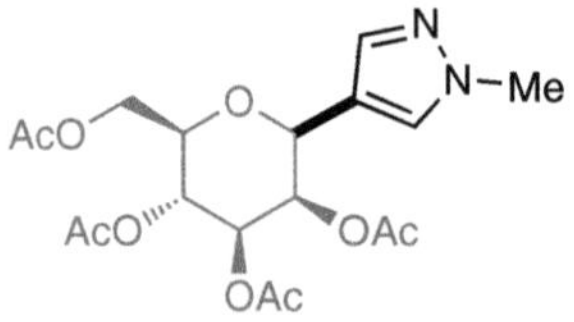

P4

(0%)

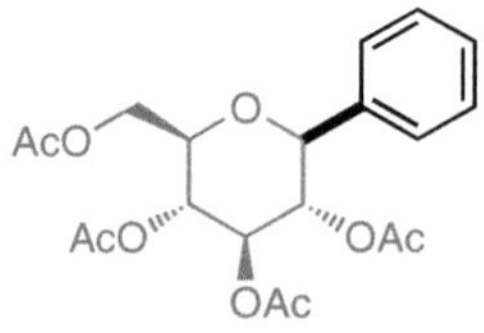

Q1

(96%, α/β = 3:1)

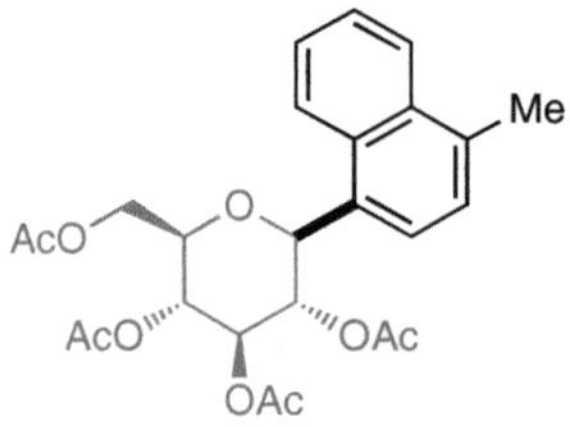

Q2

(34%, α/β = 1.3:1)

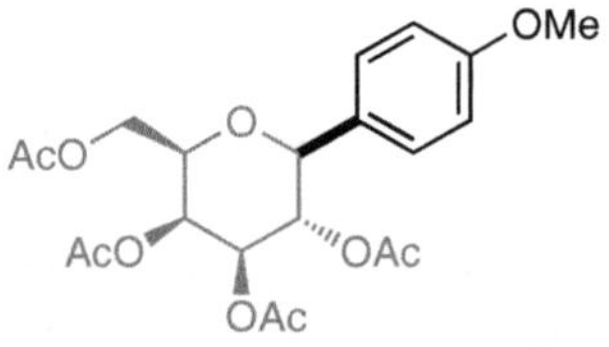

R

(82%, α/β = >9:1)

Fig. 4.38: Cobalt-catalyzed cross-coupling of aromatic Grignard reagents with 1-bromo glycosides.

Following their interest in carbohydrates chemistry, Cossy *et al.* extended their method to the cobalt-catalyzed cross-coupling of *C*-bromo-furanosides with aromatic Grignard reagents using the same catalytic system (Fig. 4.39) [39]. When the C2-hydroxy and C3-hydroxyl of the furanoside derivative were protected as acetates, the cross-coupling product was obtained in low yield due to the instability of the starting material. With the *tert*-butyldiphenylsilyl protected *C*-bromo furanoside, the reaction is diastereoselective in favour of the 1,4-*cis*-furans giving the corresponding *C*-phenyl furanoside in high yield, and the *trans*-diastereoisomer is not isolated. When the configuration at the C2- and C3-stereocenters was inverted, only the 1,4-*trans* *C*-phenyl furans was obtained. It is noteworthy that in absence of any substituent at C2, the diastereoselectivity is very poor (α/β = 1.3:1). While substituents are tolerated on the phenyl group of the Grignard reagent and give the product in excellent yields, the amount of the catalyst has to be increased with heteroaromatic Grignard reagents [($Co(acac)_3$ (10 mol %)/TMEDA (15 mol %)], and in this case, moderate yields were obtained in the coupling product.

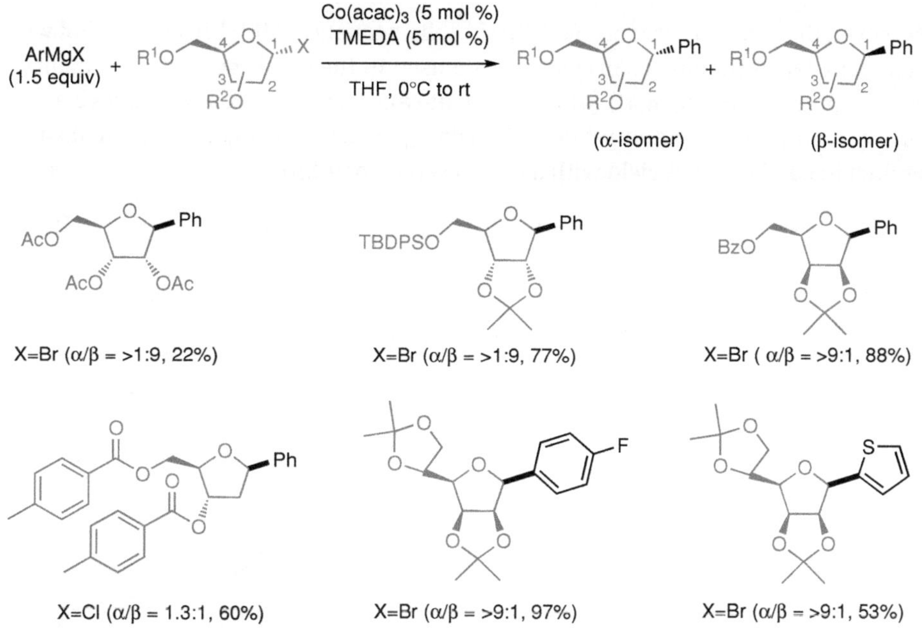

Fig. 4.39: Cobalt-catalyzed cross-coupling of aromatic Grignard reagents with 1-halogeno furanosides.

This method has been applied to the synthesis of (–)-isoaltholactone, a natural product which shows antitumoral, antifungal and antibacterial activities (Fig. 4.40).

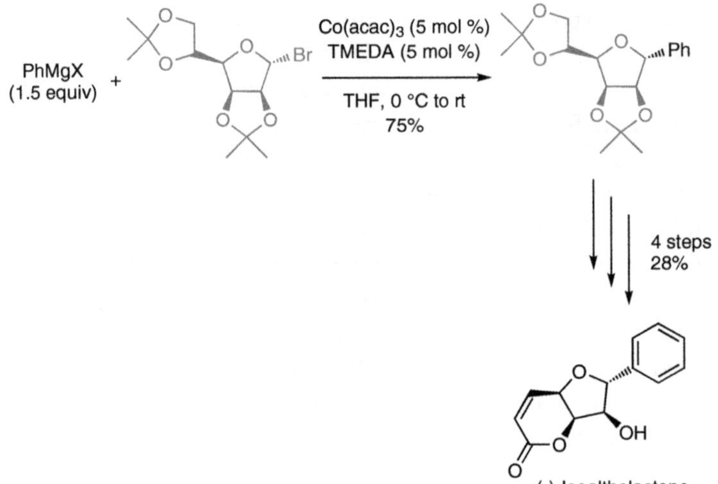

(-)-Isoaltholactone

Fig. 4.40: Synthesis of (–)-isoaltholactone.

Moreover, the catalytic system cobalt(III) acetylacetonate / TMEDA in association with hexamethylenetetramine (1:5:1) can also be used in the synthesis of the antidepressant (±)-paroxetine (Fig. 4.41) [40] The cobalt-catalyzed cross-coupling between the corresponding bromide and *p*-fluorophenylmagnesium bromide affords the desired product in good isolated yield with a *trans*/*cis* ratio of 90:10.

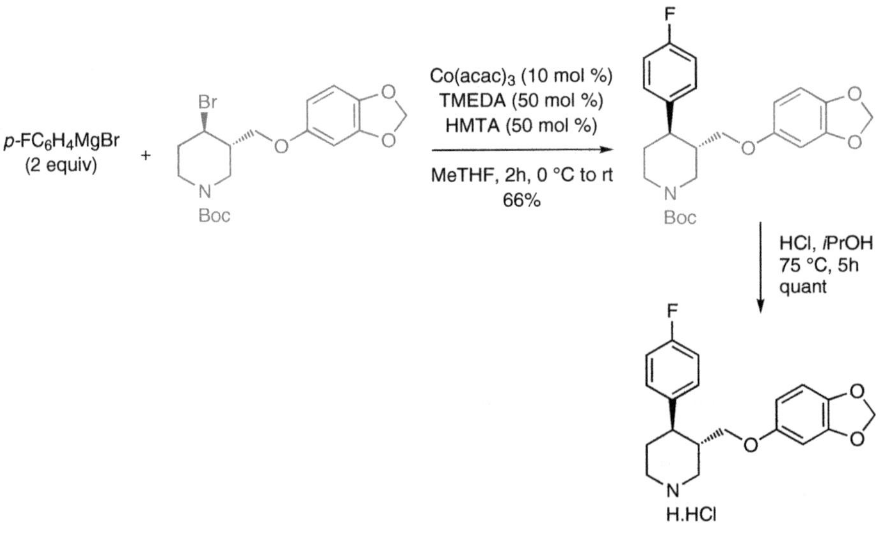

(±)-Paroxetine hydrochloride

Fig. 4.41: Synthesis of (±)-paroxetine.

Very recently, Cossy *et al.* reported the cobalt-catalyzed arylation of saturated nitrogen heterocycles derivatives such as iodo-azetidines, pyrrolidines and piperidines with aromatic Grignard reagents [41]. The reaction conditions are very similar to those used by Oshima *et al.* [34]. Indeed, the reaction is performed in the presence of cobalt(II) chloride and *N,N,N',N'*-tetramethylcyclohexane-1,2-diamine (1:1.2) in THF and only alkyl iodides were used. Selected examples are shown in Fig. 4.42. Different iodo substituted azetidines, pyrrolidines and piperidines were studied. As a rule, good to excellent yields were obtained when different electron-withdrawing or electron-donating substituents were attached to the Grignard reagents. The reaction is not subject to steric hindrance and *o*-tolylmagnesium bromide was successfully coupled. Heteroaromatic Grignard reagents can also be efficiently coupled. To the best of our knowledge, this is the first synthesis of 3-arylazetidines by cobalt catalysis.

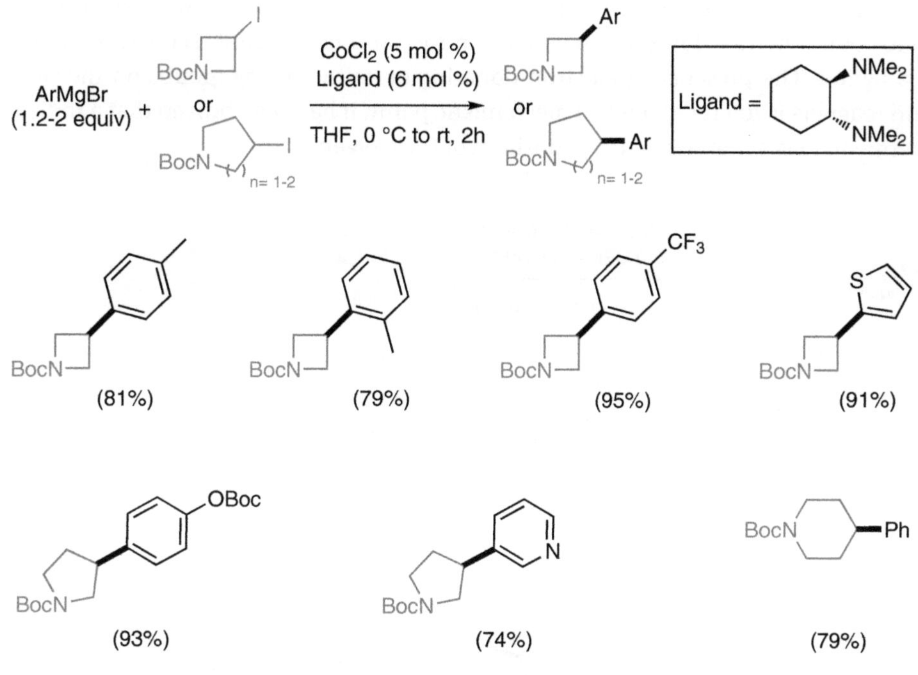

Fig. 4.42: Cobalt-diamine-catalyzed cross-coupling reactions of *N*-heterocycles iodides with aromatic Grignard reagents.

When a mixture of 2,3-disubstituted iodo-azetidines (*cis/trans* = 75:25) was treated with phenylmagnesium bromide, the reaction is diastereoselective in favour of the *trans*-isomer and the product was isolated in excellent yield (Fig. 4.43).

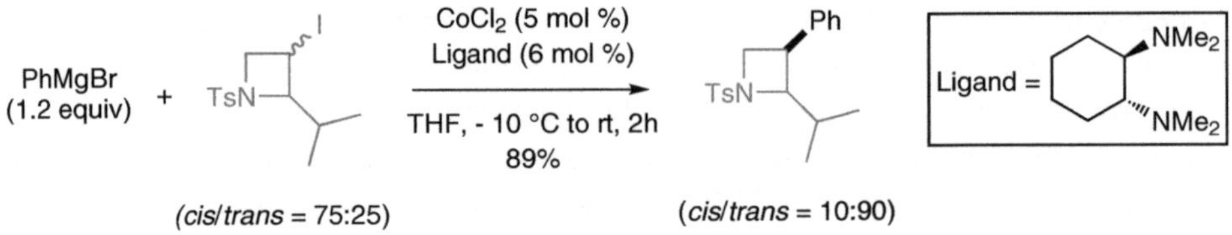

Fig. 4.43: Cobalt-diamine-catalyzed cross-coupling reaction of 2,3-disubstituted azetidines with aromatic Grignard reagent.

Several aryl-substituted NHC complexes of cobalt(II) have been used in the cross-coupling of aryl Grignard reagents with alkyl halides, but very low yields were obtained [9,42].

Very recently, Knochel *et al.* have shown that cyclic or heterocyclic TBS-protected halohydrins (bromo and iodo) underwent a highly diastereoselective cobalt-catalyzed cross-coupling with various aryl- and heteroarylmagnesium reagents (Fig. 4.44) [43]. The *trans* α-arylated cyclic alcohol derivatives were obtained in good yield with a dr up to >99:1. The reaction was carried out in the presence of $CoCl_2{\cdot}2LiCl$ and TMEDA as a ligand. The presence of electron-donating or -withdrawing groups on the Grignard reagents is tolerated. From a mechanistic point, it has been proven that a radical intermediate is generated at the α-position of the oxygen.

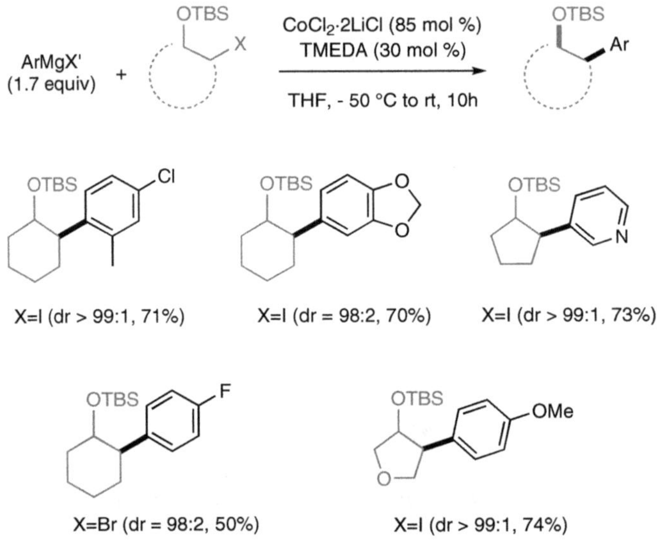

Fig. 4.44: Cobalt-catalyzed cross-coupling reaction of cyclic halohydrins with aromatic Grignard reagent.

This coupling was used to synthesize an arylated TBS-protected cyclohexanol, an intermediate in the synthesis of the NK1 antagonist hNK1 (Fig. 4.45).

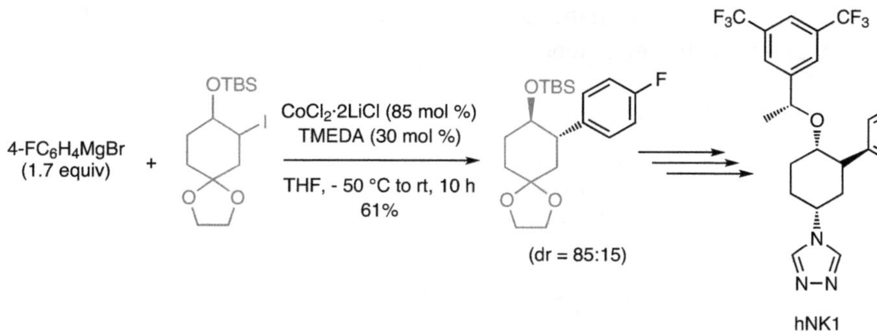

Fig. 4.45: Cobalt-catalyzed synthesis of an intermediate in the synthesis of hNK1.

4.3.3 With alkenyl halides

The first reaction between aromatic Grignard reagents and vinyl halides in the presence of cobalt(II) chloride was reported by Kharasch in 1943 (Fig. 4.46) [2]. Interestingly, vinyl bromides or vinyl chlorides give good yields with phenyl or naphthylmagnesium bromide. The presence of a substituent in a geminal position of the vinyl halide afforded very low yields. Significant amounts of biaryls and polyvinyl polymers are formed, as a result of homocoupling of alkenyl halides and of aryl Grignard reagents and the quantity is increasing when the alkenyl halide is sterically hindered.

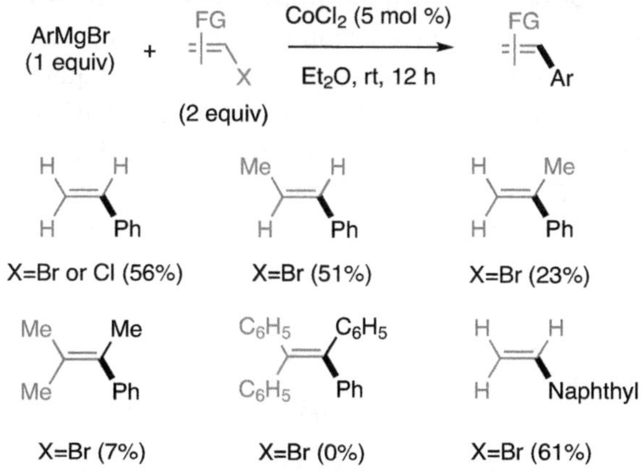

Fig. 4.46: Cobalt-catalyzed reaction of vinyl halides with aromatic Grignard reagents.

Later on, Collet and Jacques studied the cobalt-catalyzed cross-coupling between polychloroethylenes and aryl Grignard reagents in order to form diarylacetlyenes with the same catalyst [44]. As a rule, moderate yields of diarylacetylenes were obtained from electron-rich arylmagnesium bromides together with large amounts of biaryls and reduction products of Grignard reagents (Fig. 4.47). No differences were observed between tri- and tetrachloroethylene.

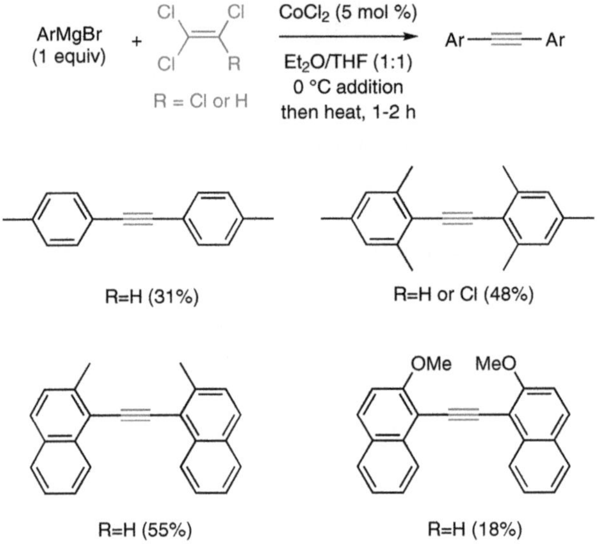

Fig. 4.47: Cobalt-catalyzed reaction of polychloroethylenes with arylmagnesium bromides.

Uemura *et al.* also investigated the cross-coupling between aromatic Grignard reagents and styryl tellurides in the presence of bis(triphenylphosphine)cobalt(II) (Fig. 4.48) [21–22]. In the case of (*Z*)-phenyl styryl telluride, the reaction with phenylmagnesium bromide is stereoselective and gives the stilbene in favour of the *cis*-isomer in quantitative yield. However the reaction is sensitive to the substituent of the styryl telluride, thus (*Z*)-ethylester styryl telluride only provides the *trans*-isomer in poor yield. Biphenyl and biphenyl telluride were produced in substantial amount, which makes the purification step delicate. The mechanism is similar to the one described in Fig. 4.19.

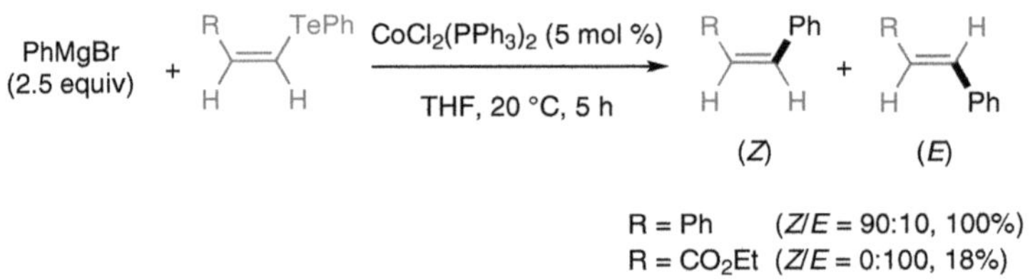

Fig. 4.48: Cobalt-catalyzed cross-coupling between aromatic Grignard reagents and biaryl tellurides.

Cahiez *et al.* reinvestigated the reaction of Kharasch in 1998 and observed that the solvent plays a crucial role in this reaction [17]. Indeed, replacement of diethyl ether by a mixture of THF/NMP improves the yields (Fig. 4.49). The reaction is more efficient with electron-rich Grignard reagents than with electron-poor Grignard reagents. For more substituted alkenyl bromides, an excess of the Grignard reagent is required as well as additional cobalt-catalyst, to afford the product in good yield.

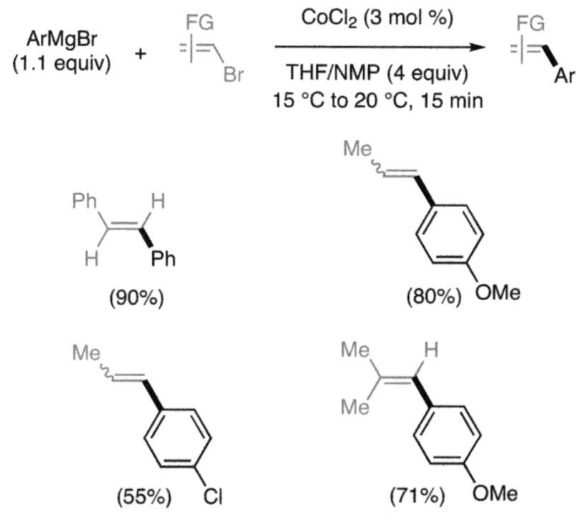

Fig. 4.49: Cobalt-catalyzed reaction of vinyl bromides with aromatic Grignard reagents in THF/NMP.

Shirakawa and Hayashi have described the first arylation of alkenyl triflates by Grignard reagents in presence of cobalt(III) acetylacetonate and triphenylphosphine (1:4) [45]. Alkenyl triflates can be easily prepared from ketones or aldehydes and are known to be highly reactive with cobalt complexes. Hence, the reaction is compatible with heteroaromatic or aromatic Grignard reagents bearing electron-donating or electron-withdrawing groups or bulky substituents. Products were obtained in better yields compared to those obtained by Cahiez *et al.* (Fig. 4.50) [17]. Interestingly, cyclic alkenyl triflates can also undergo this reaction.

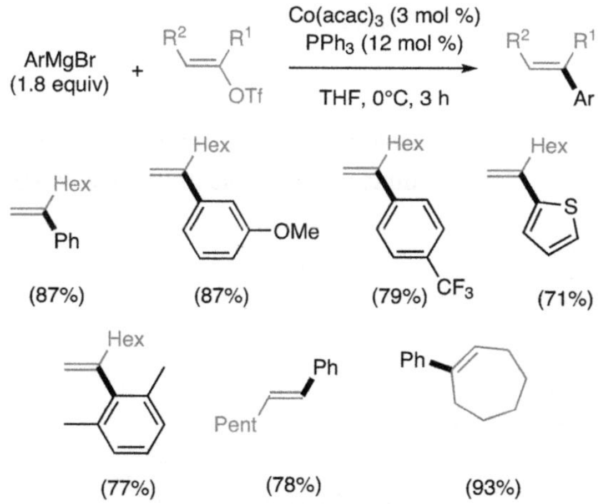

Fig. 4.50: Cobalt-catalyzed reaction of alkenyl triflates with aromatic Grignard reagents.

A plausible mechanism is proposed in Fig. 4.51. The first step is the reduction of cobalt(III) acetylacetonate by the Grignard reagent followed by the oxidative addition of alkenyl triflates to $Ar_2Co(0)$ di-ate complex **S**. The coupling product is obtained by reductive elimination of **T** to give ArCo(0) ate complex **U** which reacts with Grignard reagent to regenerate di-ate complex **S**.

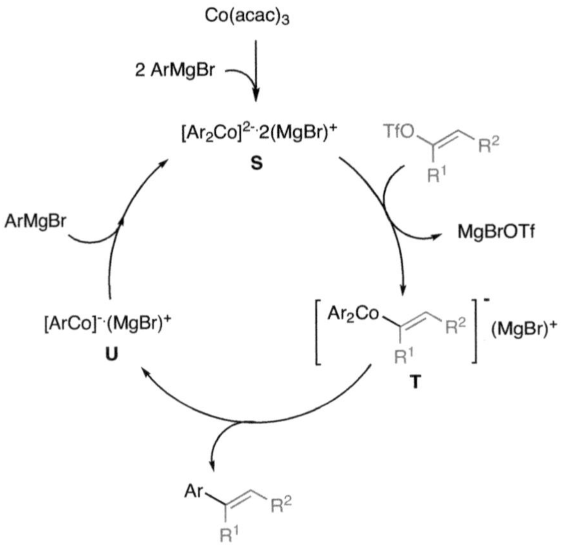

Fig. 4.51: Plausible mechanism for the reaction of alkenyl triflates with aromatic Grignard reagents by cobalt-catalysis.

4.3.4 With allylic ethers

The only cobalt-catalyzed allylation of an aromatic Grignard reagent by allylic ethers and allylic acetals was described by Oshima *et al.* in 2004 [23]. Cobalt(II) chloride associated to 1,5-bis-(diphenylphosphino)pentane is the most effective complex for this reaction. Treatment of 1- and 3-phenyl-2-propenyl methyl ethers with phenylmagnesium bromide affords (*E*)-1,3-diphenyl-prop-1-ene in good yields. However, the reaction is not regioselective and (*E*)-2-octenyl methyl ether gives a mixture of regioisomers (Fig. 4.52).

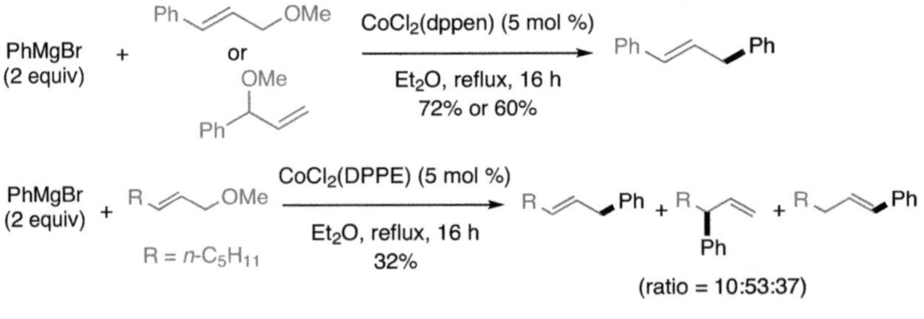

Fig. 4.52: Cobalt-catalyzed cross-coupling between allylic methyl ether and phenylmagnesium bromide.

The reaction between allylic acetals with substituents at the terminal olefinic position and phenyl Grignard reagent afforded the monophenylated product (Fig. 4.53).

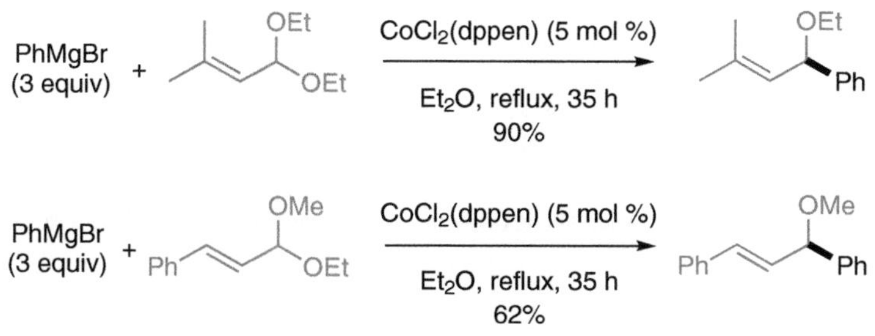

Fig. 4.53: Cobalt-catalyzed cross-coupling between allylic acetals and phenylmagnesium bromide.

4.4 Coupling reactions of allylmagnesium reagents

4.4.1 With aryl halides

Oshima *et al.* were interested, in 2002, in the cross-coupling between aryl halides and allylic Grignard reagents catalyzed by cobalt(II) chloride and triphenylphosphine [10]. However, only poor yields were obtained (Fig. 4.54). In presence of both aryl and alkyl bromides, the reaction is highly selective and only allylation of the alkyl bromides was observed.

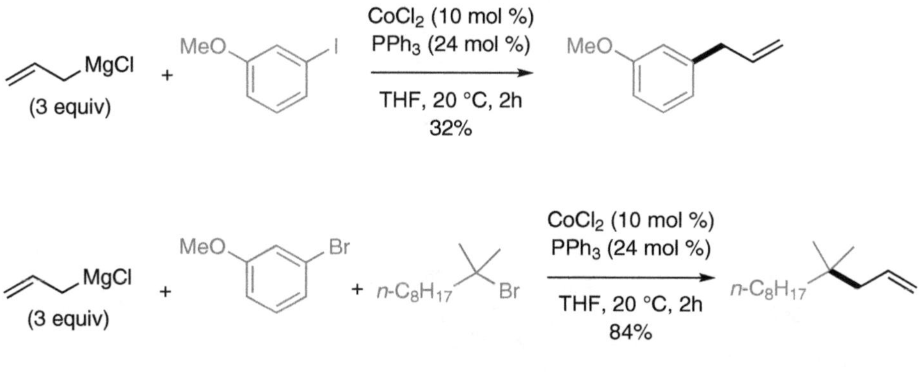

Fig. 4.54: Cobalt-catalyzed cross-coupling between allylic Grignard reagents and aryl halides.

The same year, Oshima *et al.* reported the allylation of an activated *N*-heterocyclic chloride with cobalt(II) acetylacetonate in dioxane in good yield (Fig. 4.55) [7]. Whereas the reaction between allylmagnesium chloride and 2-chloropyridine failed even at −20 °C, 2-chloropyridine bearing an alkene moiety is a good substrate and could be allylated under these conditions.

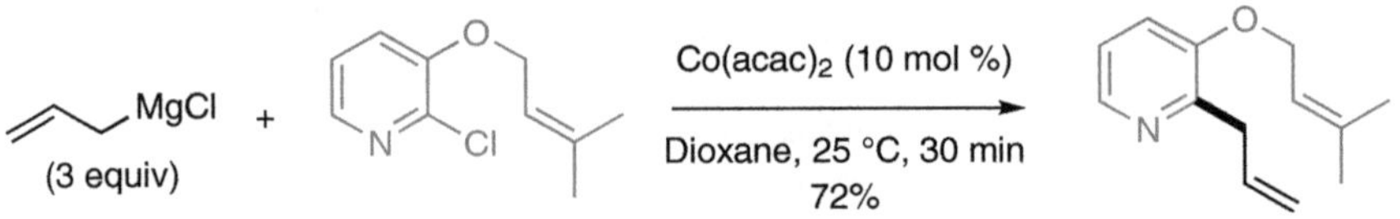

Fig. 4.55: Cobalt-catalyzed cross-coupling between allylic Grignard reagents and *N*-heterocyclic chlorides.

4.4.2 With alkyl halides

Oshima *et al.* also extended the cross-coupling between aryl halides and allylic Grignard reagents to the coupling between alkyl halides and allyl Grignard reagents in the presence of cobalt(II) chloride and 1,3-bis(diphenylphosphino)propane (Fig. 4.56) [10, 46]. Generally, good yields were obtained from tertiary alkyl bromides, whereas secondary and primary alkyl bromides were less reactive even with a large excess of Grignard reagents (3 equiv). In this case, alkyl iodides have to be used. Alkoxy groups were well tolerated contrary to esters, amides and carbonates.

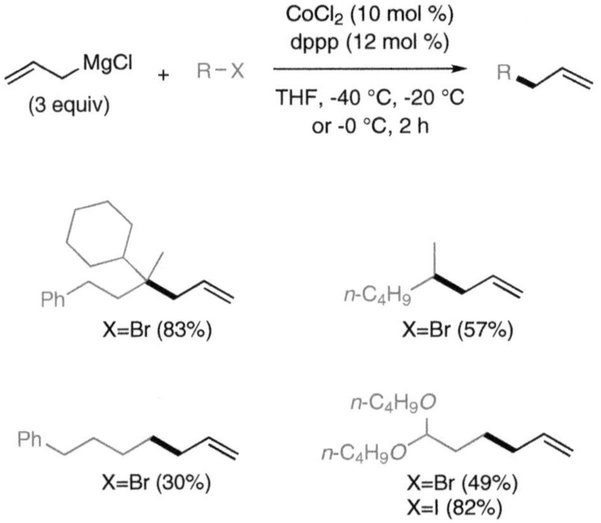

Fig. 4.56: Cobalt-catalyzed cross-coupling between allylic Grignard reagents and alkyl halides.

Unfortunately, reactions with substituted allylic Grignard reagents are not regioselective, as alkylation can also occur at the more substituted side of the allylic Grignard reagent (Fig. 4.57).

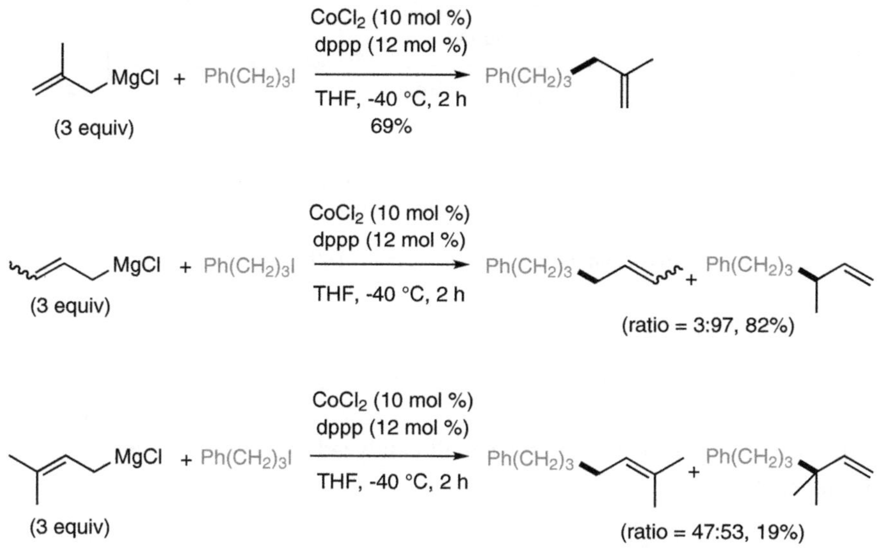

Fig. 4.57: Cobalt-catalyzed cross-coupling between substituted allylic Grignard reagents and alkyl halides.

Asymmetric allylation of racemic alkyl bromides is also possible in the presence of $[CoCl_2\{(-)\text{-chiraphos}\}]$. The enantiomeric excess was low, but could be increased at a very low temperature (Fig. 4.58).

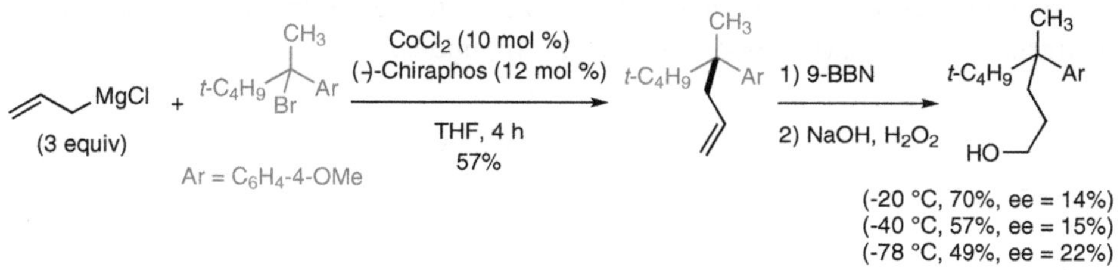

Fig. 4.58: Cobalt-catalyzed asymmetric cross-coupling between allylic Grignard reagents and alkyl halides.

4.4.3 With allylic ethers

Allylation of allylic Grignard reagents was also described by Oshima *et al.* in 2006 [47]. The regioselectivity depends on the presence of the ligand. With no ligand, cobalt(II) chloride led only to the linear substitution product in good yield. Nevertheless, the addition of phosphine ligands affords the branched substitution product in significant amount (Fig. 4.59).

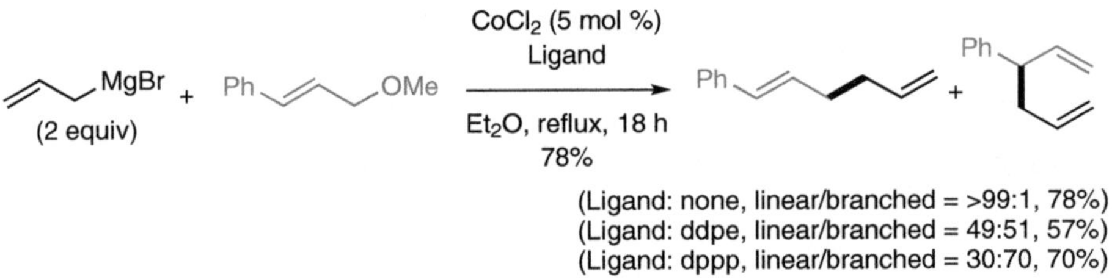

Fig. 4.59: Cobalt-catalyzed cross-coupling between allylic methyl ether and allylic Grignard reagents.

4.5 Coupling reactions of alkynylmagnesium reagents

Few results are reported concerning the cobalt-catalyzed cross-coupling with alkynyl Grignard reagents.

4.5.1 With alkyl halides

The first alkylation of acetylenic Grignard reagents in the presence of a catalytic amount of $Co(acac)_3$ was reported in 2006 by Okamoto *et al.* [48]. They have developed a new method for cobalt-catalyzed benzyl–alkynyl coupling under mild conditions (Fig. 4.60). With an alkynyl Grignard reagent substituted by an alkyl such as a butyl group, benzyl bromide is required. This method is not general. However, the reaction gives good yields with an alkynyl Grignard reagent functionalized by a trimethylsilyl with benzyl chlorides and benzyl bromides.

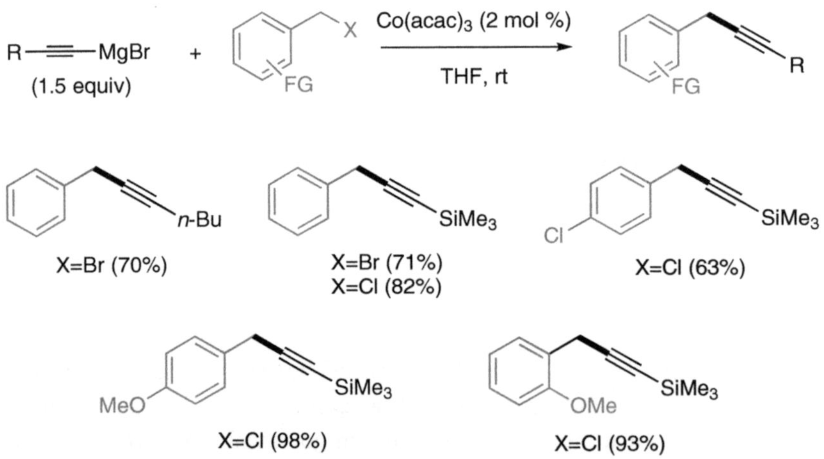

Fig. 4.60: Cobalt-catalyzed benzylation of alkynyl Grignard reagents

The same year, Oshima *et al.* extended the reaction to primary and secondary alkyl halides [49]. 2-(Trimethylsilyl)ethynylmagnesium bromide reacts in tetramethylethylenediamine (TMEDA) as the solvent, with various alkyl iodides and bromides in good yields in the presence of $Co(acac)_3$ as a catalyst in simple conditions. The cobalt-mediated cross-coupling proceeds via a radical pathway. The single electron transfer allows the use of unactivated secondary alkyl halides in the cross-coupling.

More recently, Wu's group described a synthesis of an enantiopure pyrrolidine derivative using a cobalt-catalyzed carbon–carbon bond formation previously mentioned [37]. The coupling reaction of (*S*)-2(iodomethyl)pyrrolidine with 2-(trimethylsilyl)ethynylmagnesium bromide was carried out in the presence of the catalytic system $CoCl_2$/IPr·HCl in excellent yield. The product was used as an intermediate in the synthesis of (+)-(*S*)-tylophorine, known to be effective against drug-sensitive and multidrug-resistant cancer cells (Fig. 4.61).

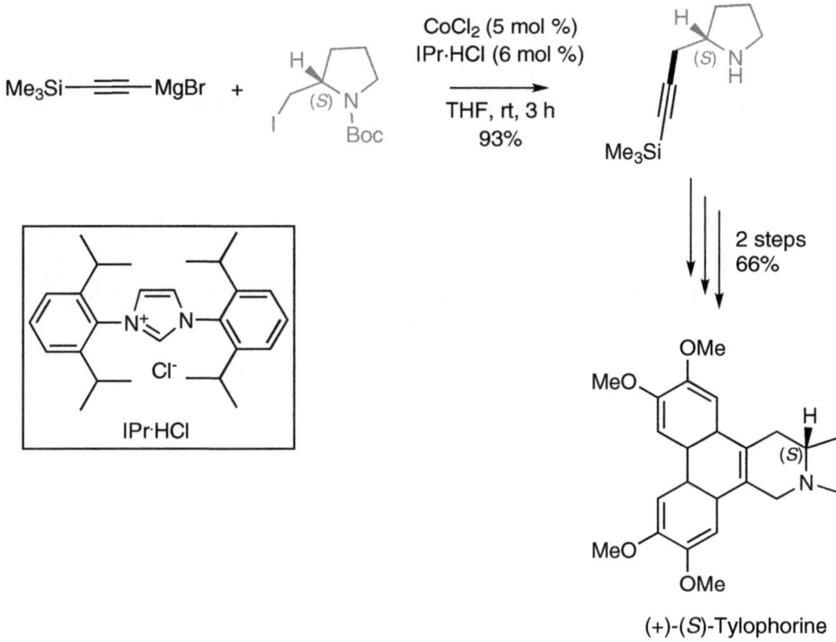

Fig. 4.61: Cobalt-catalyzed cross-coupling between alkynyl Grignard reagent and enantiopure pyrrolidine derivatives for the synthesis of (+)-(*S*)-tylophorine.

4.5.2 With alkenyl triflates

Conjugated enynes are important motifs in natural products and precursors for the synthesis of pharmaceuticals. These compounds can be obtained by a Sonogashira reaction or alternatively by a cross-coupling of alkynyl-metals with an alkenyl (pseudo)halide in the presence of a catalyst. The first cobalt-catalyzed sp–sp$_2$ coupling was reported by Shirakawa and Hayashi [50]. They showed that alkynyl Grignard reagents reacted with alkenyl triflates or nonaflates in the presence of Co(acac)$_3$ in THF at 20 °C. Selected examples are reported in Fig. 4.62.

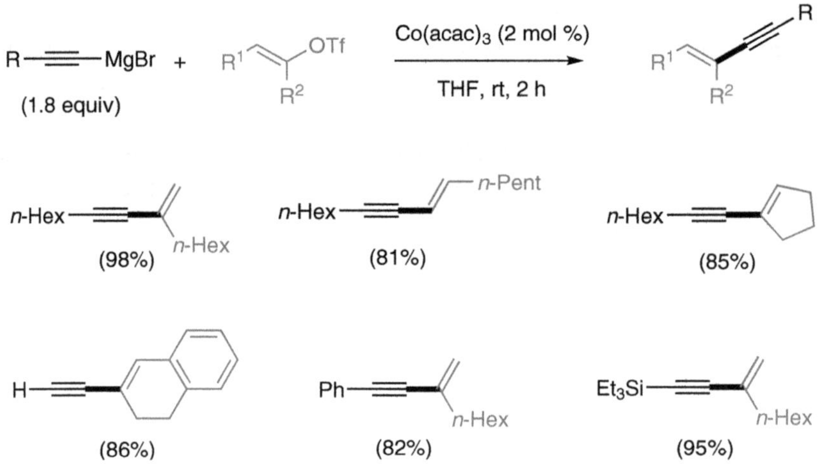

Fig. 4.62: Cobalt-catalyzed cross-coupling of alkenyl triflates with alkynyl Grignard reagents.

A catalytic cycle was proposed based on Oshima *et al.* mechanism (Fig. 4.63).

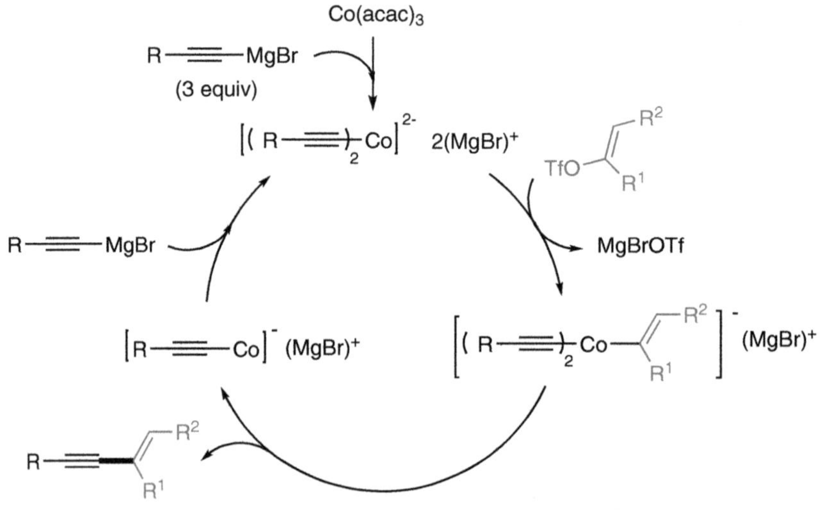

Fig. 4.63: Mechanism for the cobalt-catalyzed cross-coupling of alkenyl triflates with alkynyl Grignard reagents.

4.5.3 With alkynyl halides

The first access to symmetrical diacetylenes compounds was obtained by reaction of phenylethynyl iodide with a bromomagnesium acetylene derivative in the presence of a cobalt salt catalyst. Under these conditions, a satisfactory yield of 75 % was obtained [51]. It was one of the first cobalt-catalyzed cross-coupling involving a Grignard reagent and an alkynyl halide.

Three years later, this result was discounted by Weedon *et al.* to form unsymmetrical diacetylenes. They obtained only low yields (around 30 %) for the coupling product of an acetylenic Grignard reagent and a 1-haloalkyne, in the presence of cobalt chloride as the catalyst (Fig. 4.64) [24].

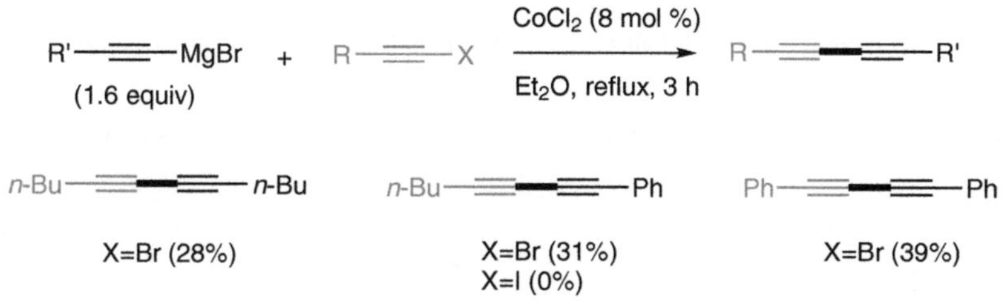

Fig. 4.64: Cobalt-catalyzed cross-coupling of alkynyl Grignard reagents with alkynyl halides.

4.6 Coupling reactions of alkenylmagnesium reagents

4.6.1 With alkyl halides

As reported previously with 2-(trimethylsilyl)ethynylmagnesium bromides, cobalt-mediated cross-couplings of 1-(trimethylsilyl)ethenylmagnesium bromides with alkyl bromides and iodides proceed smoothly [49]. With optically pure bromoalkyls, the stereochemistry is maintained without suffering from β-elimination. The choice of the solvent has a significant impact on the yield and TMEDA is the best additive. However, it should be noted that good yields were obtained only with a trimethylsilyl substituted Grignard reagents (R' = TMS).

Cossy *et al.* extended the cross-coupling of alkenyl Grignard reagents to glycosides which afforded *C*-alkenyl glycosides when cobalt(III) acetylacetonate and TMEDA (1:1) were used [38]. Selected results are shown in Fig. 4.65. The reaction is highly diastereoselective in favour of the α-isomer and *C*-vinyl glycosides were isolated in good yields. However, the reaction seems to work only with substituted alkenyl Grignard reagents, as vinylmagnesium bromide did not react.

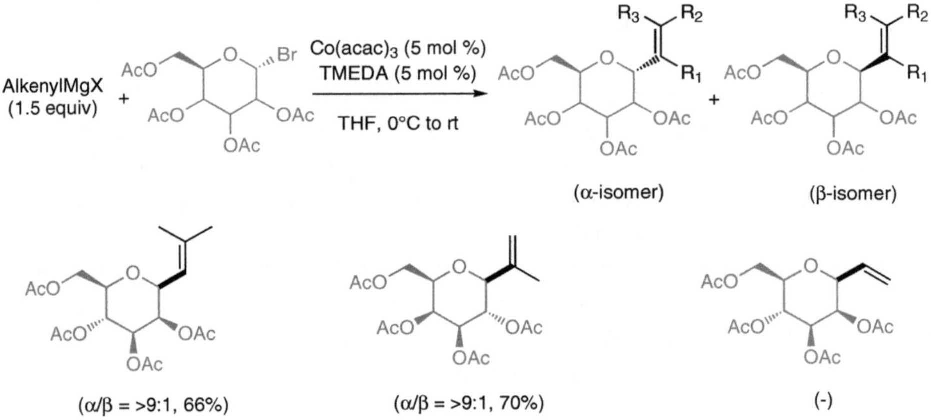

Fig. 4.65: Cobalt-catalyzed cross-coupling of alkenylmagnesium halides with 1-bromo glycosides.

These conditions can also be applied for the synthesis of *C*-vinyl furanosides [39].

4.6.2 With alkenyl species

4.6.2.1 Alkenyl halides

When the Cahiez's group reported the cobalt-catalyzed alkenylation of organomagnesium reagents, one example of a coupling with an alkenyl Grignard reagent was described. However, the yield was lower compared to the cross-coupling with alkyl Grignard reagents [17]. In addition, higher temperatures were required in this case (Fig. 4.66).

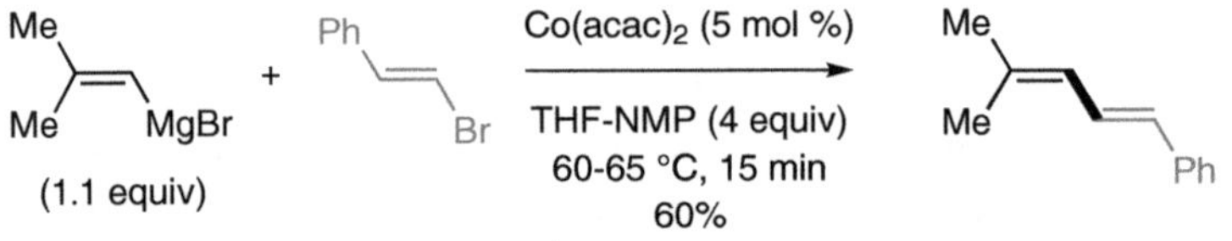

Fig. 4.66: Cobalt-catalyzed cross-coupling of alkenyl Grignard reagents with alkenyl bromide.

4.6.2.2 Alkenyl triflates

Later, Shirakawa, Imazaki and Hayashi applied the cobalt catalysis to the coupling of alkenyl triflates with alkenylmagnesium bromides to access conjugated dienes [45]. The effective catalyst is $Co(acac)_3$ associated with PPh_3 and good to excellent yields were obtained (Fig. 4.67). Moreover, the stereochemistry of the Grignard reagents was retained.

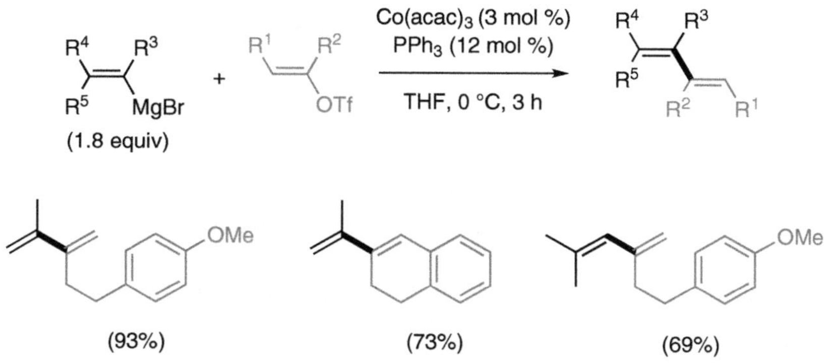

Fig. 4.67: Cobalt-catalyzed cross-coupling of alkenyl Grignard reagents with alkenyl triflates.

4.7 Tandem cyclization/cross-coupling

Some cobalt-catalyzed cross-coupling reactions of alkyl halides with Grignard reagents involving preliminary intramolecular cyclizations, have mainly been developed by Oshima *et al.* These various reactions suggested that some radical species were formed from alkyl halides. Various Grignard reagents were used.

4.7.1 With arylmagnesium halides

Oshima *et al.* reported, in 2001, the synthesis of benzyl-substituted tetrahydrofuran derivatives. The cobalt-catalyzed reaction involves a tandem radical cyclization and arylation by the Grignard reagent of 5-hexenyl halide derivatives [52]. Different substrates were studied in the presence of $CoCl_2$(dppe) affording various nitrogen- or oxygen-containing heterocyclic and carbocyclic compounds in good to excellent yields (Fig. 4.68). Iodo- and bromo-acetals, bearing a terminal double bond, react in the same way whereas with chloro-acetals the reaction does not proceed. Stereochemistry and yield highly depend on the number and the position of the substituents. Heteroaromatic or aromatic Grignard reagents can be successfully employed, however, the reaction is sensible to steric hindrance, thus *ortho*-subsituted aryl Grignard reagents can only be coupled in poor yields. It is worthy of note that when the cyclization is intramolecular, arylation is not possible.

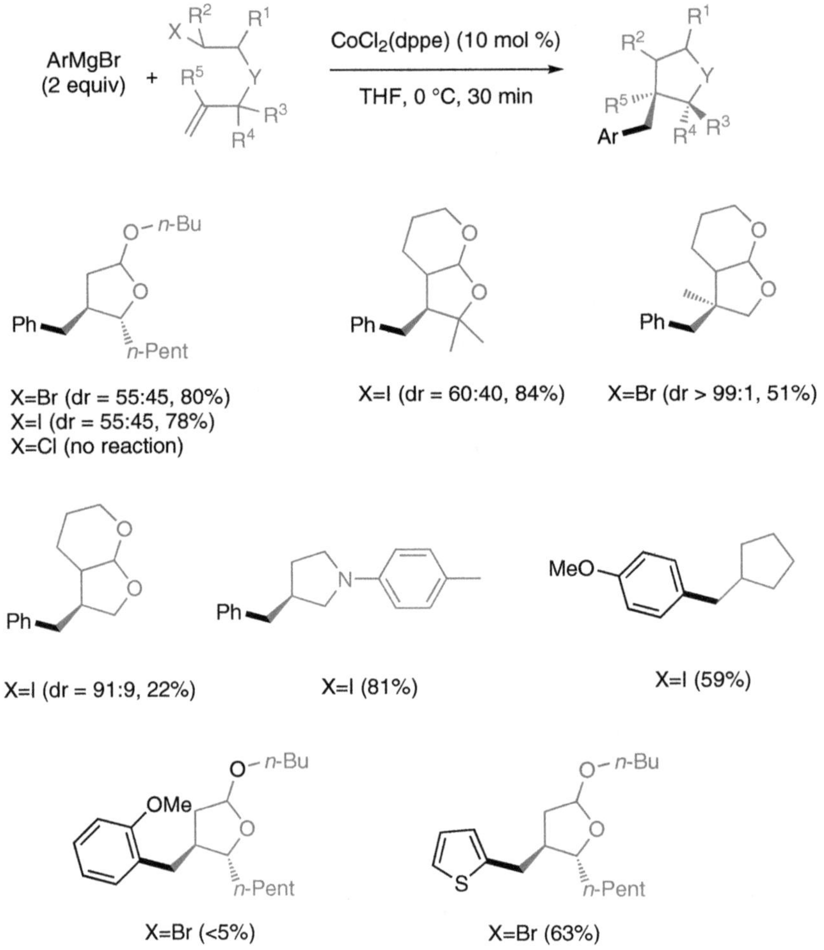

Fig. 4.68: Cobalt-catalyzed radical cyclization and cross-coupling reaction of 5-hexenyl halide derivatives with aromatic Grignard reagents.

A possible mechanism is shown in Fig. 4.69. $CoCl_2$(dppe) reacts with four equivalents of phenylmagnesium bromide to give cobalt(0) complex **V** and one equivalent of biphenyl. Single electron transfer to 5-hexenyl halide substrate from complex **V** yields cobalt(I) complex **W** and a radical anion is generated. Loss of the halide present on the substrate and subsequent cyclization lead to a cyclopentylmethyl radical, which after recombination with complex **W** gives divalent cobalt complex **X**. Product and complex **Y** are obtained after reductive elimination, and complex **V** is regenerated after insertion of phenylmagnesium bromide on **Y**.

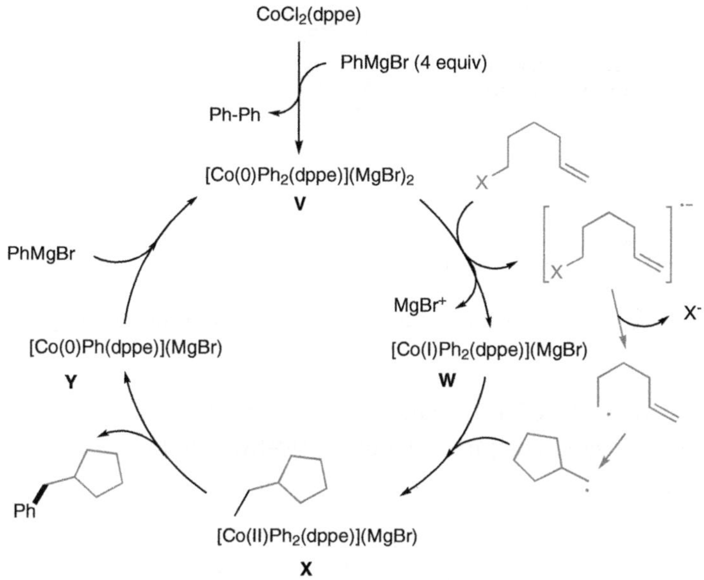

Fig. 4.69: Mechanism for cobalt-catalyzed radical cyclization and cross-coupling reaction of 5-hexenyl halide derivatives with aromatic Grignard reagents.

Oshima *et al.* have replaced 1,2-bis(diphenylphosphino)ethane by chiral diamine *N,N,N',N'*-tetramethylcyclohexane-1,2-diamine as a ligand (Fig. 4.70) [34]. With this ligand alkyl iodides are more reactive than alkyl bromides and cyclopentane derivatives were obtained in good to excellent yields. Moreover, sequential cyclization/arylation was possible on a substrate bearing an internal double bond.

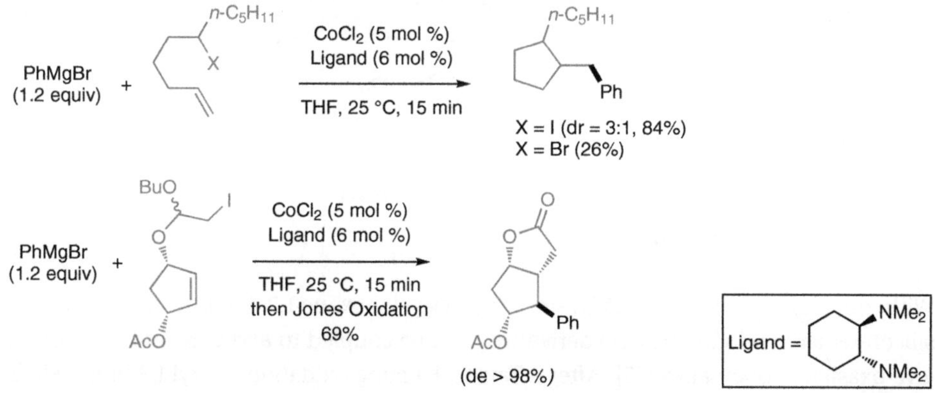

Fig. 4.70: Cobalt-diamine-catalyzed cross-coupling reactions of alkyl halides with aromatic Grignard reagents.

In addition, under these conditions δ-olefinic iodo-pyrrolidines can also be cyclized in good yields (Fig. 4.71) [41].

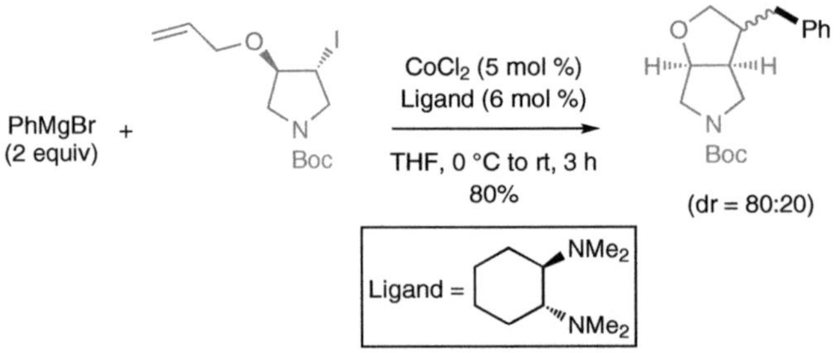

Fig. 4.71: Cobalt-diamine-catalyzed cross-coupling reactions of alkyl halides with aromatic Grignard reagents.

This method was successfully utilized as the key reaction in the synthesis of the EP_2-receptor agonist, AH-13205, a synthetic prostaglandin derivative that lowers intraocular pressure (Fig. 4.72) [34].

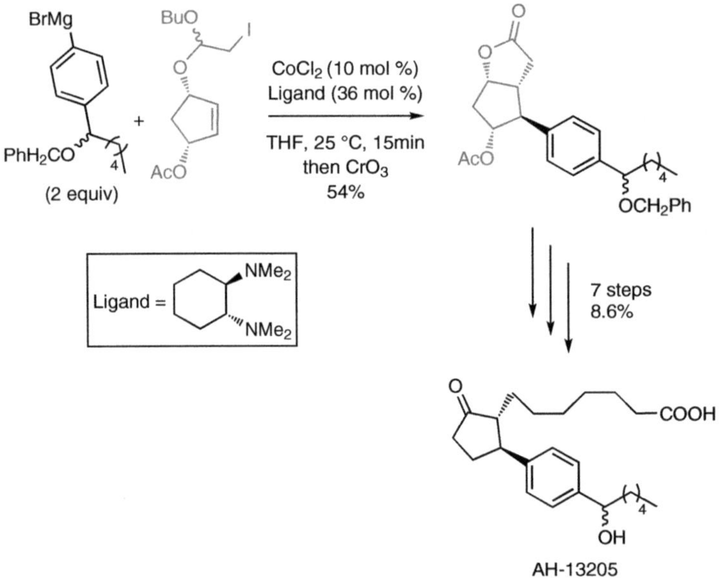

Fig. 4.72: Total synthesis of AH-13205.

With the same diamine *N,N,N',N'*-tetramethylcyclohexane-1,2-diamine as the ligand, silicon-tethered 6-iodo-1-hexene derivatives can be coupled to aryl Grignard reagents to give oxasilacyclopentanes [53]. After a Tamao-Fleming oxidation, 4-aryl-1,3-butanediols are obtained in moderate to good yields, most of the time, as a 50:50 mixture of diastereomers (Fig. 4.73). Different electron-poor or electron-rich Grignard reagents can be used; however, the reaction is sensitive to steric hindrance and highly substituted aryl Grignard reagents cannot be successfully coupled. It is noteworthy that the corresponding silicon-tethered 6-bromo-1-hexene derivatives give the desired products in very low yields.

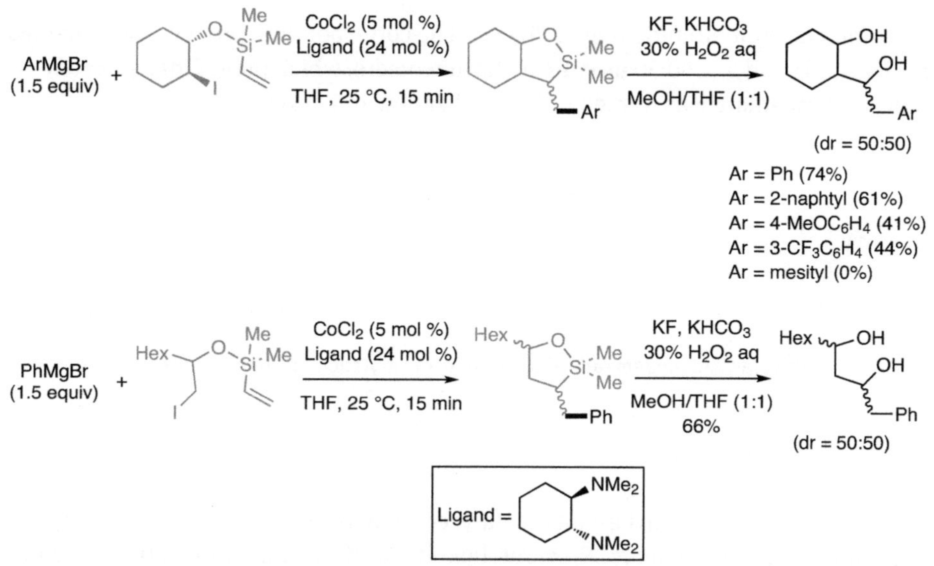

Fig. 4.73: Cobalt-diamine-catalyzed cyclization/cross-couplings of siloxy-tethered 6-iodo-1-hexene derivatives with aromatic Grignard reagents followed by a Tamao-Fleming oxidation.

As 1,3-diols are important units in biologically active compounds, this method was utilized to synthesize an antagonist of human CCR5 receptor (Fig. 4.74) [53].

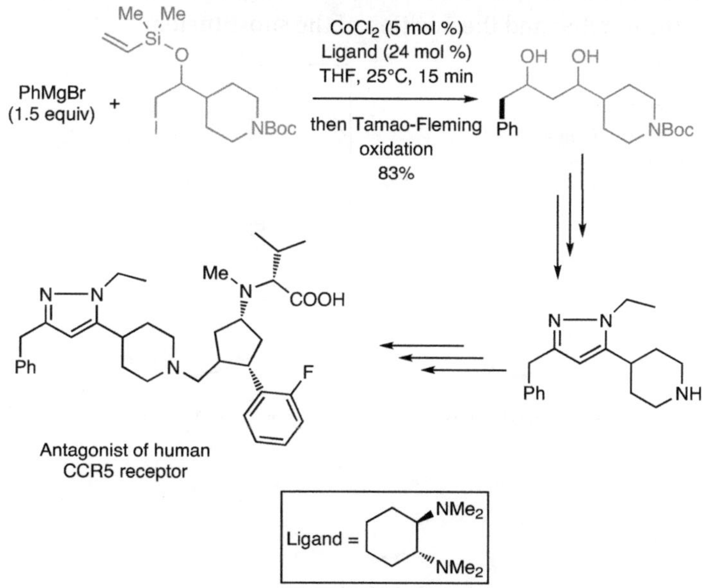

Fig. 4.74: Cobalt-diamine-catalyzed synthesis of an antagonist of a human CCR5 receptor.

Cossy *et al.* used cobalt(III) acetylacetonate and the inexpensive TMEDA (1:1) for the cyclization of a δ-olefinic 1-bromo glycoside followed by a cross-coupling with phenylmagnesium bromide (Fig. 4.75) [38]. The corresponding product was obtained in very good yield in a 50:50 mixture of diastereomers, which proves that an anomeric radical intermediate was formed before cyclization and cross-coupling.

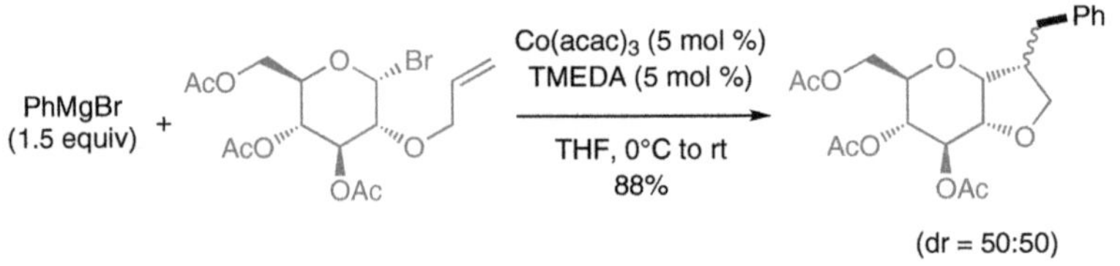

Fig. 4.75: Cobalt-catalyzed tandem radical cyclization and arylation of 1-bromo glycosides.

4.7.2 With alkylmagnesium halides

Oshima *et al.* also reported the cobalt-catalyzed sequential cyclization/cross-coupling with alkyl Grignard reagents. Previous ligands used for tandem reactions such as dppe, TMEDA and *N,N,N',N'*-tetramethylcyclohexane-1,2-diamine, resulted in very low yields of the cyclic products, whereas NHC ligands were much more efficient. Among them, SIEt·HCl proved to be the best ligand [54]. Selected reaction of various 6-iodo- or 6-bromo-1-hexene substrates with allyl- or phenyl-dimethylsilylmethyl-magnesium chlorides are shown in Fig. 4.76. As a rule, carbocyclic and oxygen- or nitrogen-containing heterocyclic compounds were obtained in good yields. The stereochemistry depends on the number and the position of the substituents.

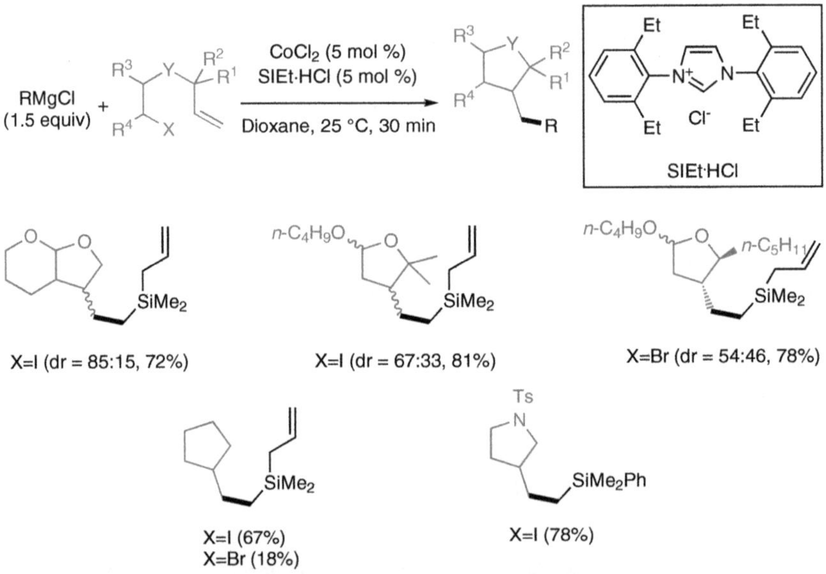

Fig. 4.76: Cobalt/NHC-catalyzed sequential cyclization/cross-couplings with 6-halo-1-hexene derivatives and alkyl Grignard reagents.

The cyclization/cross-coupling reaction was also performed on siloxy-tethered 6-iodo-1-hexene derivatives with phenyl-dimethylsilylmethylmagnesium chloride with SIEt·HCl ligand (Fig. 4.77) [54–55]. The corresponding products were easily converted to diols after a Tamao-Fleming oxidation in good yield. These results were similar to the ones obtained by using *N,N,N',N'*-tetramethylcyclohexane-1,2-diamine ligand, aryl Grignard reagents and siloxy-tethered 6-iodo-1-hexene derivatives [53].

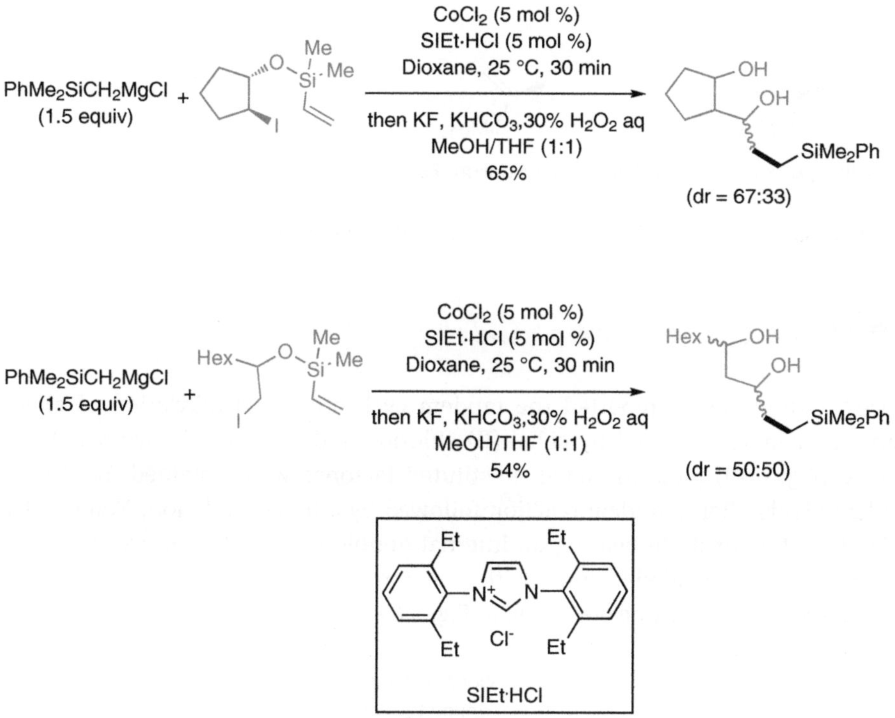

Fig. 4.77: Cobalt/NHC-catalyzed cross-couplings of siloxy-tethered 6-iodo-1-hexene derivatives with alkyl Grignard reagents followed by a Tamao-Fleming oxidation.

4.7.3 With alkenylmagnesium halides

Oshima *et al.* have also investigated the cyclization followed by an alkenylation of 6-iodo-1-hexene substrates, catalyzed by cobalt(III) acetylacetonate in TMEDA (Fig. 4.78) [49]. To obtain good yields, a large excess of Grignard reagents is necessary (4 equiv) as well as a high catalyst loading (40 mol %). Moreover, under these conditions only trimethylsilyl-substituted alkenylmagnesium reagents gave promising results. The mechanism proposed in this case is similar to the one described in Fig. 4.69.

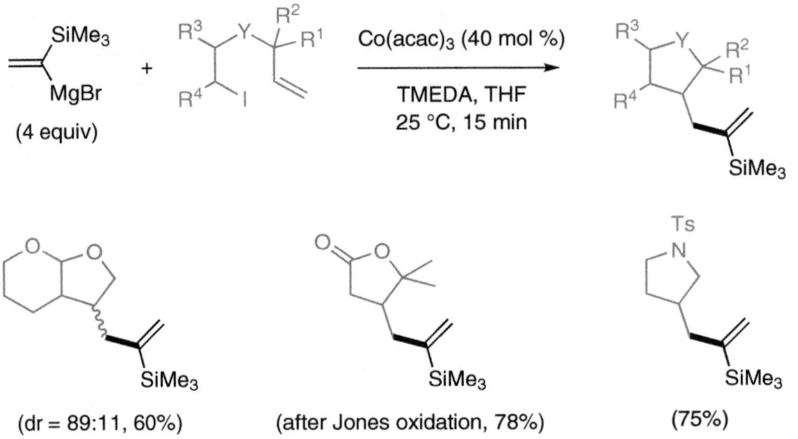

Fig. 4.78: Cobalt-catalyzed tandem radical cyclization and alkenylation of 6-iodo-1-hexenes.

4.7.4 With allylmagnesium halides

In 2002, Oshima *et al.* reported the tandem cyclization and alkylation of allylic Grignard reagents catalyzed by cobalt(II) chloride and 1,3-bis(diphenylphosphino) propane (Fig. 4.79) [46]. 3-Butenyl-substituted lactones were obtained in good to excellent yields after a tandem reaction followed by a Jones oxidation. Noteworthy, cyclization of a substrate bearing an internal double bond, followed by allylation, was possible and provided lactones with a quaternary center. The mechanism of the reaction is similar to the one described in Fig. 4.69.

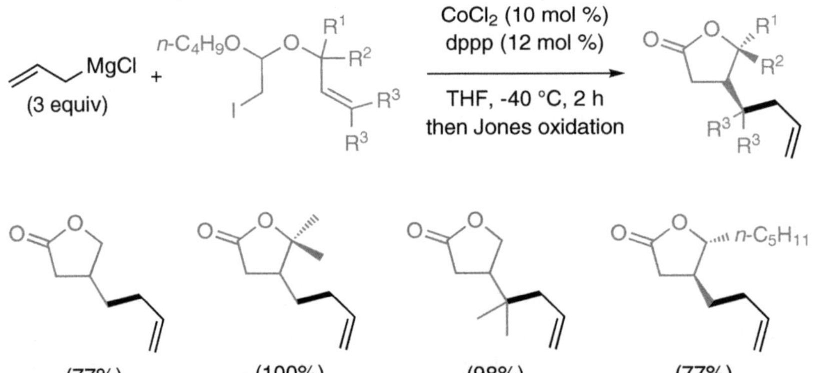

Fig. 4.79: Cobalt-catalyzed tandem radical cyclization and cross-coupling reaction between allylic Grignard reagents and 5-hexenyl iodo derivatives.

However, when the reaction was performed in diethyl ether, cyclopropane derivatives were obtained (Fig. 4.80) [10]. Allyl-, prenyl- and crotylmagnesium bromides afforded the corresponding cyclopropanes in good yield.

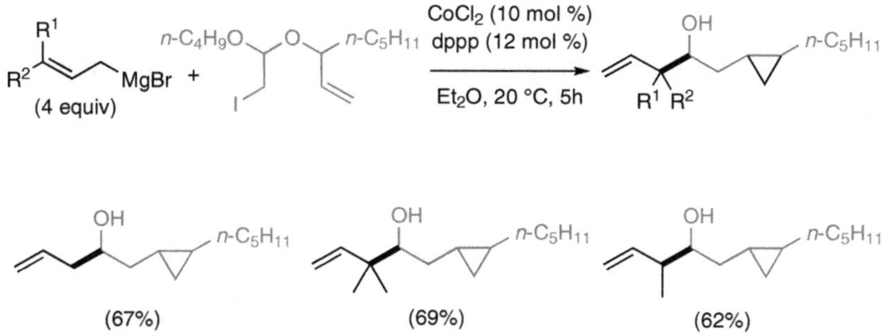

Fig. 4.80: Cobalt-catalyzed reaction between allylic Grignard reagents and 5-hexenyl iodide derivatives in diethyl ether.

A plausible mechanism is proposed in Fig. 4.81. $CoCl_2$(dppe) is first reduced by the allylmagnesium bromide to an active [Co^n] species. Single electron transfer to 5-hexenyl iodide **AA** from [Co^n] affords the radical anion of the substrate. Loss of the iodide and subsequent 5-*exo* radical cyclization lead, after recombination with a cobalt complex, to (4-butoxy-3-oxacyclopentyl)methylcobalt and/or magnesium species **AB**. In diethyl ether **AB** undergoes an intramolecular cyclopropanation with loss of n-C_4H_9O-M (M = Co or Mg) to give the cyclopropane derivative **AC**. Diethyl ether, which is less coordinating than THF, enhances the complexation of oxygen atoms of **AB** with cobalt and/or magnesium and this results in an elimination process. The resulting aldehyde **AC** then reacts with allylmagnesium bromide by carbonyl allylation to afford alcohol **AD**.

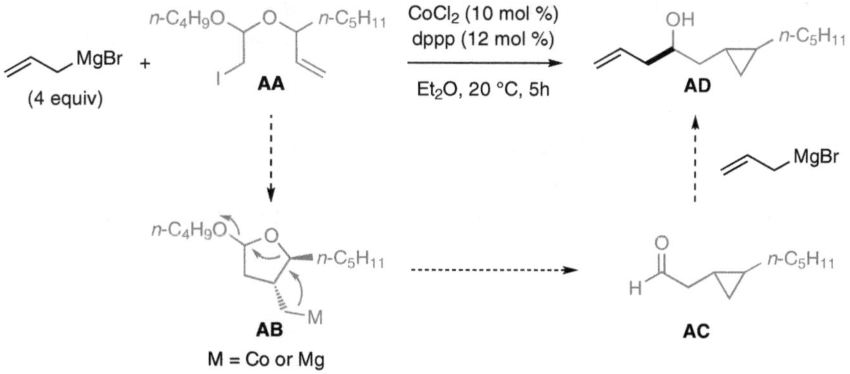

Fig. 4.81: Mechanism of the cobalt-catalyzed formation of cyclopropane from allylmagnesium bromides and 5-hexenyl iodide derivatives.

4.7.5 With alkynylmagnesium halides

Oshima *et al.* reported also a cyclization process followed by alkynylation of 6-iodo-1-hexene substrates catalyzed by cobalt(III) acetylacetonate in TMEDA (Fig. 4.82) [49]. Yields are slightly lower than with alkenyl Grignard reagents [49]. It is worth noting that under these conditions only trimethylsilyl-substituted alkynylmagnesium reagents gave satisfactory results.

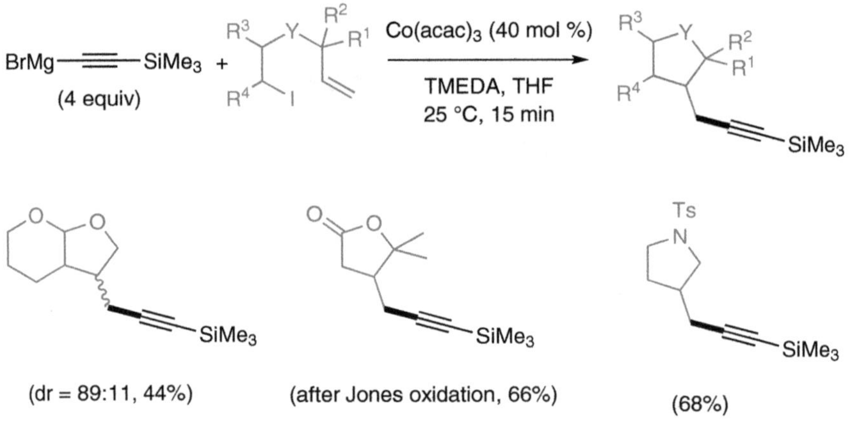

Fig. 4.82: Cobalt-catalyzed tandem radical cyclization and alkenylation of 6-iodo-1-hexenes.

The use of NHC ligands is essential and enables the sequential cyclization / cross-coupling of 6 iodo-1-hexene derivatives with a scope of various alkyl-substituted alkynyl Grignard reagents [54]. However, a large excess of alkynyl Grignard reagents (5 equiv) is necessary to provide alkynylated cyclic products in good yields. Whereas SIEt·HCl was the best ligand for the reaction of siloxy-tethered 6-iodo-1-hexene derivatives with phenyl-dimethylsilylmethyl-magnesium chloride [54, 55], no coupling products were obtained with 6-iodo-1-hexene derivatives. In this case, IMes·HCl ligand is very effective (Fig. 4.83). The reaction is even possible with sterically hindered alkynes and with alkynyl Grignard reagents bearing a siloxy group. Unfortunately, 2-trimethylsilyl-ethynyl or phenylethynyl Grignard reagents did not react and gave a mixture of the starting 6-iodo-1-hexenes and non-alkynylated cyclic products.

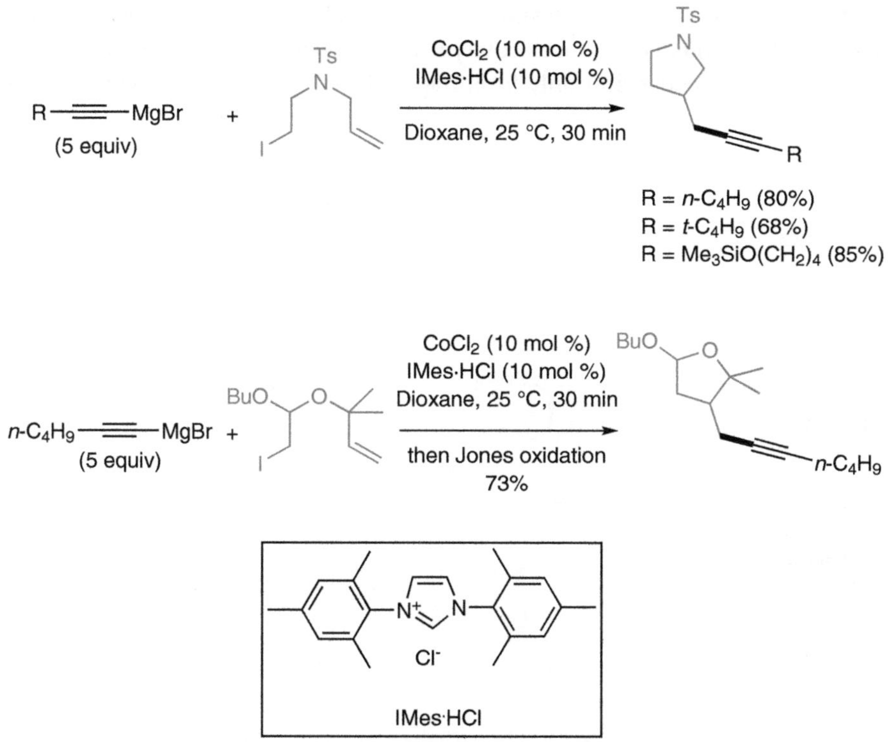

Fig. 4.83: Cobalt/NHC-catalyzed cross-couplings of 6-iodo-1-hexene derivatives with alkynyl Grignard reagents.

4.8 Three-component reactions

Oshima *et al.* reported, in 2003, a three-component reaction for the synthesis of homoallylsilanes from alkyl halides, 1,3-dienes and trimethylsilylmethylmagnesium chloride [56]. This reaction is catalyzed by cobalt(II) chloride and 1,6-bis(diphenylphosphino) hexane (1:1).To the best of our knowledge, this is the only reported example of cobalt-catalyzed three-component coupling involving a Grignard reagent. Selected examples are summarized in Fig. 4.84. Primary, secondary and tertiary alkyl halides gave homoallylsilanes in good to excellent yields. Alkyl bromides provided higher product yields than alkyl iodides, and alkyl chlorides did not react. Only trimethyl- or phenyldimethyl-silylmethylmagnesium chlorides could be implied in the reaction, other alkyl or aryl Grignard reagents were not effective. The reaction is working with 1-aryl-1,3-dienes bearing electron-withdrawing or electron-donating groups on the benzene ring. Coupling is also possible when groups in *ortho*-position are present.

Also, substituents such as chloride and bromide are tolerated. With 2,4-pentadienylcyclohexane, no problem of regioselectivity was observed. Whereas 1-aryl-1,3-dienes were highly stereoselective and only form the (*E*)-isomer, 1-alkyl or vinyl-1,3-dienes gave a mixture of (*E*)- and (*Z*)-isomers.

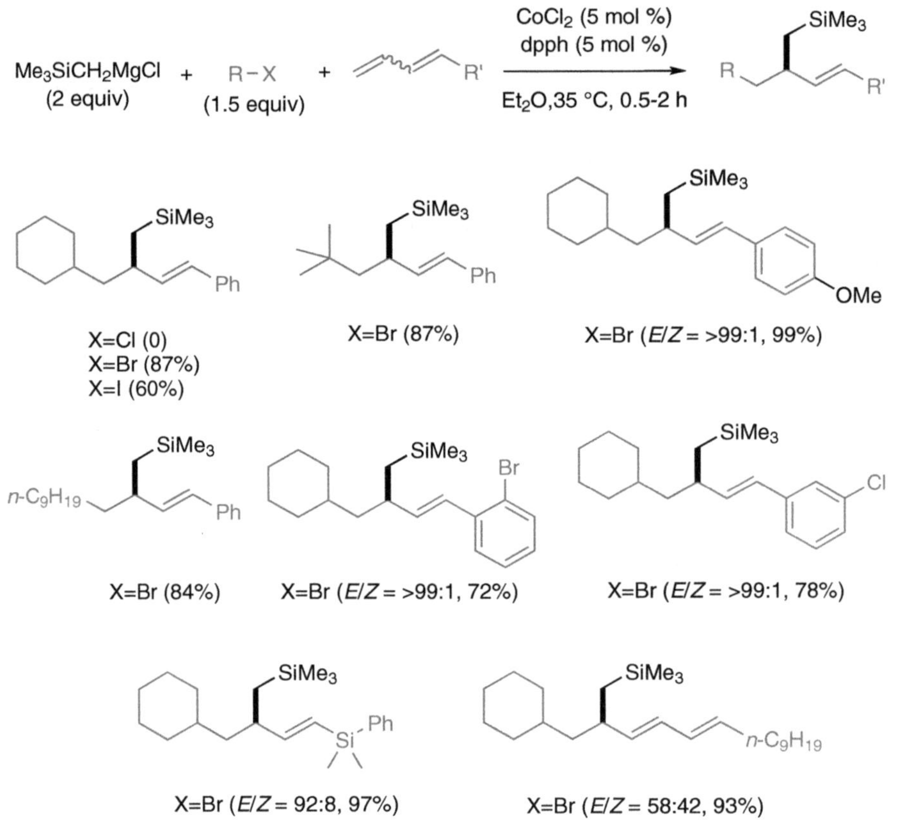

Fig. 4.84: Cobalt-catalyzed three-component coupling reaction between alkyl halides, 1,3-dienes and trimethylsilylmethyl Grignard reagent.

The following reaction mechanism is reported in Fig. 4.85. Cobalt(II) chloride is first reduced by the Grignard reagent to provide an active cobalt(0) complex **AE** which after a single electron transfer with the alkyl halide led to L_nCoX complex **AF** and an alkyl radical, which then attacks the 1-aryl-1,3-diene to form an allylic radical. After recombination of the cobalt(I) species **AF** and the radical, Π-allylcobalt complex **AG** is formed and then alkylated with the Grignard reagent to give **AH**. After reductive elimination, homoallylsilane is obtained and cobalt complex **AE** is regenerated.

4.9 Conclusion

Since the preliminary results of Kharash *et al.*, a wide range of new cobalt-catalyzed cross-coupling reactions has been studied for the last 30 years by combination of alkyl, aryl, allyl, alkenyl and alkynyl Grignard reagents with halides and pseudohalides. Generally, depending on the nature of the ligand, satisfying yields were obtained, which make cobalt-catalyzed reactions a promising way for future development. Moreover, simple ligands are generally used with cobalt salts compared to bulky, expensive and less commercially available analogues often associated with palladium and nickel catalysts. The use of cobalt salts is an interesting alternative to other economical and toxic transition metals. Some functional groups could be present both on the Grignard reagent and the halide even if functionalized Grignard reagents synthesis can be challenging. Most of the reactions described in this chapter proceed through a radical mechanism. Radical intermediates enable the tandem cyclization/cross-coupling reactions, and combination of radical species with cobalt species opens new synthetic possibilities. However, at the present time, Grignard reagents are the most commonly used organometallic species by several groups for performing cobalt-catalyzed cross-coupling reactions. Moreover, new cobalt catalytic systems have to be developed to decrease the catalyst loading. It is worth pointing out that some simple cobalt-catalyzed cross-coupling alternatives were developed using organozinc compounds instead of Grignard reagents to increase the scope of the functional groups [57, 58]. The scope of cobalt-catalyzed cross-coupling reactions has been significantly expanded and efforts have to be particularly made for asymmetric transformations.

References

[1] Wakefield BJ. Organomagnesium Methods in Organic Chemistry. London, UK, Academic Press; 1995.
[2] Kharasch MS, Fuchs CF. J Am Chem Soc 1943; 65(4): 504–7.
[3] Krasovskiy A, Knochel P. Angew Chem Int Ed 2004; 43(25): 3333–6.
[4] Kharasch MS, Fuchs CF. J Org Chem 1945; 10(4): 292–7.
[5] Kharasch MS, Lambert FL, Urry WH. J Org Chem 1945; 10(4): 298–306.
[6] Davies DI, Done JN, Hey DH. J Chem Soc C Org 1969; (15): 2019–21.
[7] Ohmiya H, Yorimitsu H, Oshima K. Chem Lett 2004; 33(10): 1240–1.
[8] Hamaguchi H, Uemura M, Yasui H, Yorimitsu H, Oshima K. Chem Lett 2008; 37(11): 1178–9.
[9] Przyojski JA, Arman HD, Tonzetich ZJ. 2013; 32(3): 723–32.
[10] Ohmiya H, Tsuji T, Yorimitsu H, Oshima K. Chem-Eur J 2004; 10(22): 5640–8.
[11] Ohmiya H, Yorimitsu H, Oshima K. Angew Chem Int Ed 2005; 44(22): 3488–90.
[12] Cahiez G, Chaboche C, Duplais C, Giulliani A, Moyeux A. Adv Synth Catal 2008; 350(10): 1484–8.
[13] Zhou W, Napoline JW, Thomas CM. Eur J Inorg Chem 2011; 2011(13): 2029–33.
[14] Ren P, Stern L-A, Hu X. Angew Chem Int Ed 2012; 51(36): 9110–3. and references cited herein.

[15] Iwasaki T, Takagawa H, Singh SP, Kuniyasu H, Kambe N. J Am Chem Soc 2013; 135(26): 9604–7.
[16] Iwasaki T, Takagawa H, Okamoto K, Singh S, Kuniyasu H, Kambe N. Synthesis 2014; 46(12): 1583–92.
[17] Cahiez G, Avedissian H. Tetrahedron Lett 1998; 39(34): 6159–62.
[18] Cahiez G, Marquais S. Tetrahedron Lett 1996; 37(11): 1773–6.
[19] Kamachi T, Kuno A, Matsuno C, Okamoto S. Tetrahedron Lett 2004; 45(24): 4677–9.
[20] Affo W, Ohmiya H, Fujioka T, Ikeda Y, Nakamura T, Yorimitsu H, Oshima K, Imamura Y, Mizuta T, Miyoshi K. J Am Chem Soc 2006; 128(24): 8068–77.
[21] Uemura S, Fukuzawa S-I, Patil SR. J Organomet Chem 1983; 243(1): 9–18.
[22] Uemura S, Fukuzawa S. Tetrahedron Lett 1982; 23(11): 1181–4.
[23] Mizutani K, Yorimitsu H, Oshima K. Chem Lett 2004; 33(7): 832–3.
[24] Black HK, Horn DHS, Weedon BCL. J Chem Soc Resumed 1954; (0): 1704–9.
[25] Korn TJ, Cahiez G, Knochel P. Synlett 2003; (12): 1892–4.
[26] Xi Z, Liu B, Lu C, Chen W. Dalton Trans 2009; (35): 7008–14.
[27] Hatakeyama T, Hashimoto S, Ishizuka K, Nakamura M. J Am Chem Soc 2009; 131(33): 11949–63.
[28] Matsubara K, Sueyasu T, Esaki M, Kumamoto A, Nagao S, Yamamoto H, Koga Y, Kawata S, Matsumoto T. Eur J Inorg Chem 2012; 2012(18): 3079–86.
[29] Gülak S, Stepanek O, Malberg J, Rezaei Rad B, Kotora M, Wolf R, Von Wangelin AJ. Chem Sci 2013; 4(2): 776–84.
[30] Kuzmina OM, Steib AK, Markiewicz JT, Flubacher D, Knochel P. Angew Chem Int Ed 2013; 52(18): 4945–9.
[31] Li J, Li X, Sun H. J Organomet Chem 2013; 743: 114–22.
[32] Madhupriya S, Elango KP. Spectrochim Acta A Mol Biomol Spectrosc 2014; 118: 337–42.
[33] Zeng J, Liu KM, Duan XF. Org Lett 2013; 15(20): 5342–5.
[34] Ohmiya H, Yorimitsu H, Oshima K. J Am Chem Soc 2006; 128(6): 1886–9.
[35] Ohmiya H, Wakabayashi K, Yorimitsu H, Oshima K. Tetrahedron 2006; 62(10): 2207–13.
[36] Cahiez G, Chaboche C, Duplais C, Moyeux A. Org Lett 2009; 11(2): 277–80.
[37] Hsu S-F, Ko C-W, Wu Y-T. Adv Synth Catal 2011; 353(10): 1756–62.
[38] Nicolas L, Angibaud P, Stansfield I, Bonnet P, Meerpoel L, Reymond S, Cossy J. Angew Chem Int Ed Engl 2012; 51(44): 11101–4.
[39] Nicolas L, Izquierdo E, Angibaud P, Stansfield I, Meerpoel L, Reymond S, Cossy J. J Org Chem 2013; 78(23): 11807–14.
[40] Despiau CF, Dominey AP, Harrowven DC, Linclau B. Eur J Org Chem 2014; 2014(20): 4335–41.
[41] Barré B, Gonnard L, Campagne R, Reymond S, Marin J, Ciapetti P, Brellier M, Guérinot A, Cossy J. Org Lett 2014; 16(23): 6160–3.
[42] Matsubara K, Kumamoto A, Yamamoto H, Koga Y, Kawata S. J Organomet Chem 2013; 727: 44–9.
[43] Hammann JM, Steib AK, Knochel P. Org Lett 2014; 16(24): 6500–3.
[44] Collet A, Jacques J. Synthesis 1972; 1972(01): 38–9.
[45] Shirakawa E, Imazaki Y, Hayashi T. Chem Lett 2008; 37(6): 654–5.
[46] Tsuji T, Yorimitsu H, Oshima K. Angew Chem Int Ed 2002; 41(21): 4137–9.
[47] Yasui H, Mizutani K, Yorimitsu H, Oshima K. Tetrahedron 2006; 62(7): 1410–5.
[48] Kuno A, Saino N, Kamachi T, Okamoto S. Tetrahedron Lett 2006; 47(15): 2591–4.
[49] Ohmiya H, Yorimitsu H, Oshima K. Org Lett 2006; 8(14): 3093–6.
[50] Shirakawa E, Sato T, Imazaki Y, Kimura T, Hayashi T. Chem Commun 2007; (43): 4513–5.
[51] Schlubach HH, Franzen V. Über Justus Liebigs Ann Chem 1951; 572(1): 116–21.
[52] Wakabayashi K, Yorimitsu H, Oshima K. J Am Chem Soc 2001; 123(22): 5374–5.
[53] Someya H, Kondoh A, Sato A, Ohmiya H, Yorimitsu H, Oshima K. Synlett 2006; 2006(18): 3061–4.
[54] Someya H, Ohmiya H, Yorimitsu H, Oshima K. Org Lett 2007; 9(8): 1565–7.

[55] Someya H, Ohmiya H, Yorimitsu H, Oshima K. Tetrahedron 2007; 63(35): 8609–18.
[56] Mizutani K, Shinokubo H, Oshima K. Org Lett 2003; 5(21): 3959–61.
[57] Gosmini C, Moncomble A. Isr J Chem 2010; 50(5–6): 568–76.
[58] Molander GA. Science of Synthesis: Cross Coupling and Heck-Type Reactions Vol. 1: C-C Cross Coupling Using Organometallic Partners. Georg Thieme Verlag; 2014; 623–715.

Gérard Cahiez, Alban Moyeux

5 Grignard reagents and Manganese

5.1 Introduction

Since the discovery of organomagnesium reagents by Grignard, about one century ago [1], organometallic reagents have had a considerable impact in organic synthesis, especially for the creation of C–C bonds. The use of transition metals as catalysts, begun in the 1940s, increasing the field of application of organometallic chemistry for the elaboration of organic molecules. During a long time, palladium and nickel were the most popular transition metals [2]. However, the need for the development of more economical and eco-friendly procedures compatible with sustainable development has led to the study of other transition metals, like iron and manganese, that have not been extensively studied until recently [3]. Thus, the chemistry of organomanganese reagents and manganese-catalyzed reactions were only developed from the 1970s [3h,j].

This chapter begins with a brief description of the preparation of organomanganese reagents and their reactivity since it is essential in order to have a good understanding of manganese-catalyzed reactions involving Grignard reagents.

5.2 Preparation and reactivity of organomanganese reagents [3h,i]

Organomanganese reagents are generally prepared by the transmetallation from Grignard reagents or organolithium compounds and manganese halides. Manganese iodide or bromide are used in ether. From manganese bromide, it is better to prepare the soluble ate complex $MnBr_2$•LiBr since the magnesium or lithium exchange is very fast (Fig. 5.1).

$$MnBr_2 + LiBr \xrightarrow[20\ ^\circ C]{Ether} \underset{(Soluble)}{MnBr_2 \cdot LiBr} \xrightarrow[\substack{20\ ^\circ C \\ \approx 100\%}]{RMgX} RMnBr$$

Fig. 5.1: Preparation of organomanganese bromide from manganese bromide.

These last years, manganese chloride has been more frequently used than manganese iodide and bromide since it is a very cheap salt. To facilitate the transmetallation, the soluble ate complexes $MnCl_2 \bullet LiCl$ or $MnCl_2 \bullet 2LiCl$ were prepared in THF by mixing manganese chloride with one or two equivalents of lithium chloride (Fig. 5.2).

$$\mathrm{MnCl_2 + 2\,LiCl \xrightarrow[20\,^{\circ}C]{THF} MnCl_2{\cdot}2LiCl\ (Soluble)}$$

Fig. 5.2: Preparation of the ate complex $MnCl_2 \bullet 2LiCl$.

The manganese/magnesium or manganese/lithium exchange then occurs instantaneously at room temperature. According to the ratio $RMgX/MnCl_2$ or $RLi/MnCl_2$, several organomanganese reagents can be prepared (Fig. 5.3, eq 1–4). It should be noted that lithium organomanganates can be prepared in ether or THF whereas magnesium organomanganates are only obtained in THF.

$$\mathrm{RMgX + MnCl_2{\cdot}2LiCl \xrightarrow{THF} RMnX\ (MgClX + 2\,LiCl)} \qquad \text{(eq 1)}$$

R= Aryl, Alkenyl, Alkynyl...
X= Cl, Br

$$\mathrm{\underset{(2\ equiv)}{RMgX} + MnCl_2{\cdot}2LiCl \xrightarrow{THF} R_2Mn\ (2\,MgClX + 2\,LiCl)} \qquad \text{(eq 2)}$$

R= Aryl, Alkenyl, Alkynyl...
X= Cl, Br

$$\mathrm{\underset{(3\ equiv)}{RMgX} + MnCl_2{\cdot}2LiCl \xrightarrow{THF} R_3MnMgX\ (3\,MgClX + 2\,LiCl)} \qquad \text{(eq 3)}$$

R= Aryl, Alkenyl, Alkynyl...
X= Cl, Br

$$\mathrm{\underset{(4\ equiv)}{RMgX} + MnCl_2{\cdot}2LiCl \xrightarrow{THF} R_4Mn(MgX)_2\ (4\,MgClX + 2\,LiCl)} \qquad \text{(eq 4)}$$

R= Aryl, Alkenyl, Alkynyl...
X= Cl, Br

Fig. 5.3: Preparation of organomanganese reagents in THF.

In solution, the structure of the organomanganese reagents described above are not known. In all cases, the transmetallation probably leads to an equilibrium between a complex mixture of organomanganese species like RMnX, R_2Mn, R_3MnM (M=Li or MgX), R_4MnM_2, etc. The formation of polymeric forms or agregates with metallic salts present in the reaction mixture is also very likely. The nature of the compounds formed by reacting a manganese halide with a large excess of Grignard reagents (or RLi) is not known. A mixture of high order manganates $R_xMn_y(MgX)_z$, like for organocuprates, is probably obtained. Accordingly, the nature of the intermediates involved in the manganese-catalyzed reactions is currently not obvious. It is important to note that the transmetallation Mg or Li/Mn takes place very quickly. This point is essential to understand the chemoselectivity of the manganese-catalyzed procedures since the transmetallation Mg/Mn is the first step of all manganese-catalyzed reactions of Grignard reagents. Organomanganese reagents were also obtained by metallation using TMP_2Mn (Fig. 5.4) [4].

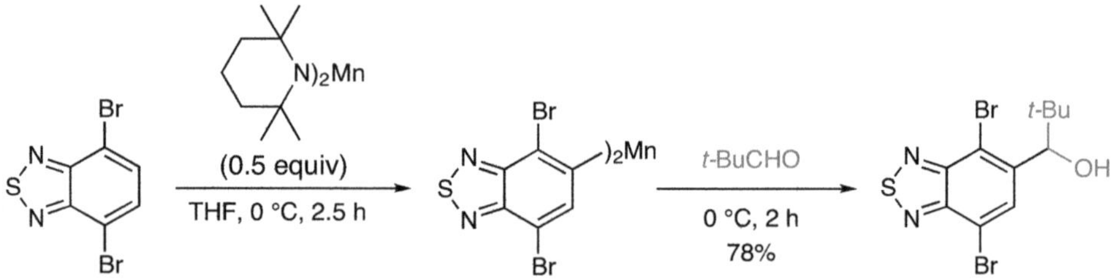

Fig. 5.4: Preparation of organomanganese reagents by metallation with TMP_2Mn.

Organomanganese reagents can also be prepared by direct insertion of manganese metal from organic halides. Thus, manganese-mediated Reformatsky or Barbier reactions were described (Fig. 5.5) [5].

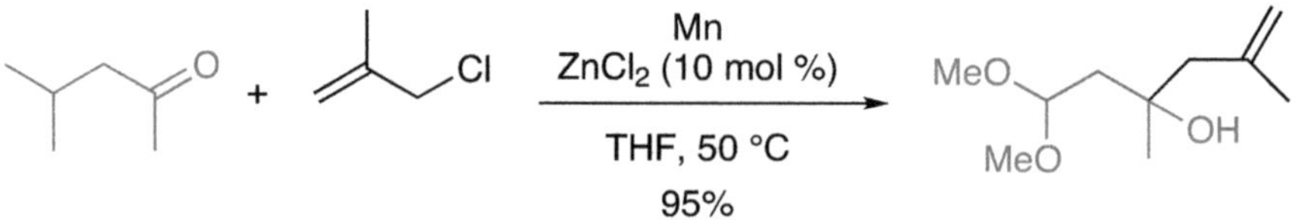

Fig. 5.5: Manganese-mediated Barbier reaction.

However, these methods have, until now, a limited scope by comparison with the transmetallation of RMgX or RLi with manganese salts. Moreover, it is not pertinent to develop them in the framework of a chapter devoted to manganese-catalyzed reactions of Grignard reagents.

As the reactivity of organomanganese reagents was already reviewed [3h,j], only some reactions are summarized in Fig. 5.6.

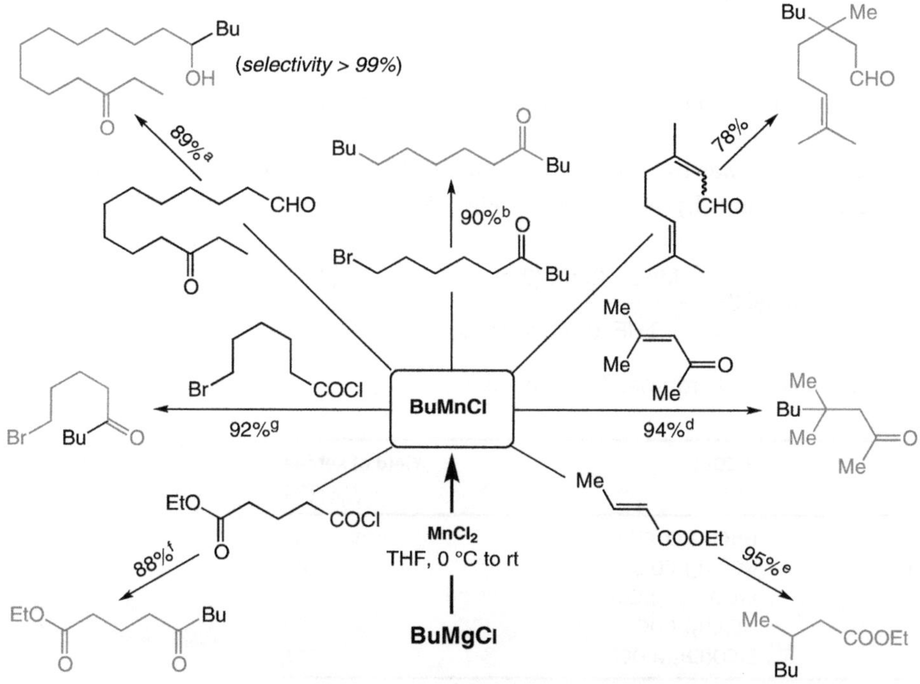

a/ rt, 2h; **b**/ CuCl (3 mol %), THF/NMP, 20°C, 1 h; **c**/ CuCl (1 mol %), -30°C, 30 min; **d**/ CuCl (1 mol %), 0°C, 3 h; **e**/ 1 equiv Me_3SiCl, CuCl (1 mol %), -0°C, 1 h; **f**/ -10 °C to rt, 1.5 h; **g**/ CuCl (1 mol %), -10°C to 20 °C, 30 min.

Fig. 5.6: Main reactions of organomanganese reagents.

5.3 Reactions of alkyl Grignard reagents

5.3.1 With acyl chlorides

As shown above (Fig. 5.6), organomanganese reagents are easily acylated by acyl chlorides or related acylating agents. The reaction is efficient, highly chemoselective and the scope of its application is very large.

Organomanganates can also be acylated. From a di-ate complex $R_4Mn(MgX)_2$, the four alkyl groups are transferred. However, it is interesting to note that the acylation of the first alkyl group occurs almost instantaneously (Fig. 5.7), even at low temperature (–78 °C).

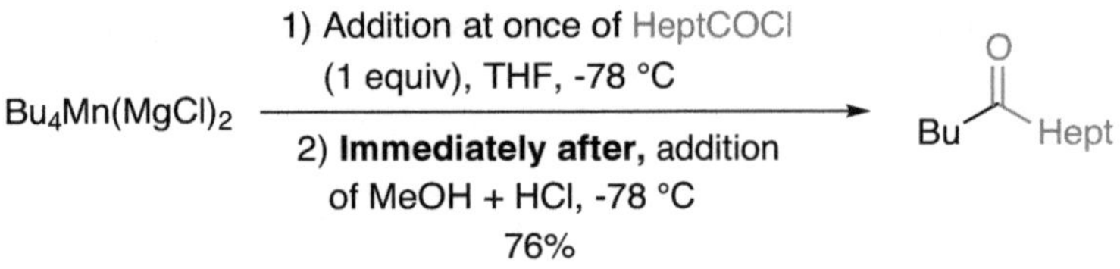

Fig. 5.7: Acylation of the first R group of the organomanganate $R_4Mn(MgX)_2$.

This observation was the starting point to develop a manganese-catalyzed acylation of Grignard reagents by acyl chlorides (Fig. 5.8) [6].

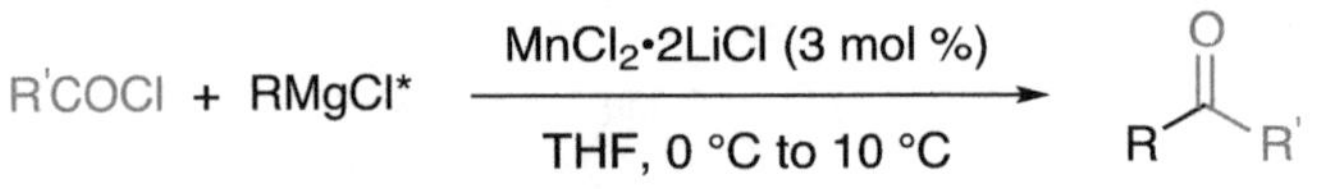

* RMgCl was added dropwise for 30 to 45 min.

R	R'COCl	Yield of ketone (%)
Bu	$Br(CH_2)_3COCl$	86
*i*Pr	$NC(CH_2)_6COCl$	84
Bu	$EtOOC(CH_2)_4COCl$	83
"	$NC(CH_2)_6COCl$	84
"	$EtCO(CH_2)_5COCl$	91

Fig. 5.8: Manganese-catalyzed acylation of Grignard reagents by carboxylic acid chlorides.

Excellent yields in ketones were obtained by adding a Grignard reagent in 30–45 min to a mixture of the carboxylic acid chloride with a catalytic amount of manganese chloride ($MnCl_2$•2LiCl) in THF, between 0 °C and 10 °C. The reaction is highly chemoselective, thus, functionalized carboxylic acid chlorides led to the corresponding ketones in good yields. Even a keto-carboxylic acid chloride can be used.
The catalytic cycle reported in Fig. 5.9 was proposed. This mechanism is consistent with our observations about the rapid transmetallation $R_3MnMgX + RMgX \rightarrow R_4Mn(MgX)_2$ and the fast acylation of the first alkyl group of $R_4Mn(MgX)_2$ by acyl chlorides (Fig. 5.7). It is important to note that on one hand, the rate of addition of the Grignard reagent to the reaction mixture has to be carefully controlled to maintain the concentration of $R_4Mn(MgX)_2$ and, on the other hand to avoid the presence of an excess of Grignard reagent.

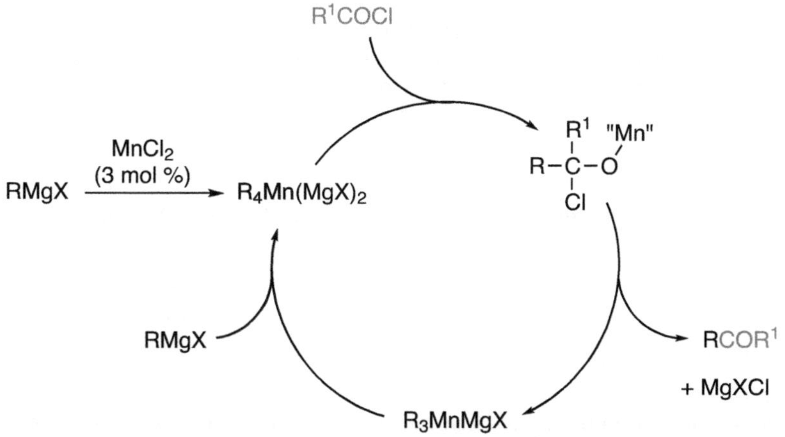

Fig. 5.9: Manganese-catalyzed acylation of Grignard reagents by acyl chlorides: putative mechanism.

The chemoselective reaction of the manganate $R_4Mn(MgX)_2$ with the acyl chloride in the presence of a ketone is not surprising. Indeed, such a selectivity was observed by reacting an organomanganese reagent with a keto-carboxylic acid chloride (Fig. 5.6).

The yields drop when hindered reagents were employed. In these cases, the use of a copper-manganese catalysis improves the yields (Fig. 5.10).

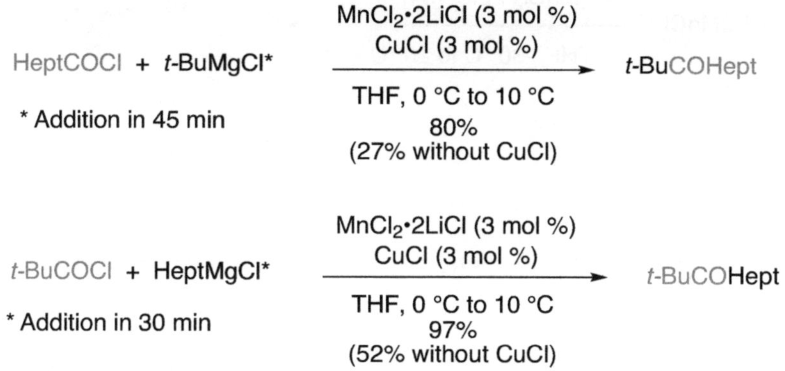

Fig. 5.10: Preparation of hindered ketones by manganese-copper-catalyzed acylation of alkyl Grignard reagents by acyl chlorides.

A similar metal-catalyzed acylation was reported using iron acetylacetonate as a catalyst. However, the manganese-catalyzed procedure is more suitable for large scale applications since it is possible to work at higher concentration (Fig. 5.11) [3h].

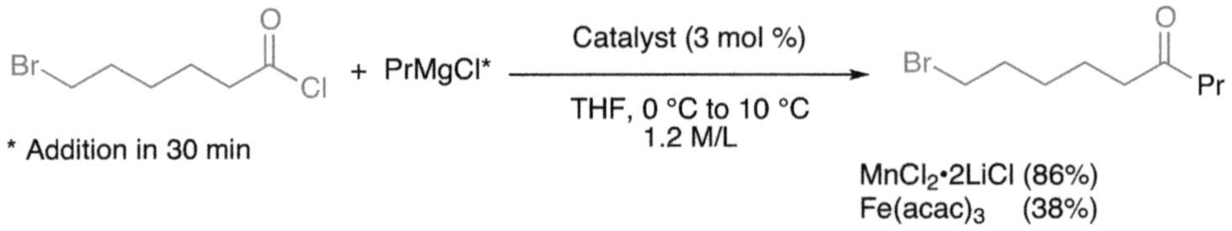

Fig. 5.11: Comparison between iron- and manganese-catalyzed acylation of alkyl Grignard reagents.

5.3.2 With enones

Grignard reagents react with conjugated enones in the presence of manganese salts to give a mixture of 1,2- and 1,4-addition with a variable amount of β-reductive dimerization product. In some cases it is possible to obtain satisfactory yields of β-reductive dimerization product (Fig. 5.12) [7].

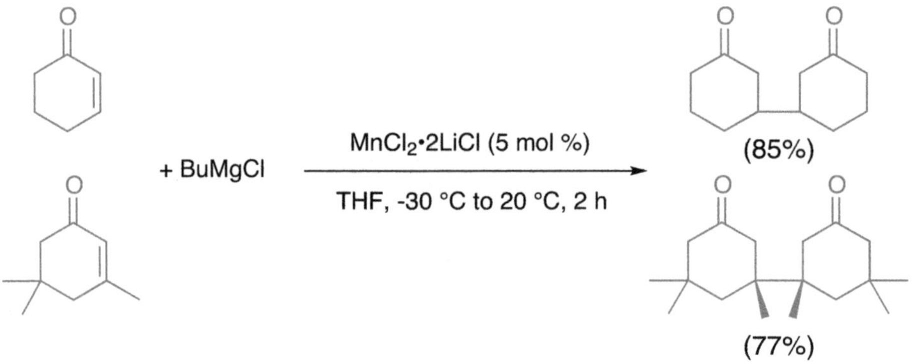

Fig. 5.12: β-Reductive dimerization of enones by treatment with alkyl Grignard reagents in the presence of $MnCl_2$.

Under manganese-copper catalysis, Grignard reagents react with enones to give the conjugate addition product very efficiently (Fig. 5.13) [7].

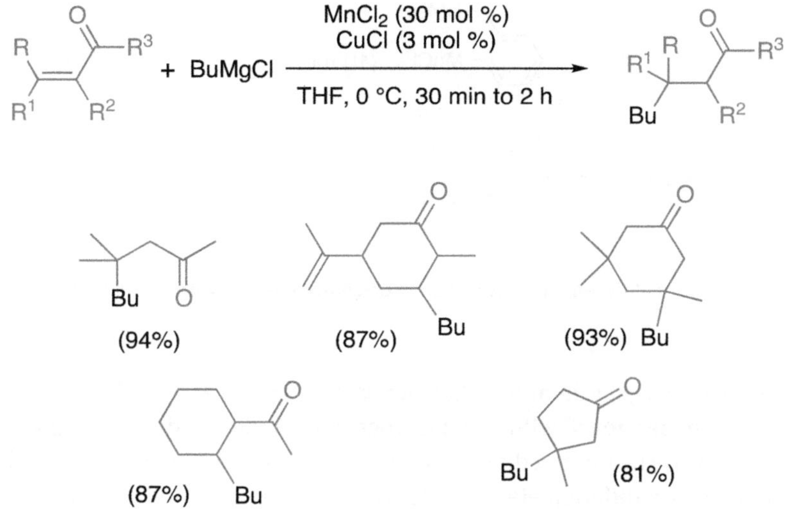

Fig. 5.13: Manganese-copper catalyzed 1,4-addition of alkyl Grignard reagents to enones.

The scope of the reaction is very large and the yields are excellent, even from enones bearing two substituents in the β-position.

5.3.3 With alkyl halides

Organomanganese reagents couple very efficiently with alkyl bromides in the presence of copper salts (Fig. 5.6). Similar results were obtained with Grignard reagents under manganese–copper catalysis. The manganese co-catalyst is a tridentate diamino arylmanganese(II) complex: 2,6-$(CH_2NMe_2)_2C_6H_3MnCl$ (pincer ligand) [8]. The reaction is very chemoselective since an ester or a keto group are tolerated (Fig. 5.14).

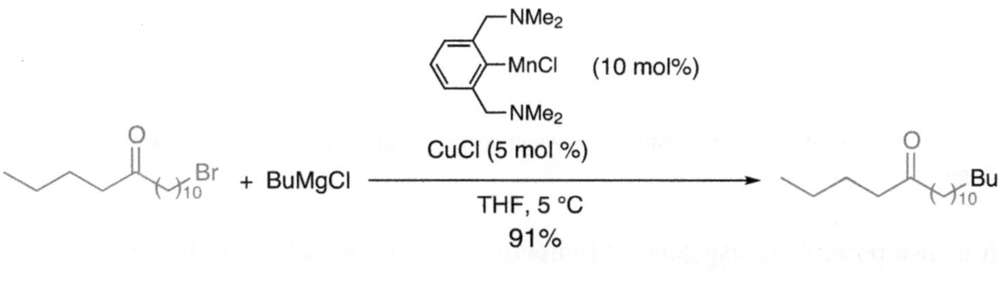

Fig. 5.14: Chemoselective manganese-copper-catalyzed alkylation of alkyl Grignard reagents.

Moreover, secondary alkyl bromides can be used sucessfully (Fig. 5.15).

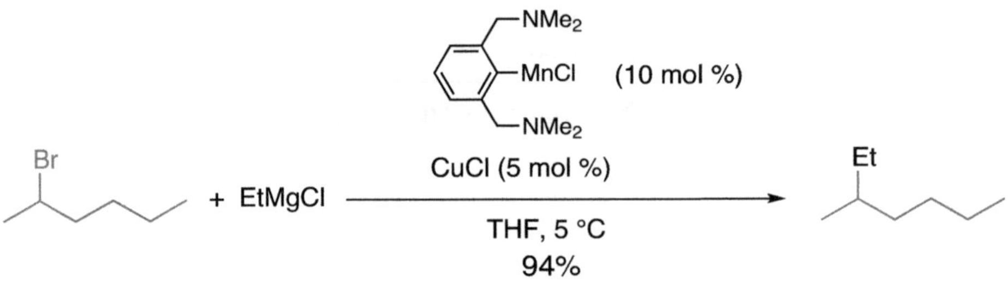

Fig. 5.15: Manganese-copper-catalyzed alkylation of alkyl Grignard reagents with secondary alkyl bromides.

The reactions with various *gem*-di- and tri-halogenoalcanes were also studied. Thus, *gem*-dibromocyclopropanes react with alkylmagnesium halides under manganese catalysis (Fig. 5.16) [9,10]. The intermediate cyclopropylmanganese compound formed *in situ* can be trapped with different electrophiles (H_2O, AllylBr, etc.) and the reaction affords moderate to good yields of the final product.

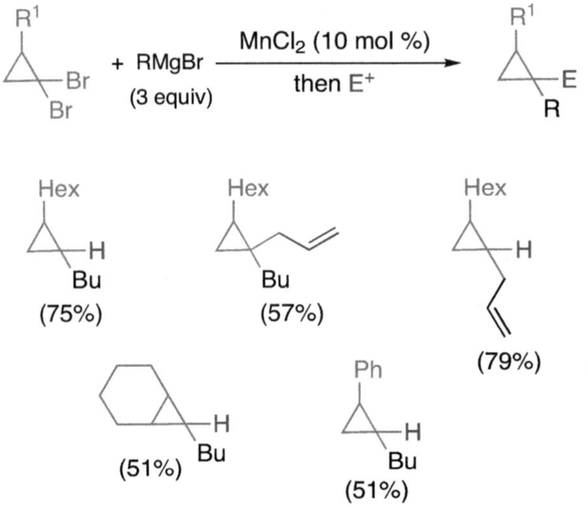

Fig. 5.16: Manganese-catalyzed reaction between *gem*-dicromocyclopropanes and alkyl or allyl Grignard reagents.

It is also possible to use *gem*-dibromoethylsilanes (Fig. 5.17) [10]. In this case, the intermediate organomanganese compound undergoes a hydride b-elimination to give selectively the corresponding *(E)*-olefin thanks to the influence of the bulky R_3Si group.

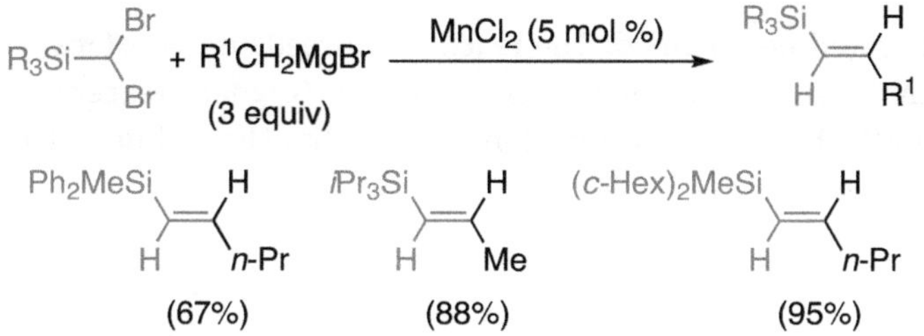

Fig. 5.17: Manganese-catalyzed reaction between *gem*-dibromoethylsilanes and alkyl Grignard reagents.

From γ-siloxy-1,1-dibromoalkanes, it is also possible to obtain mainly the *(E)*-olefins owing to the Mn-O interaction which influences the stereochemistry of the β-hydride elimination (Fig. 5.18).

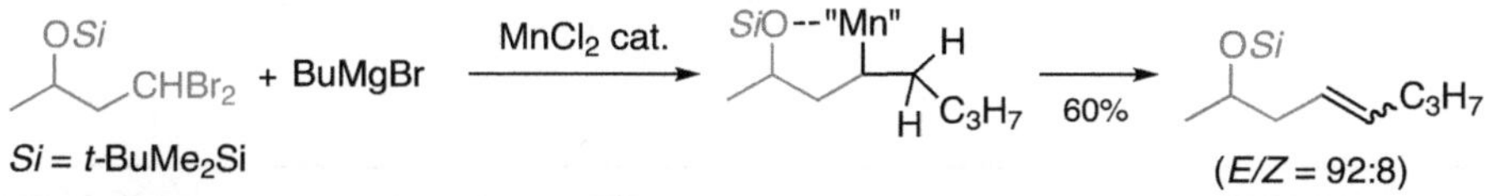

Fig. 5.18: Manganese-catalyzed reaction between γ-siloxydibromocompounds and alkyl Grignard reagents.

Tribromomethyltrialkylsilanes can also be used (Fig. 5.19).

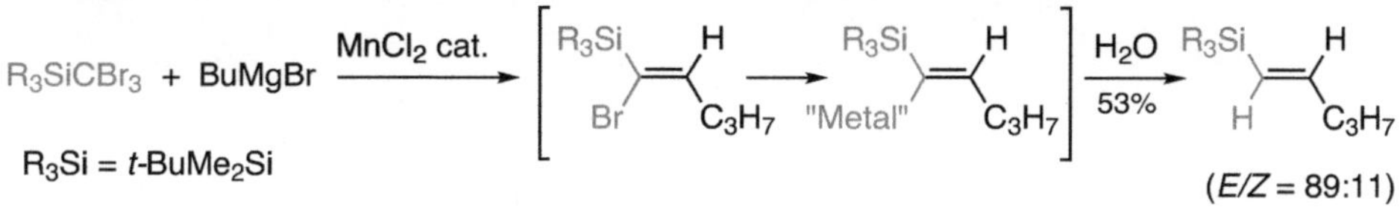

Fig. 5.19: Manganese-catalyzed reaction between tribromomethyltrialkylsilanes and alkyl Grignard reagents.

In addition, Oshima *et al.* [11] described the manganese-catalyzed reaction between 2,2-dibromo-*N*,*N*-diethylpropanamide and aliphatic Grignard reagents (Fig. 5.20).

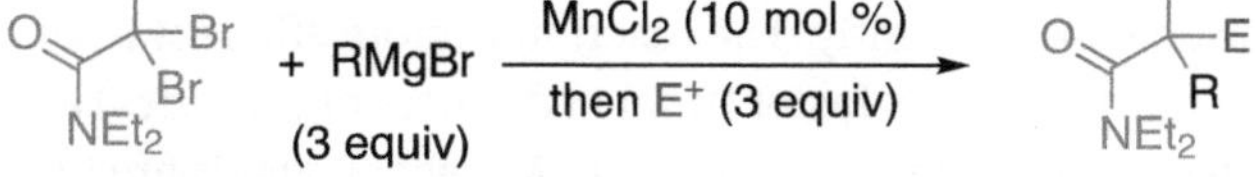

AlkMgBr	Electrophile	Yield (%)
BuMgBr	PhCHO	95
EtMgBr	H_2O	74

Fig. 5.20: Manganese-catalyzed cross-coupling of 2,2-dibromo-*N*,*N*-diethylpropanamide with Grignard reagents.

Moderate to good yields could be obtained. The reaction proceeds in two-steps: at first the 2,2-dibromo-*N,N*-diethylpropanamide reacts with the Grignard reagent, to afford a manganese enolate which can be trapped in a second time by addition of an electrophile to the reaction mixture.

5.3.4 With aryl halides

With activated aryl halides, primary or secondary alkyl Grignard reagents led to the cross-coupling product in excellent yields (Table 5.1) [12].

Table 5.1: Manganese-catalyzed coupling with activated aryl halides.

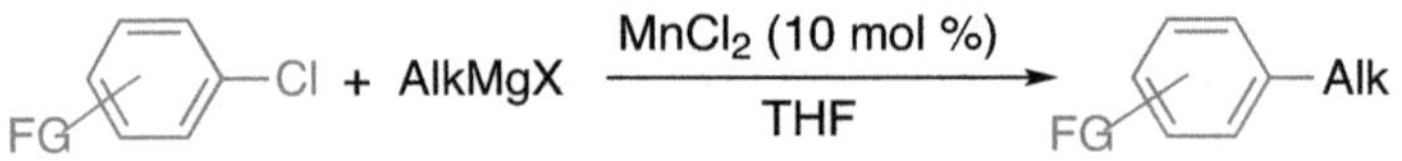

Entry	Aryl Halide		Grignard Reagent	Yield (%)
1	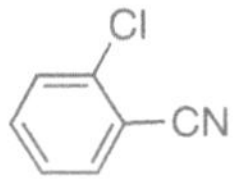		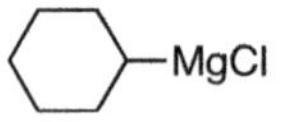	75
2	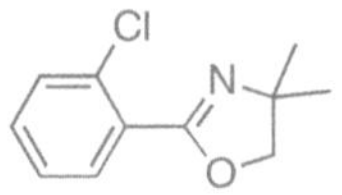		BuMgCl	67
3	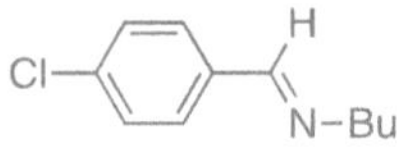		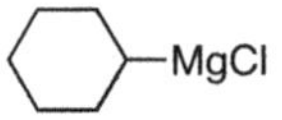	88
4	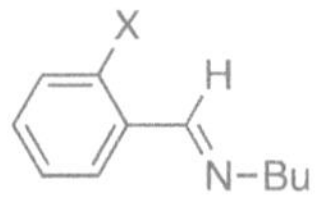	X = Br	BuMgCl	92
5		X = Cl	"	93
6		X = F	"	90
7		X = OMe	"	91

It is interesting to note that activated aryl bromides, chlorides, fluorides or even ethers (Table 5.1, entries 4–7) can be used indifferently. The reaction is more efficient than a nucleophilic aromatic substitution since it is faster (from aryl ethers or fluorides) and its scope is large (aryl chlorides and bromides can be used). Moreover, yields are high whatever the nature of the leaving group.

With *ortho*-chloroaryl ketones the cross-coupling product is formed in moderate yields since the 1,2-addition to the carbonyl group takes place competitively (Fig. 5.21) [13].

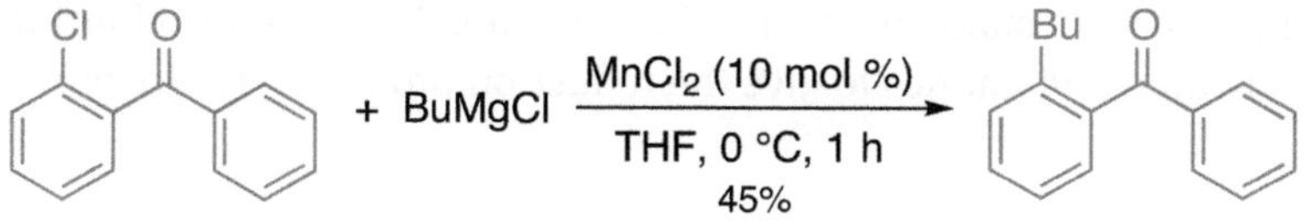

Fig. 5.21: Manganese-catalyzed coupling between PhMgCl and *o*-chloroaryl ketones.

In 2010, Yuan *et al.* [14] described the tandem acylation/cross-coupling reaction of ortho-halobenzoyl chloride with aliphatic symmetrical diorganomagnesium compounds in the presence of manganese salts (Fig. 5.22).

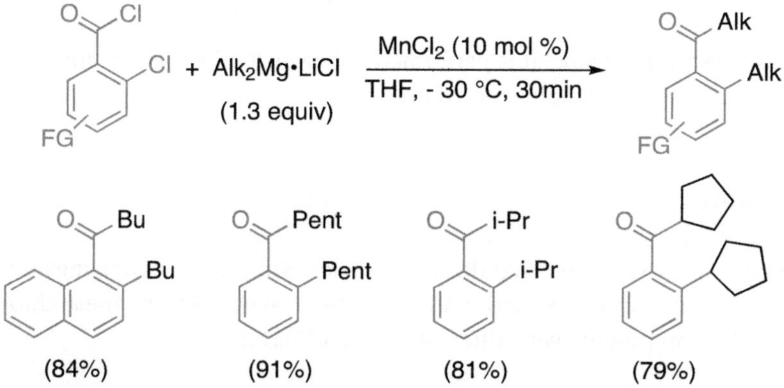

Fig. 5.22: Manganese-catalyzed tandem acylation/cross-coupling with *o*-chloro aromatic carboxylic acid chlorides.

However, the scope of this reaction is very limited since it is only possible to transfer twice the same R group.

The manganese-catalyzed coupling reaction has been extended to chloroquinolines by Rueping *et al.* (Fig. 5.23) [15]. Satisfactory yields of substitution products are generally obtained.

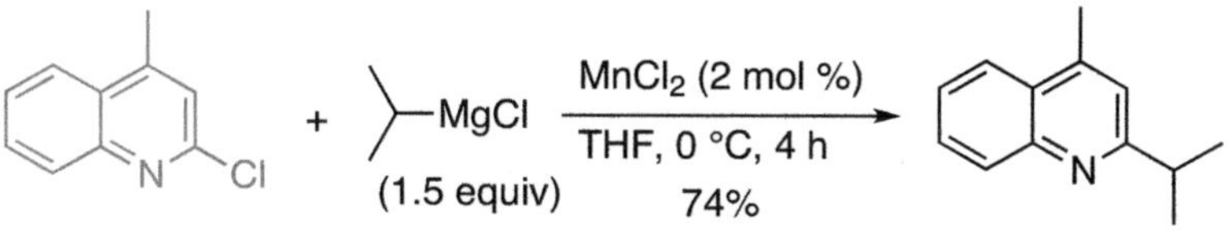

Fig. 5.23: Manganese-catalyzed coupling of chloroquinolines with Grignard reagents.

In THF, in the presence of manganese chloride, isopropylmagnesium chloride reacts with non activated aryl halides to give the reduction product very efficiently (Fig. 5.17) [16].

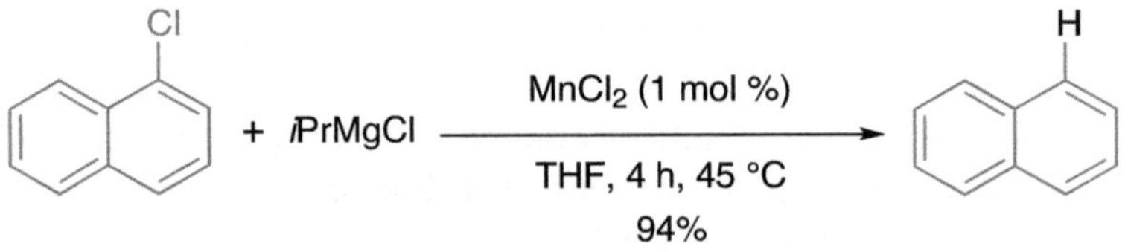

Fig. 5.24: Manganese-catalyzed reduction of aryl halides by *i*PrMgCl.

The reaction has a very large scope. It is performed between 20 °C and 45 °C and generally gives excellent yields of dehalogenated product.

5.3.5 With alkenyl halides

As in the case of non-activated aryl halides (see Section 3.4.), isopropylmagnesium chloride reacts with non-activated alkenyl halides in the presence of manganese chloride to give the reduction product very efficiently (Fig. 5.25) [16].

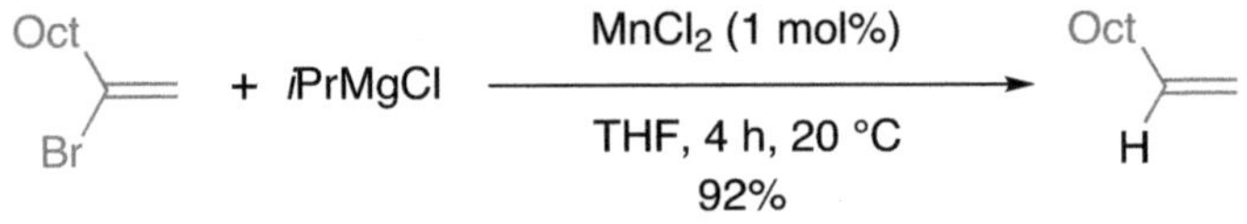

Fig. 5.25: Manganese-catalyzed reduction of alkenyl halides by iPrMgCl.

Manganese-catalyzed cross-coupling between activated alkenyl halides and Grignard reagents was also reported. Thus, chlorodienes (Table 5.2, entry 5) and chloroenynes (Table 5.2, entries 1–4) are stereoselectively alkylated by using a mixture THF/DMPU as a solvent [17].

Table 5.2: Manganese-catalyzed coupling between aliphatic Grignard reagents and chloroenynes or chlorodienes.

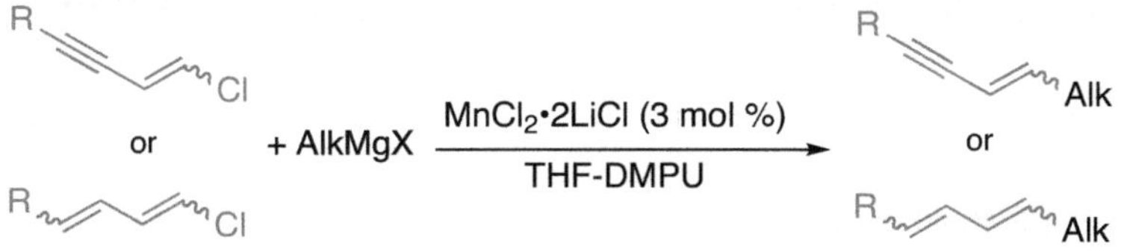

Entry	Alkenyl Halide	Grignard Reagent	Yield (%)
1		*t*-BuMgX	75
2	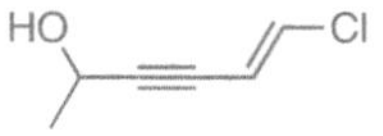	BuMgX	53
3	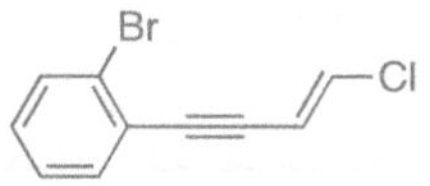	BuMgX	88
4	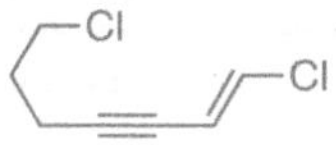	*c*-HexMgX	80
5	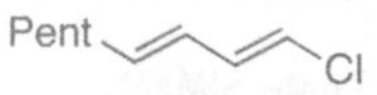	*c*-HexMgX	80

Primary, secondary and tertiary alkylmagnesium compounds afford good yields of coupling products. This procedure favorably compares with the classical Ni- or Pd-catalyzed coupling reactions of Grignard reagents with alkenyl halides.

Oshima *et al.* [18] described few examples of cross-couplings between enol triflates and benzyl or allyl Grignard reagents in the presence of manganese salts (Fig. 5.26).

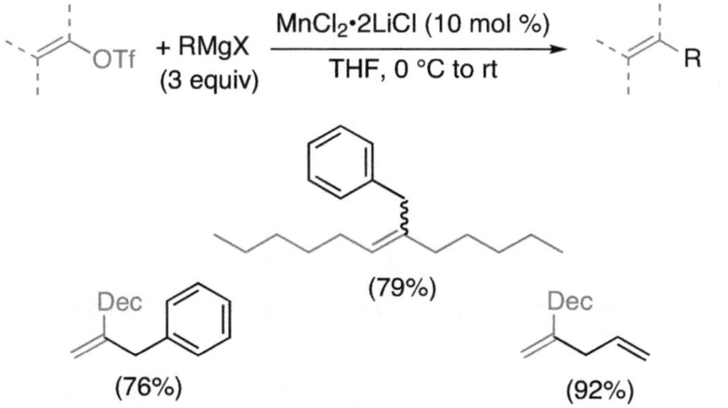

Fig. 5.26: Manganese-catalyzed coupling between enol triflates and benzyl or allyl Grignard reagents.

5.3.6 With alkynes

Oshima *et al.* described the addition of the silyl Grignard reagent $PhMe_2SiMgMe$ to alkynes in the presence of manganese salts [9, 19]. The best yields and regioselectivities were obtained with β-alkoxyalkynes (Fig. 5.27).

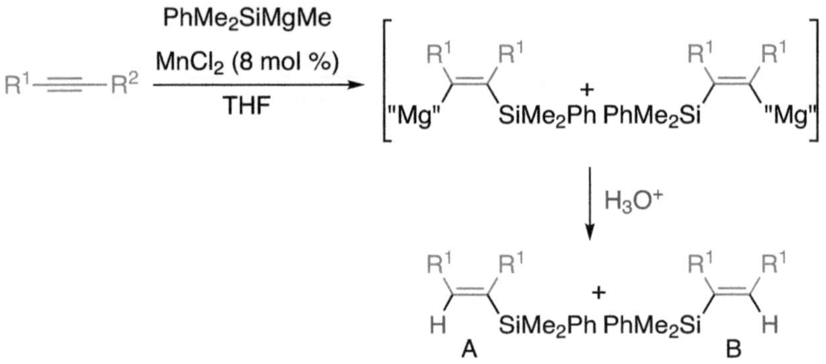

R^1	R^2	Yield (%)	Ratio A/B
Hex	H	40	80/20
$PhCH_2OCH_2CH_2$	H	90	95/5

Fig. 5.27: Manganese-catalyzed carbometallation of alkynes with $PhMe_2SiMgMe$.

It is also possible to trap the organometallic intermediate with electrophiles such as allyl bromide or benzaldehyde.

Manganese-catalyzed carbometallation of β-alkoxyalkynes can also be performed with allyl magnesium bromide (Fig. 5.28) [20, 21]. However, the yield dramatically drops when the alkoxy group becomes bulky. Moreover, alkynes having no alkoxy group are unreactive under the reaction conditions. These results highlight the dependence of the reaction to the complexation with the oxygen atom at the β-position of the triple bond.

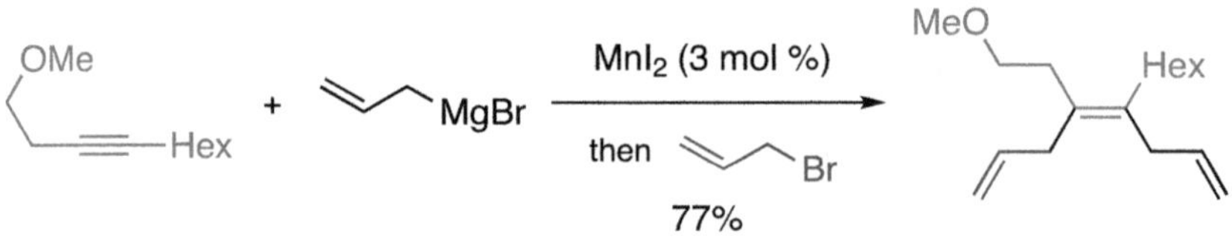

Fig. 5.28: Manganese-catalyzed carbometallation of homopropargylic ethers with allylmagnesium bromide.

Carbometallation of propargylic ethers afford allenes after elimination of the alkoxy group at the b-position of the metal (Fig. 5.29).

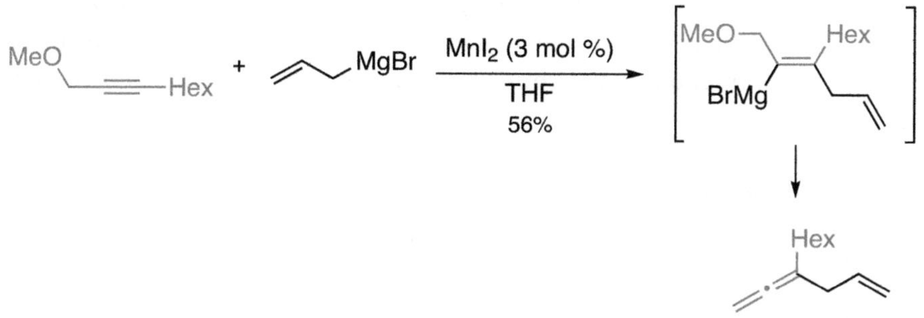

Fig. 5.29: Manganese-catalyzed carbometallation of propargylic ethers.

If the carbometallation is performed with allylmagnesium bromide in the presence of an oxidant, a 1,4,7-triene is obtained in good yields (Fig. 5.30).

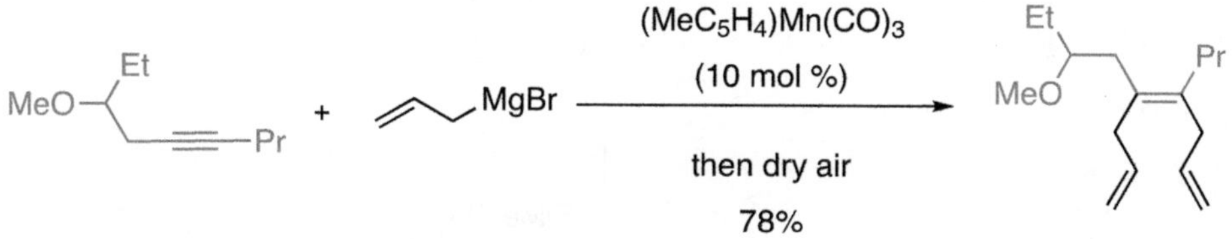

Fig. 5.30: Manganese-catalyzed carbometallation of homopropargylic ether in the presence of an oxidant.

o-Hydroxy- [22] or o-aminophenylacetylenes [23] can also be carbometallated by alkyl Grignard reagents (Fig. 5.31).

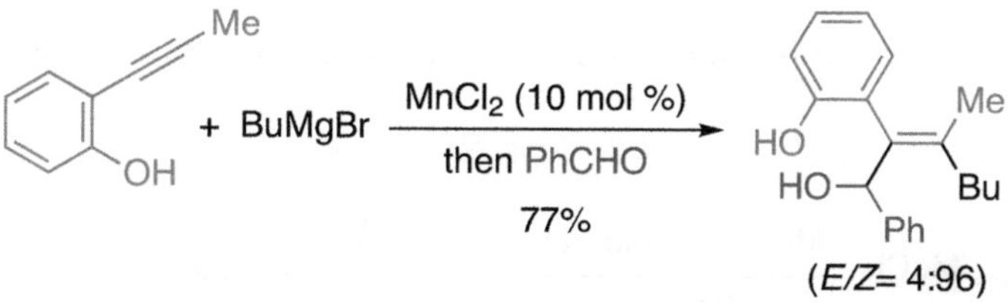

Fig. 5.31: Manganese-catalyzed carbometallation of *o*-hydroxy- or *o*-aminophenylacetylenes.

5.3.7 With 1,3-dienes

Oshima *et al.* also described the manganese-catalyzed silylmagnesiation of 1,3-dienes [19]. At –78 °C, the reaction leads to the formation of a β-silyl allylmagnesium compound. On the other hand, at room temperature, an isomerization takes place and leads to the formation of the corresponding α-silyl allylmagnesium reagent (Fig. 5.32).

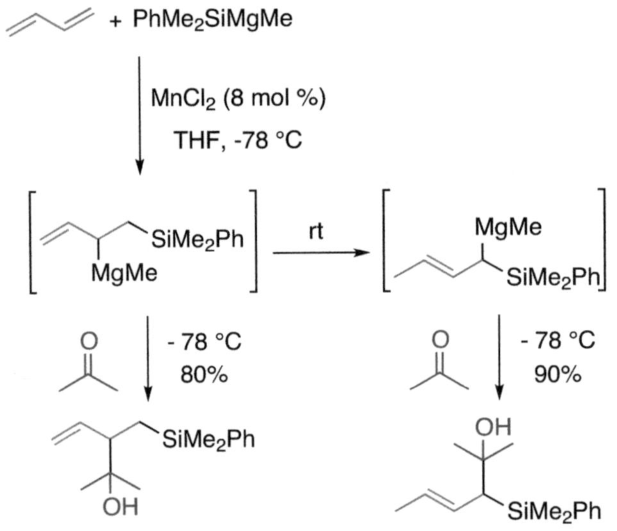

Fig. 5.32: Manganese-catalyzed silylmagnesiation of 1,3-dienes.

5.3.8 With allenes

In 2003, Oshima *et al.* described the manganese-catalyzed carbometallation of allenes [24]. Thus, addition of allylmagnesium chloride to an allene, then trapping of the intermediate with an electrophile affords various 1,5-dienes in moderate to good yields (Fig. 5.33).

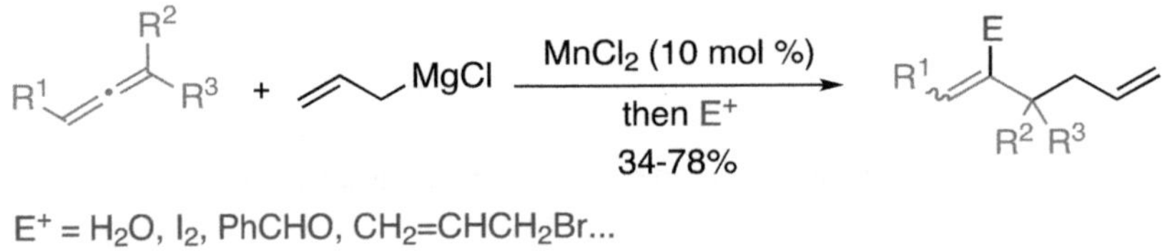

Fig. 5.33: Manganese-catalyzed carbometallation of allenes.

5.4 Reactions of aryl and heteroaryl Grignard reagents

5.4.1 With acyl halides

As already mentioned above (see Section 3.1.), alkyl Grignard reagents can be acylated in the presence of catalytic amounts of manganese. Aryl Grignard reagents [6] can also be used in this kind of coupling (Fig. 5.34).

$$\text{BuCOCl} + \text{PhMgCl} \xrightarrow[\text{THF, 0 °C, 30 min}]{\text{MnCl}_2\text{•2LiCl (3 mol \%)}} \text{BuCOPh}$$

85%

Fig. 5.34: Manganese-catalyzed acylation of phenylmagnesium chloride.

5.4.2 With alkyl halides

As previously described (see Section 3.3.), Oshima *et al.* [11] described the manganese-catalyzed reaction between 2,2-dibromo-*N,N*-diethylpropanamide and aliphatic Grignard reagents. Aromatic Grignard reagents also react under the same conditions (Fig. 5.35). Moderate yields could be obtained. The reaction proceeds in two-steps; at first the 2,2-dibromo-*N,N*-diethylpropanamide reacts with the Grignard reagent, resulting to a manganese enolate which can be trapped in a second time by addition of an electrophile to the reaction mixture.

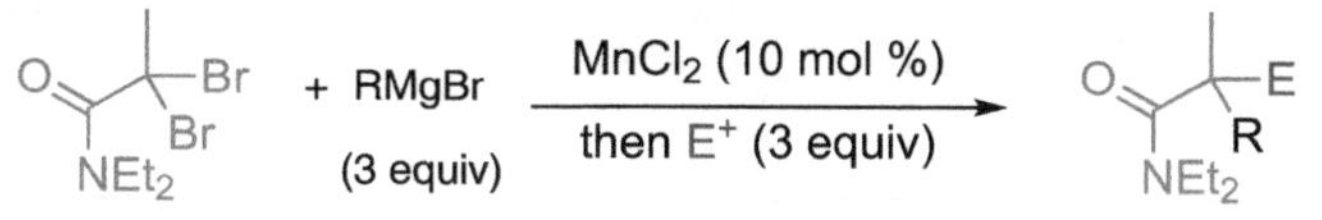

AlkMgBr	Electrophile	Yield (%)
PhMgBr	H_2O	67
PhMgBr	PhCHO	60[a]

[a] *Syn/anti* = 51:49

Fig. 5.35: Manganese-catalyzed cross-coupling of 2,2-dibromo-*N,N*-diethylpropanamide with Grignard reagents.

5.4.3 With aryl halides

Manganese-catalyzed cross-coupling reaction between Grignard reagents and aryl halides was reported [13]. Good to excellent yields could be obtained under mild conditions from various activated aryl halides and aryl Grignard reagents (Table 5.3).

Table 5.3: Manganese-catalyzed cross-coupling between aryl halides and arylmagnesium compounds.

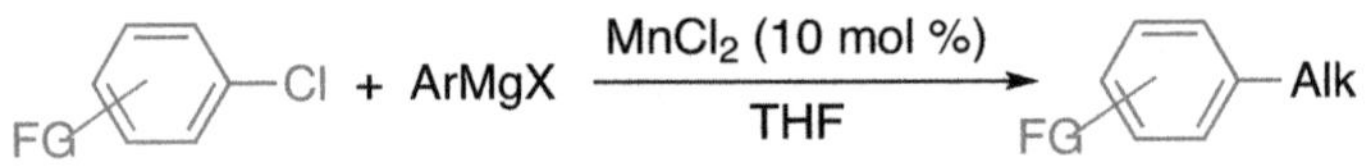

Entry	Aryl Halide	Grignard Reagent	Yield (%)
1	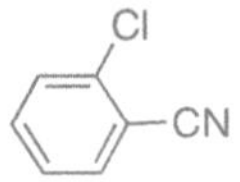	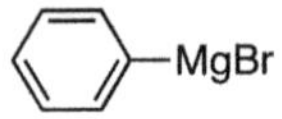	77
2	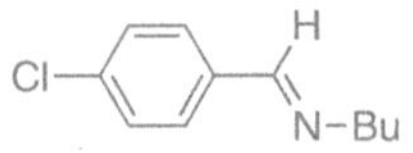	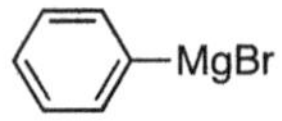	93
3	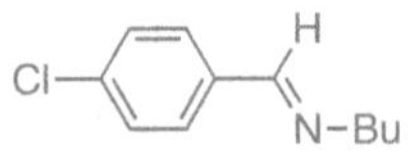	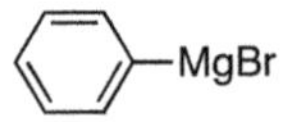	88
4	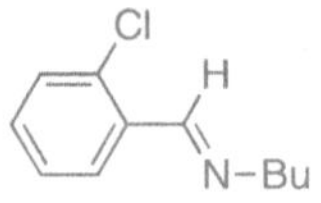		93
5	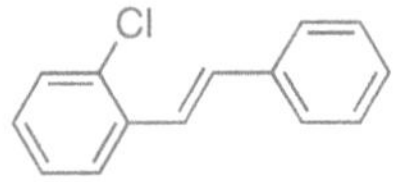	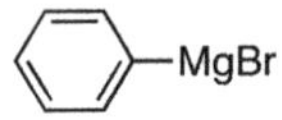	56

It is important to note that the reaction is chemoselective since various functional groups are well tolerated such as nitrile (Table 5.3, entry 1) or imine (Table 5.3, entries 2 and 3). Moreover, the cross-coupling reaction is performed with aryl chlorides which are clearly the most interesting aryl halides from an economical point of view.

This reaction is used to prepare an important intermediate of the synthesis of Irbesartan, an anti-hypertensive drug from Sanofi (Fig. 5.36).

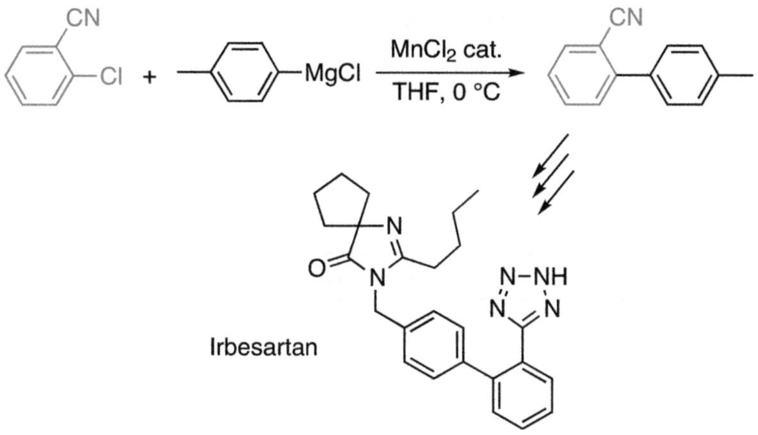

Fig. 5.36: Synthesis of Irbesartan.

The coupling reaction was further extended to aryl Grignard reagents and *o*-chloroaryl ketones (Fig. 5.37) [13].

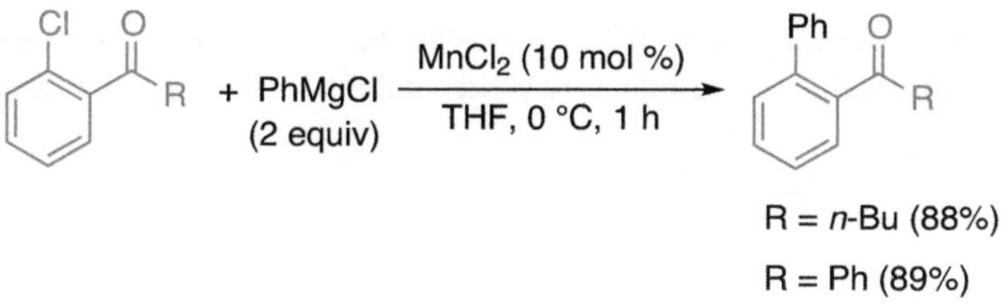

Fig. 5.37: Manganese-catalyzed cross-coupling between PhMgCl and o-chloroaryl ketones.

These substrates are clearly more challenging because of the presence of a very reactive keto group.

In 2010, Yuan *et al.* [15] described the tandem acylation/cross-coupling reaction of *o*-halobenzoyl halides with aliphatic symmetrical diorganomagnesium compounds in the presence of manganese salts (see Section 3.4). Aromatic symmetrical diorganomagnesium compounds can participate as well (Table 5.4). It should be noted that *o*-chlorobenzoyl chloride (Table 5.4, entry 1) affords lower yield than *o*-bromo or *o*-iodo benzoyl chlorides (Table 5.4, entry 2 and 3)). However, as previously mentioned, the scope of this reaction is very limited since it is only possible to transfer the same R group twice.

Table 5.4: Manganese-catalyzed tandem acylation / cross-coupling reaction.

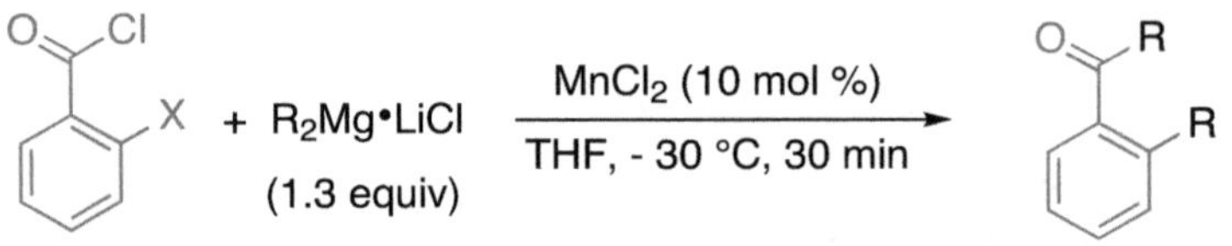

Entry	Halobenzoyl Halide	X	Coupling Product	Yield (%)
1	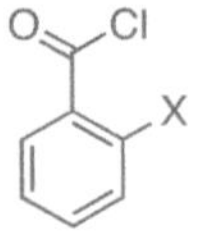	Cl	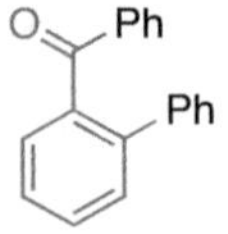	53
2		Br		90
3		I	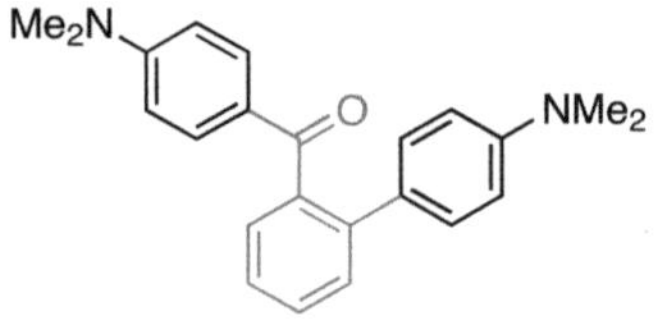	92
4		I		75

The manganese-catalyzed coupling reaction has also been extended to chloroquinolines by Rueping *et al.* (Fig. 5.38) [16]. Satisfactory yields of substitution products are generally obtained.

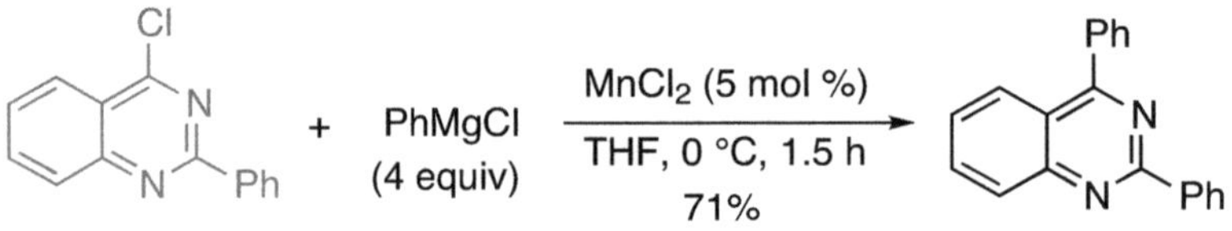

Fig. 5.38: Manganese-catalyzed cross-coupling of chloroquinolines with Grignard reagents.

5.4.4 With alkenyl halides and pseudo-halides

Aryl Grignard reagents can be efficiently coupled with simple alkenyl halides in the presence of manganese salts (Table 5.5) [25]. Moderate to good yields can be obtained under mild conditions and, as a rule, the reaction is stereoselective. Alkenyl iodides or bromides can be indifferently used (Table 5.5, entries 5 and 6). However, alkenyl chlorides, which are clearly less reactive, afford poor yields (Table 5.5, entry 4).

Table 5.5: Manganese-catalyzed coupling between aryl Grignard reagents and alkenyl halides.

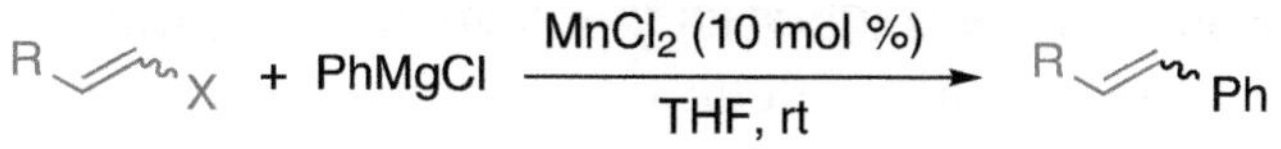

Entry	Alkenyl Halide (*E/Z*)	Yield (*E/Z*)
1	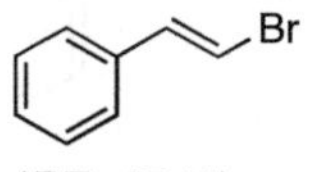 (*E/Z* = 87:13)	83% (88:12)
2	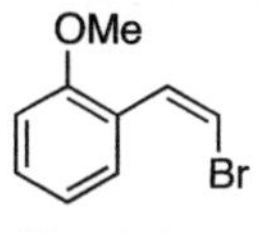 (*Z* > 99%)	79%[a] (5:95)
3	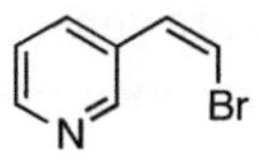 (*Z* > 99%)	71%[b] (99/1)
4	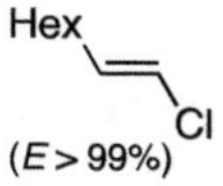 (*E* > 99%)	20%[a] (99:1)
5	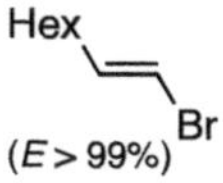 (*E* > 99%)	75% (E > 99%)
6	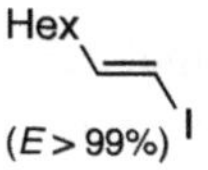 (*E* > 99%)	79% (99:1)
7	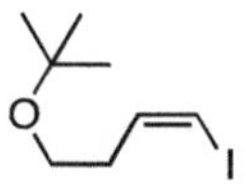 (*Z* > 99%)	75% (Z > 99%)
8	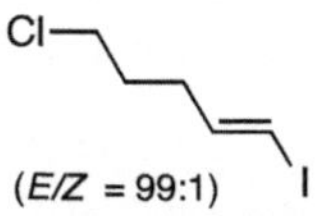 (*E/Z* = 99:1)	69% (97:3)

[a] Reaction performed at 50 °C. [b] Reaction performed at 0 °C.

Oshima *et al.* [17] described the cross-coupling between enol triflates and arylmanganates. Noteworthy, aryl, benzyl or allyl Grignard reagents can also be used in the presence of 10 mol % of the ate complex $MnCl_2$•2LiCl (Fig. 5.39). However, with the Grignard reagents the yields are lower than with the corresponding organomanganese compounds.

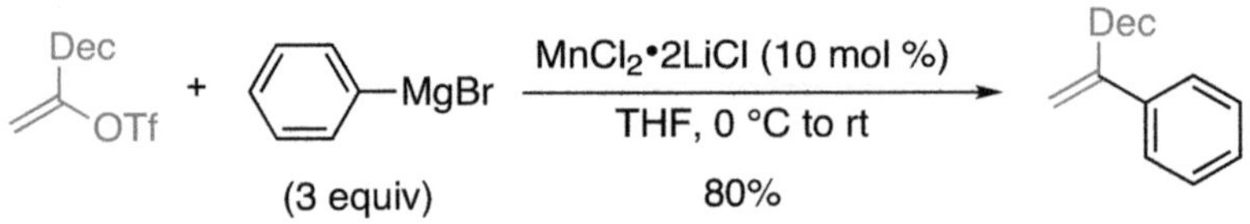

Fig. 5.39: Manganese-catalyzed cross-coupling between enol triflates and Grignard reagents.

5.4.5 Oxidative homocoupling

Homocoupling reactions of organometallic compounds are generally performed by using an oxidant in stoichiometric amounts in the presence of a metal catalyst (Fig. 5.40).

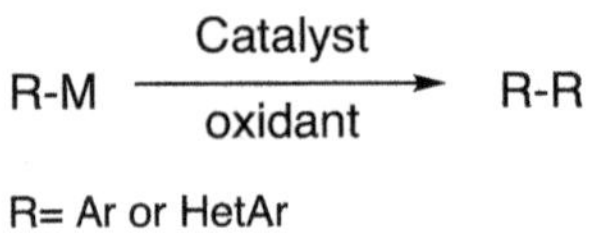

Fig. 5.40: Metal-catalyzed homocoupling reaction.

As a rule, they afford good yields under mild conditions, making them of particular interest for the preparation of di- or polyaromatic, olefinic or acetylenic conjugated compounds. Potential applications of such compounds are numerous: optical materials, molecular devices, organic conductors, etc. [26].

In 2007, Cahiez *et al.* [27] reported the manganese-catalyzed homocoupling reaction of aryl Grignard reagents using atmospheric oxygen as an oxidant (Table 5.6). Good to excellent yields can be obtained under mild conditions. Noteworthy, the reaction is chemoselective since it tolerates the presence of many functional groups (Table 5.6, entries 2, 3 and 5).

Table 5.6: Manganese-catalyzed homocoupling of aryl Grignard reagents using atmospheric oxygen as an oxidant.

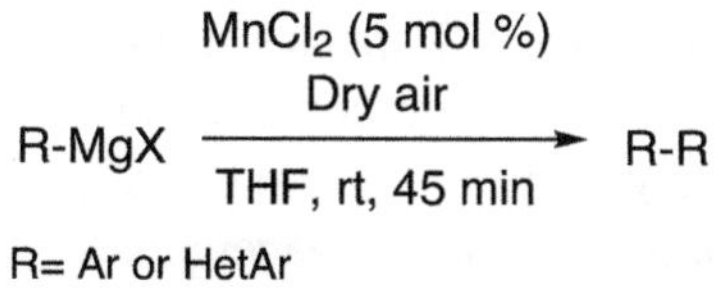

Entry	RMgX	Yield (%)
1	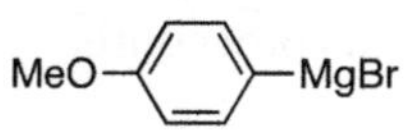	95
2	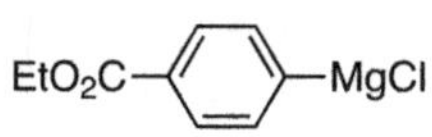	80[a]
3	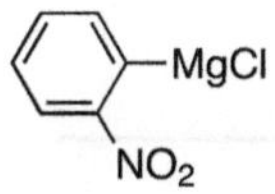	75[a]
4	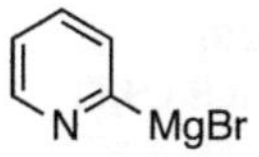	80
5	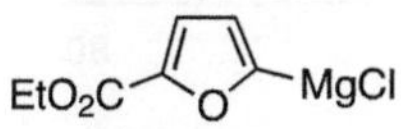	88[a]

[a] The reaction was performed at −20 °C.

A mechanism was proposed for this homocoupling reaction (Fig. 5.41).

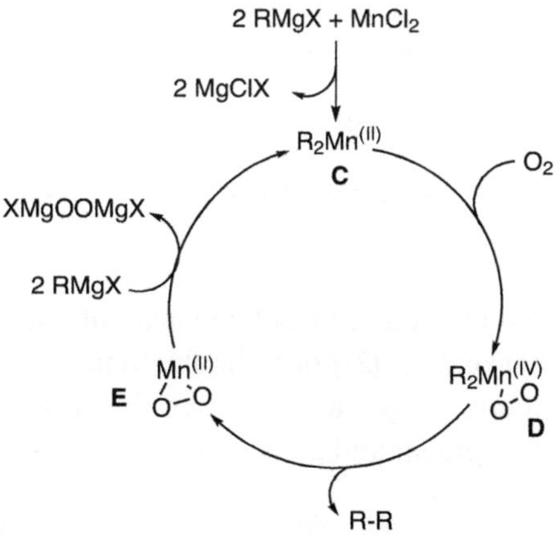

Fig. 5.41: Mechanism of the manganese-catalyzed homocoupling of Grignard reagents.

The catalytic cycle begins by a transmetallation between the Grignard reagent and manganese chloride to afford a stable diorganomanganese(II) compound **C**. Addition of dioxygen to **C** gives an unstable Mn(IV) species **D**, which, after a rapid reductive elimination leads to the formation of the peroxomanganese(II) species **E**. Further reaction of **E** with the Grignard reagent finally regenerates the diorganomanganese(II) species **C**, which can then participate again in the catalytic cycle.

Manganese-catalyzed homocoupling of aromatic Grignard reagents using 1,2-dihaloethanes as an oxidant has been also described [28]. It should be underlined that by using dihaloethanes as an oxidant instead of oxygen [27] the yields are lower (Fig. 5.42). Moreover, the scope of the reaction is clearly more limited. Thus, heteroarylmagnesium compounds cannot be used with this procedure.

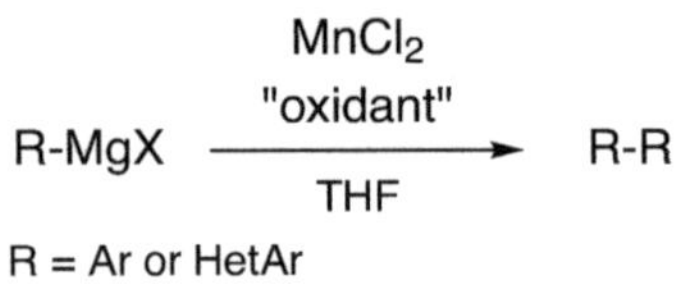

Grignard Reagent	Yield (%)	
	Method A[a] [27] oxidant = O_2 X = Br	Method B[b] [28] oxidant = $ClCH_2CH_2Cl$ X = Cl
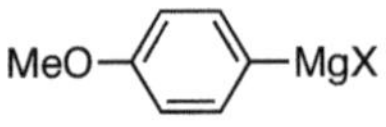	95	80
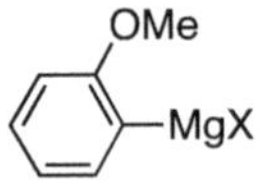	76	74
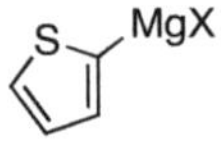	91[c]	0

[a] Method A: $MnCl_2$ (5 mol%), THF, 20 °C, 45 min. [b] Method B: $MnCl_2$ (10 mol%), THF, 20 °C, 1 h. [c] X = Cl

Fig. 5.42: Comparison between manganese-catalyzed homocoupling reaction of aryl Grignard reagents with oxygen or 1,2-dichloroethane as an oxidant.

Noteworthy, this reaction can also be performed by using iron salts instead of manganese salts as a catalyst. The oxidant is either oxygen [27] or a dihaloethane [29]. It is important to note that the use of atmospheric oxygen as an oxidant is particularly interesting in the frame of sustainable development and green chemistry since it avoids to use an organic oxidant.

Yuan *et al.* [30] reported the manganese-catalyzed homocoupling of aryl halides in the presence of manganese salts and magnesium metal (Fig. 5.43). However, the yields are in general lower than those obtained with the procedures using preformed Grignard reagents (Table 5.6) [27].

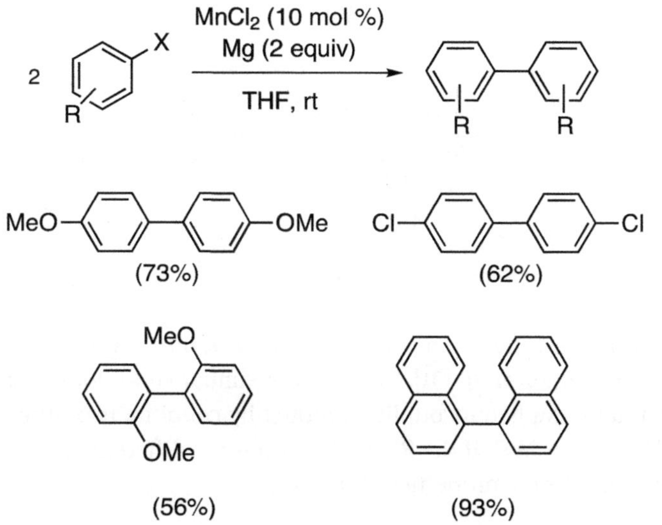

Fig. 5.43: Homocoupling of aryl halides in the presence of manganese chloride and magnesium.

The aryl halide acts both as an oxidant and a precursor for the formation of the organomagnesium species formed as an intermediate. A mechanism was proposed by the authors (Fig. 5.44).

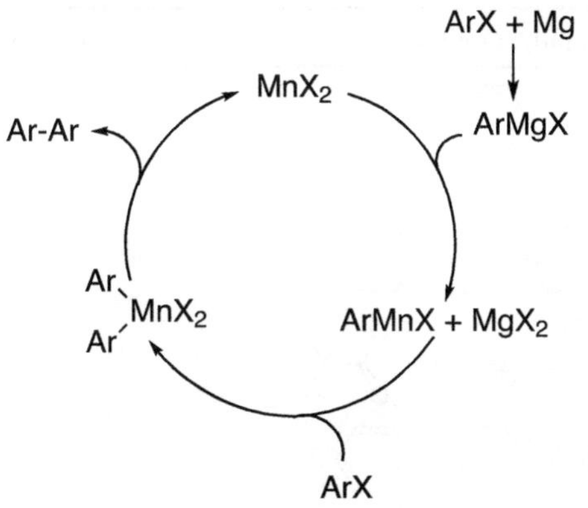

Fig. 5.44: Mechanism for the manganese-catalyzed homocoupling of aryl halides mediated by magnesium metal.

5.4.6 Oxidative heterocoupling with another Grignard reagent

The homocoupling reaction of aromatic Grignard reagents using oxygen as an oxidant [27] has been successfully applied to intramolecular heterocoupling (Table 5.6, Fig. 5.42). Thus, *N*-methylcrinasiadine, a natural product extracted from *Lapiedra Matrinezzi* was obtained in this way (Fig. 5.45).

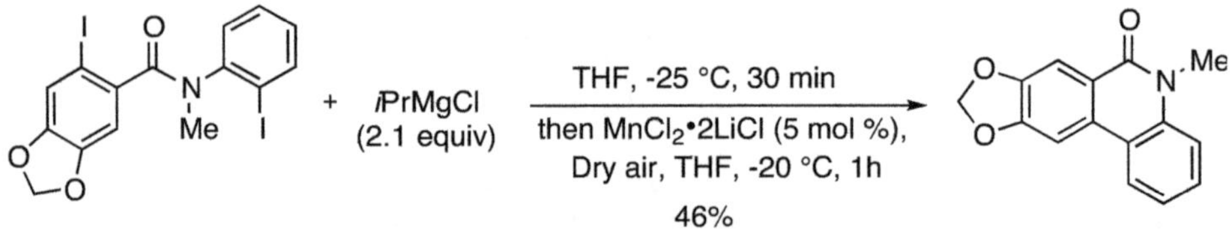

Fig. 5.45: Manganese-catalyzed synthesis of *N*-methylcrinasiadine.

Moreover, the reaction was successfully extended to the heterocoupling of two organomagnesium compounds by Cahiez *et al.* [31]. Thus, under similar conditions, it is possible to obtain preferentially the heterocoupling product by coupling two different Grignard reagents R^1MgX and R^2MgX if the R^1 and R^2 groups are different enough regarding their steric and/or electronic properties (Table 5.7).

Table 5.7: Manganese-catalyzed heterocoupling from aromatic Grignard reagents.

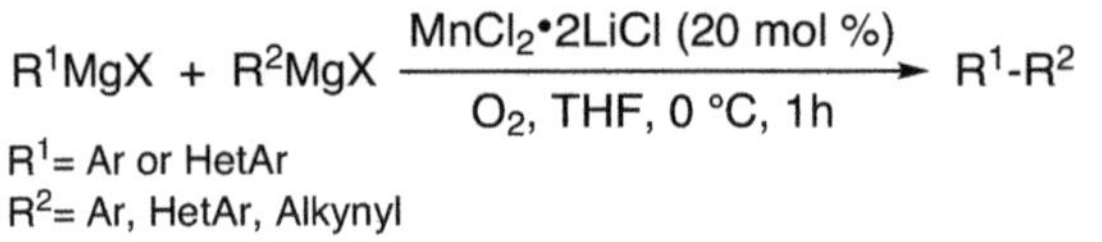

Entry	R^1MgX	Coupling product	Yield (%)
1	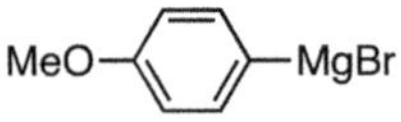	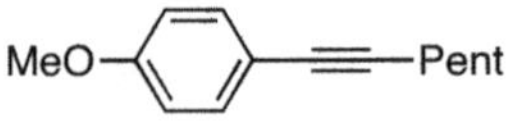	72
2	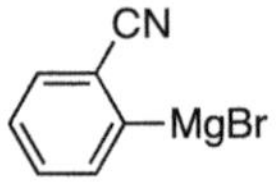	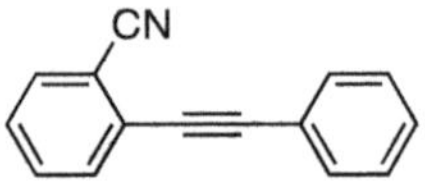	77
3	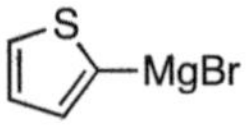	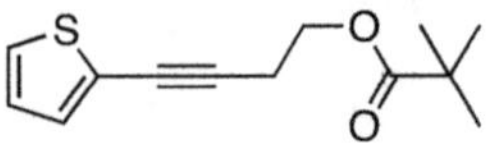	68
4	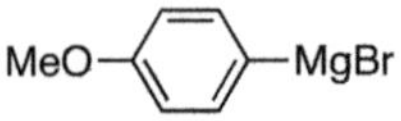	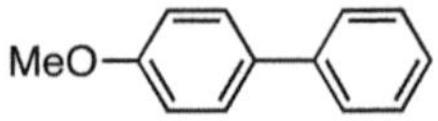	80
5	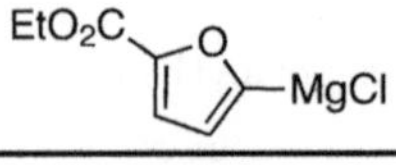	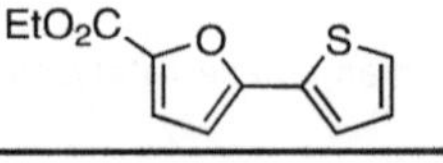	69[a]

[a] Reaction carried out at −20 °C

Interestingly, functionalized Grignard reagents (Table 5.7, entries 1, 2, 4 and 5) as well as heterocyclic Grignard reagents (Table 5.7, entries 2 and 5) can be used in this reaction.

The results were discussed on the basis of the mechanism proposed in Fig. 5.43 by considering the kinetics of the homocoupling of various Grignard reagents. Reaction of two different Grignard reagents like **F** and **G** can lead to the formation of three distinct diorganomanganese compounds **H, I** and **J**. Addition of dioxygen then affords the formation of the corresponding peroxomanganate complexes **K, L** and **M** which, by subsequent reductive elimination, can afford three different products: the two homocoupling products **N** and **P** and the heterocoupling product **O**. The formation of **N** is disfavored by steric constraints. On the other hand, reductive elimination from complex **M** is disfavored for electronic reasons. Therefore, the heterocoupling product **O** is obtained as the major product.

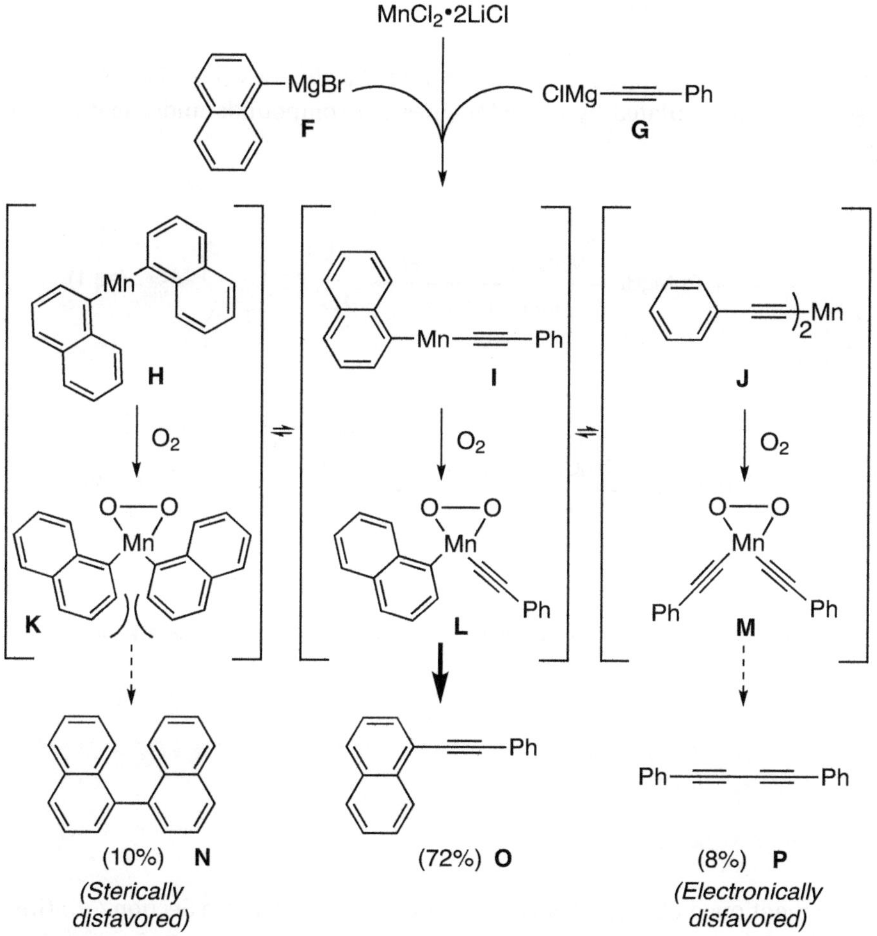

Fig. 5.46: Mechanism of the manganese-catalyzed oxidative heterocoupling of Grignard reagents.

5.4.7 Manganese-catalyzed carbometallation with aryl Grignard reagents

Various manganese-catalyzed carbomagnesiation from aliphatic Grignard reagents have been described (see Sections 3.6.–3.8).

Arylmetallation is more difficult to achieve. Thus, only few examples were described in the literature from arylmagnesium reagents. A manganese-catalyzed procedure has been described by Oshima *et al.* [22] (Fig. 5.47).

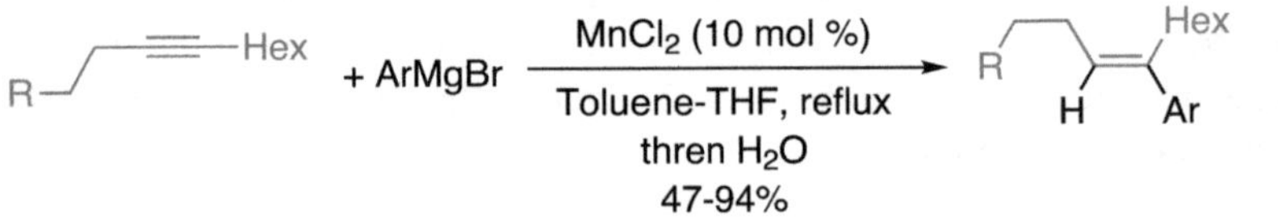

Fig. 5.47: Manganese-catalyzed arylmetallation.

o-Amino- (Fig. 5.48, eq 1) [22] or *o*-hydroxyphenylacetylenes (Fig. 5.48, eq 2) [23] can also be carbometallated by phenylmagnesium compounds under manganese catalysis.

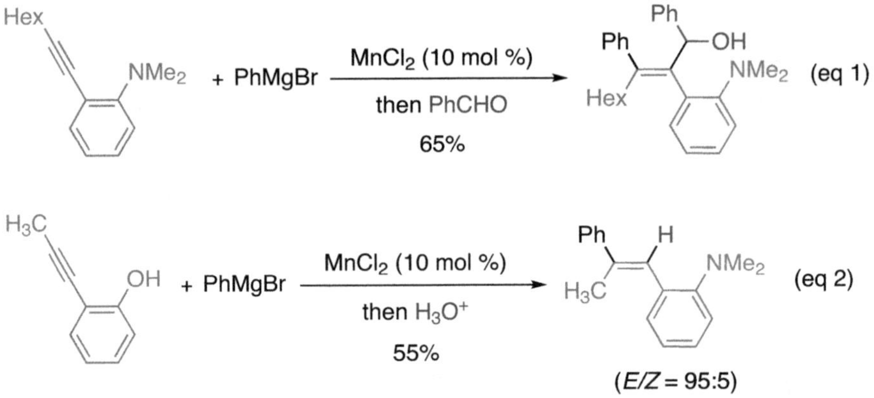

Fig. 5.48: Manganese-catalyzed carbometallation of *o*-hydroxy- or *o*-aminophenylacetylenes.

5.5 Coupling reactions of alkenyl Grignard reagents

5.5.1 With acyl halides

As already mentioned above, alkyl (see Section 3.1.) and aryl (see Section 4.1.) Grignard Reagents can be acylated in the presence of a catalytic amount of manganese. Alkenyl Grignard reagents [6] react similarly (Fig. 5.49).

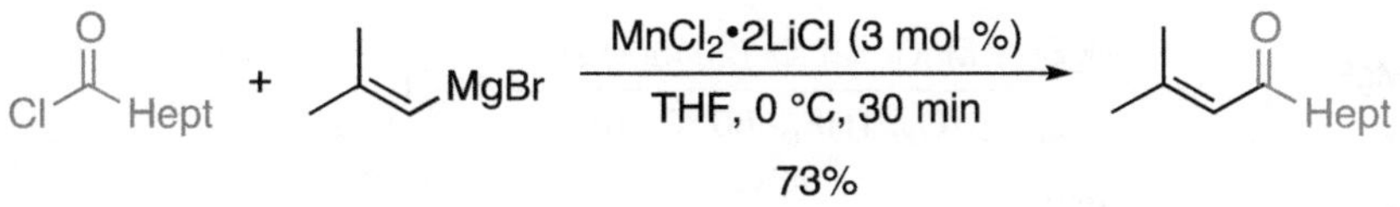

Fig. 5.49: Manganese-catalyzed acylation of phenylmagnesium chloride.

5.5.2 Oxidative homocoupling

Aryl Grignard reagents can undergo homocoupling reactions in the presence of manganese salts (see Section 4.5). Alkenyl Grignard reagents can also be used to perform such an homocoupling reaction (Fig. 5.50) [27]. A mechanism was proposed (see Fig. 5.41).

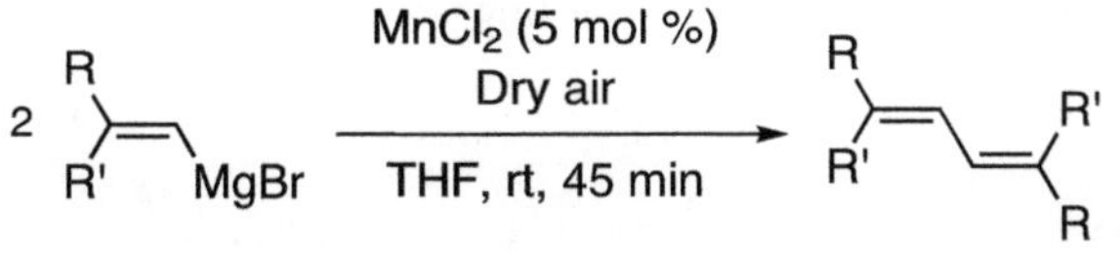

RMgX	Yield (%)
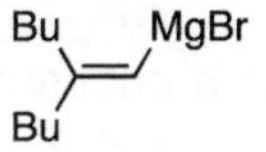	88
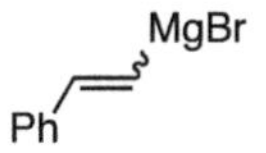	90[a]
	92[b]

[a] The reaction was performed at −20 °C. [b] The reaction was performed at −40 °C.

Fig. 5.50: Manganese-catalyzed homocoupling of alkenyl Grignard reagents using air as an oxidant.

5.5.3 Oxidative heterocoupling with another Grignard reagent

Oxidative heterocoupling from alkenylmagnesium halides has been described by Cahiez *et al.* [31] (Fig. 5.51). Interestingly, functionalized Grignard reagents can be used in this reaction. A mechanism has been proposed (see Fig. 5.46).

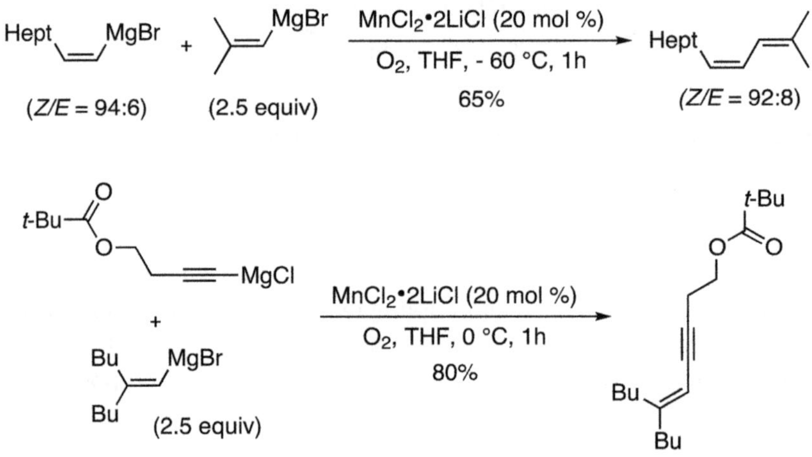

Fig. 5.51: Manganese-catalyzed oxidative heterocoupling from vinylic Grignard reagents.

5.6 Cross-coupling of alkynyl Grignard reagents

5.6.1 Oxidative homocoupling

Aryl (see Section 4.5.) and alkenyl (see Section 5.2.) Grignard reagents can undergo homocoupling reactions in the presence of manganese salts. Alkynyl Grignard reagents can also be coupled under similar conditions (Fig. 5.52) [27]. Functional groups are well tolerated. A mechanism has been proposed (see Fig. 5.41).

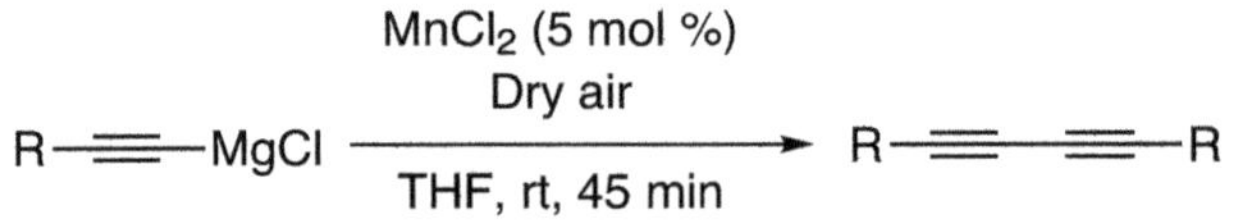

RMgX	Yield (%)
Bu—≡—MgCl	91
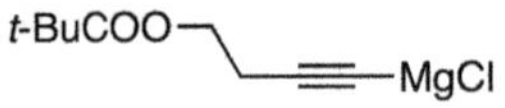	82[a]
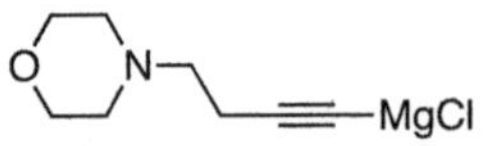	85

[a] The reaction was performed at 10 °C.

Fig. 5.52: Manganese-catalyzed homocoupling of alkenyl Grignard reagents using air as an oxidant.

Manganese-catalyzed homocoupling of ethynyl magnesium using 1,2-dihaloethanes as an oxidant has been described (Fig. 5.53) [28].

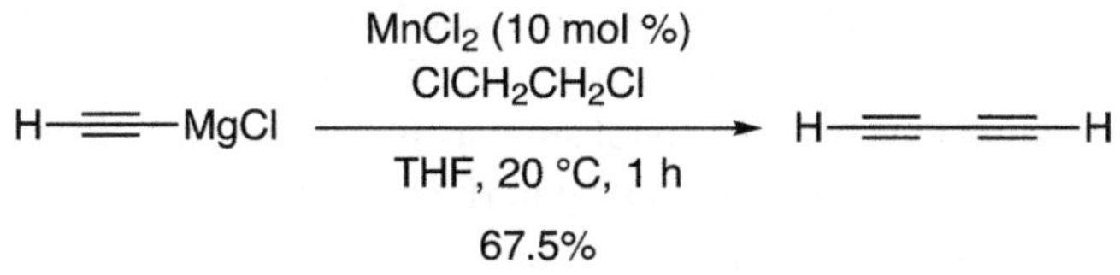

Fig. 5.53: Manganese-catalyzed homocoupling of ethynylmagnesium chloride using 1,2-dihaloethanes as an oxidant.

5.6.2 Oxidative heterocoupling with another Grignard reagent

Oxidative heterocoupling of alkynylmagnesium halides has been described by Cahiez *et al.* [31] (Fig. 5.54, see also Sections 4.6. and 5.3.).

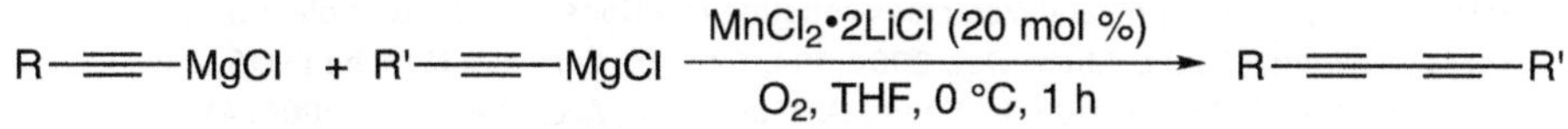

R1MgX	Coupling Product	Yield (%)
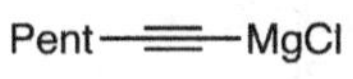	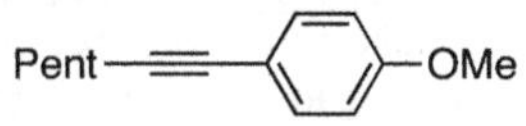	72
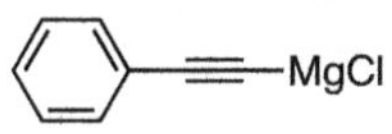	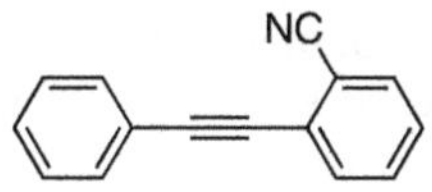	77
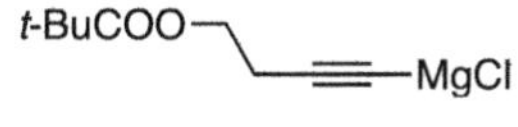	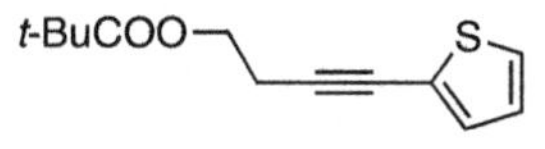	68
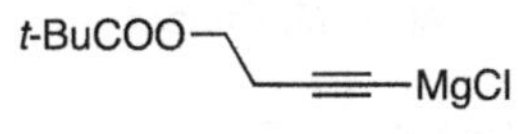	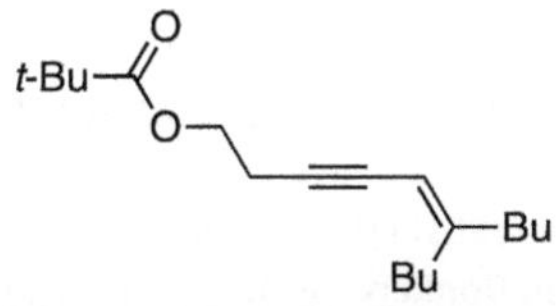	80

Fig. 5.54: Manganese-catalyzed oxidative heterocoupling from acetylenic Grignard reagents.

Interestingly, functionalized Grignard reagents can be used in this reaction. A mechanism has been proposed (see above Fig. 5.43).

5.7 Conclusion

Manganese-catalyzed reactions were developed since about forty years. Thus, manganese is the most recent transition metal which has been used to perform coupling reactions with Grignard reagents. The results already obtained show that it offers a very interesting alternative to precious or toxic metal such as palladium and nickel. The development of manganese chemistry is promising in the context of sustainable development and research of eco-friendly synthetic methods.

References

[1] Grignard V, C. R. Acad. Sci. 1900, 130, 1322.
[2] (a) Cardenas DJ, Angew. Chem. Int. Ed. 2003, 42, 384. (b) Frisch AC, Beller M, Angew. Chem. Int. Ed. 2005, 44, 674. (c) Corbet JP, Mignani G, Chem. Rev. 2006, 106, 2651.(d) Cross-Coupling Reactions: A Practical Guide, ed. Miyaura N, Topics in Current Chemistry Series 219; Berlin Heidelberg, Germany, Springer-Verlag, 2002. (e) Metal-Catalyzed Cross-Coupling Reactions, eds. de Meijere A, Diederich F, Weinheim, Germany, Wiley-VCH, 2004.
[3] (a) For reviews about iron-catalyzed cross-coupling reactions, see for example: (a) Bolm C, Legros J, Le Paih J, Zani L, Chem. Rev. 2004, 104, 6217. (b) Correa A, Mancheno OG, Bolm C, Chem. Soc. Rev. 2008, 37, 1108. (c) Sherry BD, Fürstner A, Acc. Chem. Res. 2008, 41, 1500. (d) Czaplik WM, Mayer M, Cvengros J, von Wangelin AJ, ChemSusChem 2009, 2, 396. (e) Iron Catalysis in Organic Chemistry: Reactions and Applications, ed. Plietker B, Weinheim, Germany, Wiley-VCH, 2008. (f) Cahiez G, Duplais C, Iron-Catalyzed Reactions of Grignard Reagents in: The Chemistry of Organomagnesium Compounds, eds. Rappoport Z and Marek I, Chichester, England, Wiley & sons, 2008, 595. (g) Nakamura E, Hatakeyama T, Ito S, Ishizuka K, Ilies L, Nakamura M, Iron-Catalyzed Cross Coupling Reactions in: Organic Reactions, 2013, 83, 1. For reviews about manganese-catalyzed cross-coupling reactions, see for example: (h) Cahiez G, Duplais C, Buendia J, Chem. Rev. 2009, 109, 1434. (i) Cahiez G, Gager O in: The Chemistry of Organomanganese Compounds, Chichester, England, eds. Rappoport Z and Marek I, Wiley & sons, 2011, Chap. 6, 305. (j) Layfield RA, Chem. Soc. Rev. 2008, 37, 1098.
[4] Wunderlich SH, Kienle M, Knochel P, Angew. Chem. Int. Ed. 2009, 48, 7256.
[5] (a) Cahiez G, Chavant PY, Tetrahedron Lett. 1989, 30, 7373. (b) Cahiez G, Chavant PY, Tozzolino P, Eur. Patent 323332, 1989; Chem. Abstr. 1990, 112, 38679. (c) Cahiez G, Chavant PY, Tozzolino P, Fr. Patent 2625500, 1989; Chem. Abstr. 1990, 112, 35281.
[6] Cahiez G, Laboue B, Tetrahedron Lett. 1992, 33, 4439.
[7] (a) Cahiez G, Alami M, Tetrahedron Lett. 1989, 30, 3541. (b) Marquais S, Alami M, Cahiez G, Org. Synth. 1995, 72, 135.
[8] (a) Donkervoort JG, Vicario JL, Jastrzebski JT, Gossage RA, Cahiez G, Van Koten G, Rec. Trav. Chim. Pays-Bas 1996, 115, 547. (b) Donkervoort JG, Vicario JL, Jastrzebski JT, Gossage RA, Cahiez G, Van Koten G, J. Organomet. Chem. 1998, 558, 61.
[9] Inoue R, Shinokubo H, Oshima K, Tetrahedron Lett. 1996, 37, 5377.
[10] Oshima K, J. Organomet. Chem. 1999, 575, 1.
[11] Inoue R, Shinokubo H, Oshima K, J. Org. Chem. 1998, 63, 910.
[12] Cahiez G, Lepifre F, Ramiandrasoa P, Synthesis 1999, 2138.
[13] Cahiez G, Luart D, Lecomte F, Org. Lett. 2004, 6, 4395.

[14] Zhang F, Shi Z, Chen F, Yuan Y, Appl. Organomet. Chem. 2010, 24, 57.
[15] Rueping M, Ieawsuwan W, Synlett, 2007, 247.
[16] Cahiez G, Bernard D, Normant JF, J. Organomet. Chem. 1976, 113, 107.
[17] Alami M, Ramiandrasoa P, Cahiez G, Synlett 1998, 325.
[18] Fugami K, Oshima K, Utimoto K, Chem. Lett. 1987, 16, 2203.
[19] Tang J, Shinokubo H, Oshima K, Bull. Chem. Soc. Jpn. 1997, 70, 245.
[20] Okada K, Oshima K, Utimoto K, J. Am. Chem. Soc. 1996, 118, 6076.
[21] Tang J, Okada K, Shinokubo H, Oshima K, Tetrahedron 1997, 53, 5061.
[22] Yorimitsu H, Tang J, Okada K, Shinokubo H, Oshima K, Chem. Lett. 1998, 27, 11.
[23] Nishimae S, Inoue R, Shinokubo H, Oshima K, Chem. Lett. 1998, 27, 785.
[24] Nishikawa T, Shinokubo H, Oshima K, Org. Lett. 2003, 5, 4623.
[25] Cahiez G, Gager O, Lecomte F, Org. Lett. 2008, 10, 5255.
[26] See for example a) Pu L, Chem. Rev. 1998, 98, 2405. b) Nielsen MB, Diederich F, Chem. Rev. 2005, 105, 1837.
[27] Cahiez G, Moyeux A, Buendia J, Duplais C, J. Am. Chem. Soc. 2007, 129, 13788.
[28] Zhou Z, Xue W, J. Organomet. Chem. 2009, 694, 599.
[29] Cahiez G, Chaboche C, Mahuteau-Betzer F, Ahr M, Org. Lett. 2005, 7, 1943.
[30] Yuan Y, Bian Y, Appl. Organomet. Chem. 2008, 22, 15.
[31] Cahiez G, Duplais C, Buendia J, Angew. Chem. Int. Ed. 2009, 48, 6731.

Armelle Ouali, Marc Taillefer

6 Grignard reagents and Copper

6.1 Introduction

Transition metal-catalyzed cross-coupling reactions involving organic halides or pseudohalides with organometallic Grignard reagents are among the most important method allowing for the C–C bond formation [1–4]. Since their discovery at the beginning of the last century [5], these reactions found a considerable amount of applications in organic synthesis either at the laboratory or at the industrial scale. In fact, the key to the success lied on the association of Grignard reagents with transition metal catalysts which greatly increased their application field. Since the work of Kharasch [6–8], who presented the first studies in this field, a large number of cross-coupling methods involving a variety of transition metal complexes as catalysts have been described. The Kumada-Corriu reaction [9, 10], usually corresponding to the nickel- or palladium-catalyzed coupling of electrophiles such as aryl or vinyl halides with Grignard reagents, represents one of the best examples of such a successful combination. Other transition metals are able to promote this reaction and it is also the case, in a general point of view, for the coupling of various families of Grignard reagents with different types of electrophiles which can be catalyzed by Ni-, Pd-, Fe-, Co-, Mn-, Cu-based complexes and to a less extend by Cr-, Zr- or Ti-based complexes. We will focus in this chapter on copper (Cu)-catalyzed cross-couplings of saturated and unsaturated electrophiles with Grignard reagents.

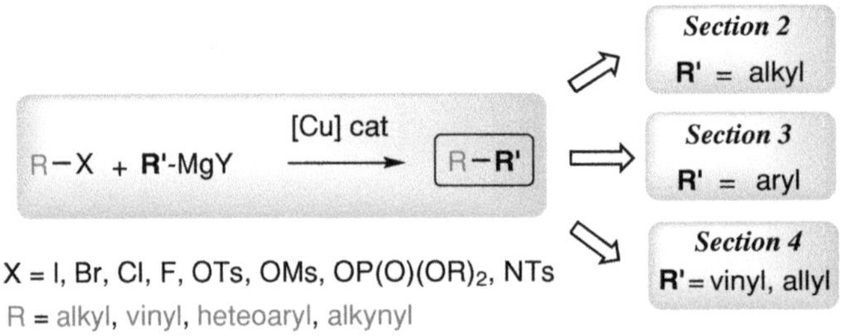

Fig. 6.1: Cu-catalyzed coupling reactions involving organic halides or pseudohalides with organometallic Grignard reagents.

The sections are organized according to the type of Grignard reagent involved: thus the first two sections are dedicated respectively to alkyl- and aryl Grignard reagents and the third and last section discusses the less widespread Grignard reagents such as vinyl, allyl and propargyl derivatives. In every case, the starting electrophile and nucleophile components are respectively highlighted in red (R) and in black bold (**R'**).

6.2 Coupling reactions from alkyl Grignard reagents (R'MgX)

6.2.1 Coupling of alkyl Grignard reagents with alkyl halides or pseudo-halides derivatives (RX)

In his seminal work, Kharasch reported, in 1943, the Co-, Cr- and to a lesser extent Cu-catalyzed cross-coupling of vinyl halides with phenylmagnesium bromide (PhMgBr) [11]. However his catalytic systems did not succeed to be extended to the cross-coupling of alkyl Grignard reagents or alkyl halides due to the formation of homo-coupling and/or disproportionation products [8]. While the possibility of efficiently performing the coupling of alkyl halides with Grignard reagents only appeared in the past decade when metals such as nickel, palladium or iron catalysts were used [4], the corresponding copper-catalyzed cross-coupling of alkyl Grignard reagents with alkyl halides was reported by Noller [12] in 1964 and in the 1970s by Tamura and Kochi (Fig. 6.2) [13–15]. This coupling employed $CuBr_2$ or Li_2CuCl_4 as precatalysts in THF at low temperature to form, in high yields, cross-coupling products from primary alkyl bromides and iodides with various Grignard reagents.

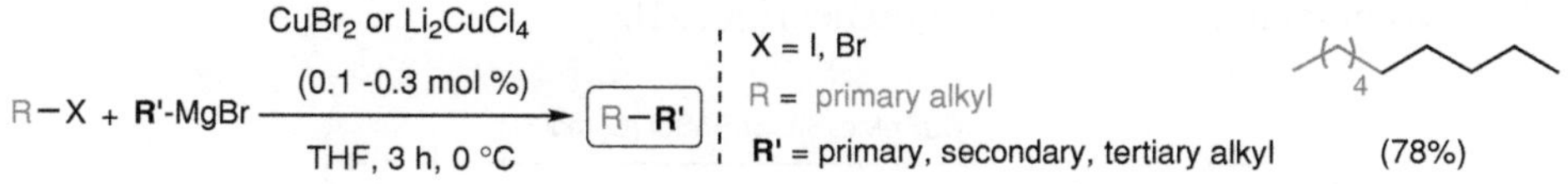

Fig. 6.2: Cu-catalyzed coupling of primary alkyl bromide or iodides with alkyl Grignard reagents [13–15].

After these breakthroughs, in 1974, Schlosser demonstrated the efficiency of low catalytic amounts of Li_2CuCl_4 (0.3–0.5 mol %) to promote, in very simple experimental conditions, the coupling of tosylate derivatives (aryl and primary alkyl tosylates) with primary and tertiary Grignard reagents (Fig. 6.3) [16, 17].

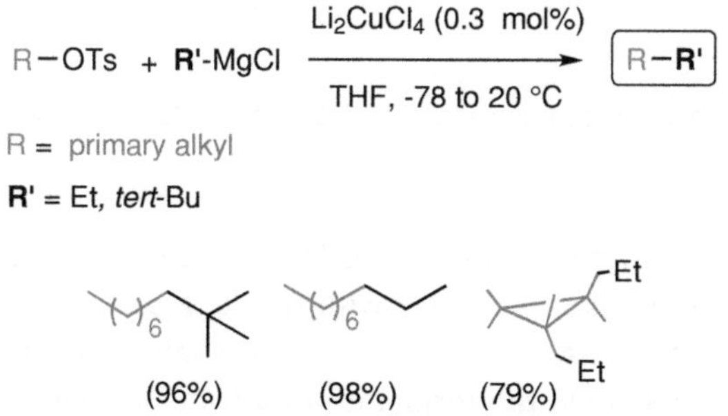

Fig. 6.3: Cu-catalyzed coupling of primary alkyl toylates with alkyl Grignard reagents [16, 17].

Later on, some interesting but isolated examples, were reported respectively by Nunomoto (1983) and Stratton (1995). The former employed catalytic amounts of Li_2CuCl_4 used by Kochi and Schlosser to couple the *n*-octyl iodide with a secondary alkyl Grignard reagent (*i*PrMgBr). This system also found a wider application with the synthesis of 2-functionnally substituted 1,3-butadienes from the corresponding coupling of 1,3-butadien-2-ylmagnesium chloride with primary functionalized alkyl iodides and bromides [18]. The second author presented a symmetrical coupling procedure combining di-halide substrates (X-R-X), magnesium turnings and Li_2CuCl_3 as the precatalyst, allowing for the synthesis of some di-haloalkanes (X-R-R-X) potentially interesting as precursors for the synthesis of insect pheromones or fatty acid-ester [19].

In 1997, Burns *et al.* discovered a new copper-based catalytic system ($CuBr.Me_2S.LiBr.PhSLi$) soluble in the reaction solvent (THF) which was efficient at 0–18°C for the coupling of primary alkyl electrophiles (X = I, Br, OTs) with primary alkyl Grignard reagents (Fig. 6.4). The presence of a thiol (Me_2S) and further addition of HMPA (hexamethylphosphoramide, 6 % v/v) stabilized the Grignard catalytic system, thus allowing the reaction to proceed at higher temperature and the scope of the coupling could be broaden. For example, secondary tosylates or challenging mesylates react with primary Grignard reagents while primary tosylates couple with secondary or tertiary Grignard reagents. Interestingly, this method extended to the C–alkyl—C–aryl bond formation has been applied by the authors to the synthesis of *exo*-metacyclophanes (homologues of exocalix[4]arenes) [20, 21].

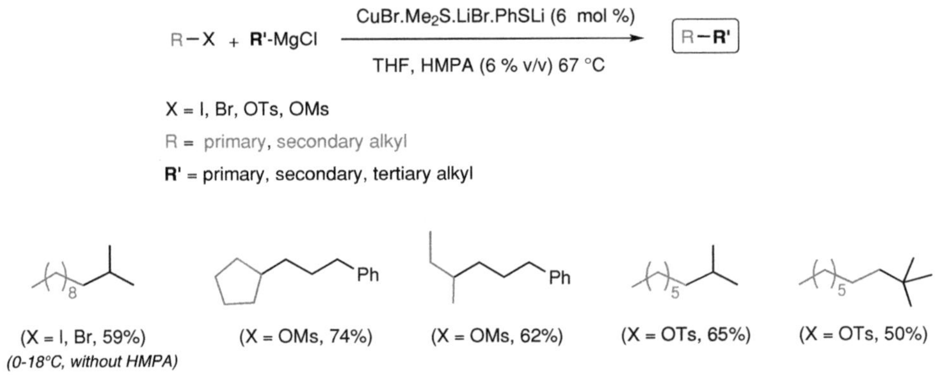

Fig. 6.4: Cu-catalyzed coupling involving primary or secondary alkyl toylates or mesylates and alkyl Grignard reagents [20, 21].

In 1998, Van Koten *et al.* described a copper/manganese (CuCl/[Mn],Y = Cl_2Li) co-catalyzed coupling of *n*-primary or secondary alkyl bromides with primary, secondary, and tertiary alkyl magnesium chlorides (Fig. 6.5) [22]. The reactions take place in a very short reaction time (15 min) probably *via* the formation of a transmetalation product intermedite ([Mn]; Y = **R'**) prior the cross-coupling product itself.

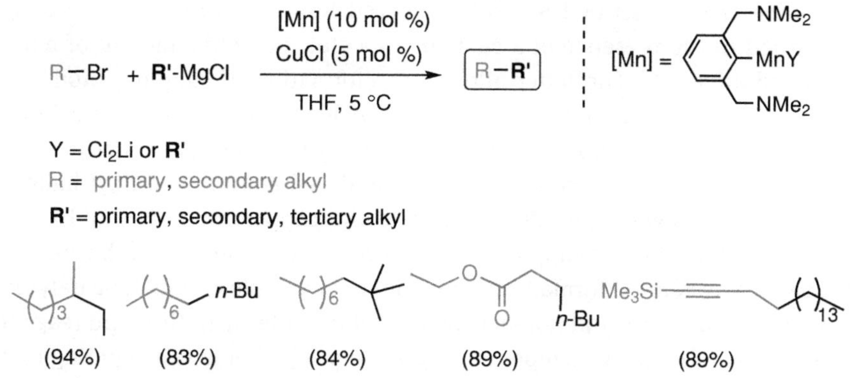

Fig. 6.5: Cu/Mn-catalyzed coupling involving primary functionalized or secondary alkyl bromides and alkyl Grignard reagents [22].

In 2000, Cahiez *et al.* reported the copper-catalyzed alkylation of alkyl Grignard reagents using the traditional CuCl or Li_2CuCl_4 salts as the precatalysts (Fig. 6.6). Under mild conditions (THF, 20 °C), primary alkyl halides (X = I, Br, OTs) functionalized by an ester, an amide, a nitrile or a keto group could react with various Grignard reagents including the most challenging tertiary alkyl ones. The key to the success of this coupling relies on the presence of NMP (*N*-methyl pyrrolidinone) as the solvent, which prevents side reactions and the restriction of functionality usually observed in cross-coupling involving organometallic compounds. The authors have shown that the addition of NMP could however not be efficient enough to allow the coupling from challenging alkyl chlorides or secondary and tertiary alkyl halides [23].

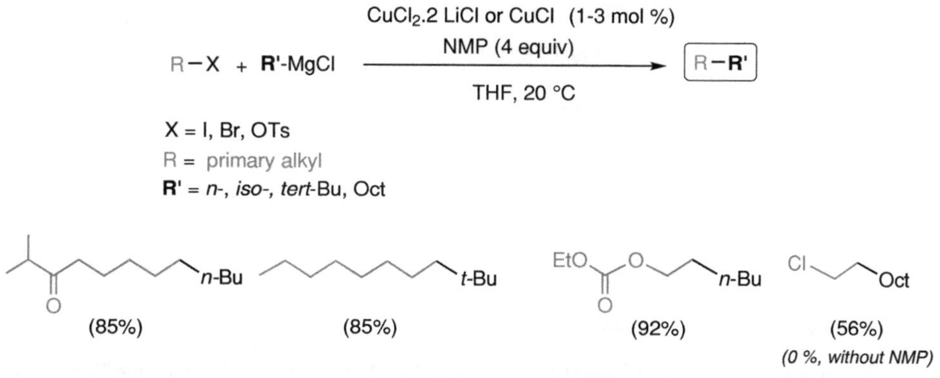

Fig. 6.6: Cu-catalyzed coupling of primary alkyl halides with alkyl Grignard reagents in the presence of the solvent NMP [23].

In 2002, Kambe *et al.* described a nickel-catalyzed Kumada-Corriu type coupling [9, 10] performed in the presence of a catalytic to a stoichiometric amount of butadiene as the additive [24]. The latter allowed the formation of coupling products, obtained from alkyl bromides or tosylates, in good yields. In the absence of the butadiene, reduction and elimination products were mostly obtained. A similar positive impact was observed for the related palladium-catalyzed coupling of alkyl halides with alkyl Grignard reagents [25, 26]. Applied to the copper catalysis ($CuCl_2$), the addition of butadiene which is supposed to stabilize an active species of the copper, also improved the system performance. Thus, the first example of a metal-catalyzed cross-coupling of unreactive and non-activated alkyl fluorides with Grignard reagent of alkyl (and also aryl) type, was reported (Fig. 6.7, eq. 1) [27]. It is worth noting that the observed reactivity order (R-Cl<R-F<R-Br) could not be reasonably explained in terms of bond energies of the alkyl halides. The same year, these authors expanded the application field of their catalytic copper/1,3-butadiene system to the cross-coupling of unactivated secondary alkyl iodides with primary alkyl Grignard reagents (Fig. 6.7, eq. 2) [28]. After the first reports of Burns and Van Koten involving the coupling of secondary alkyl halides with Grignard reagents [20–22], only few results were reported in this field.

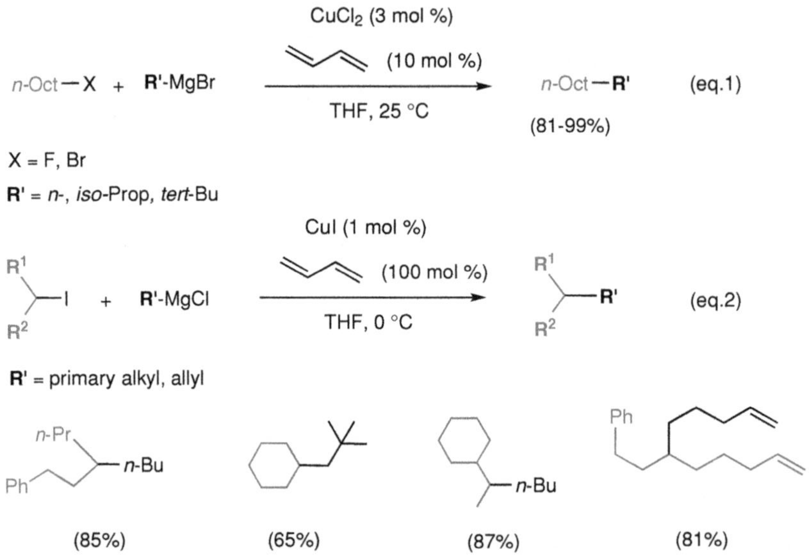

Fig. 6.7: Cu-catalyzed coupling of alkyl Grignard reagents in the presence of butadiene with primary alkyl fluorides or bromides (eq. 1) and secondary alkyl iodides (eq. 2) [27].

As a variation of their protocol, the same authors disclosed that the coupling of alkyl Grignard reagents with primary alkyl chlorides, less efficient (3 % yield) than alkyl fluorides in the presence of butadiene, proceeded efficiently when the copper catalyst was assisted by another additive, the 1-phenylpropyne (Fig. 6.8) [29, 30]. Taking advantage of the difference of reactivity between bromo- and chloroalkyl electrophiles, interesting highly site-selective sequential cross-coupling involving two different alkyl Grignard reagents was performed from a chlorobromoalkane.

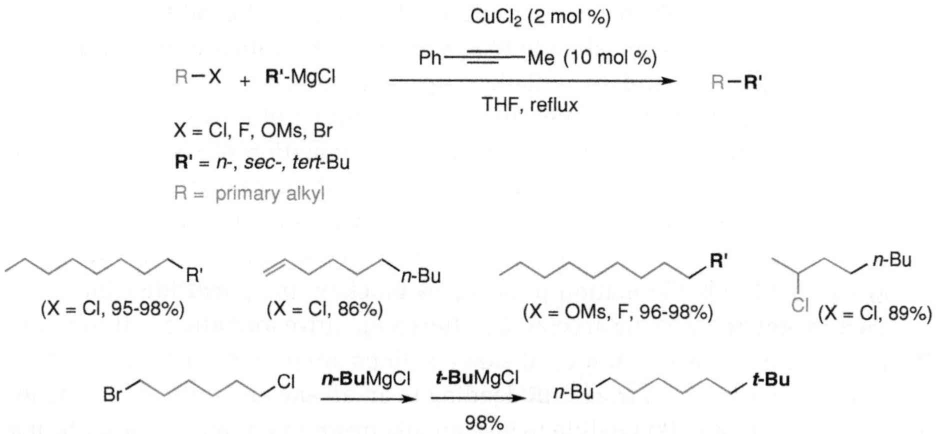

Fig. 6.8: Cu-catalyzed coupling of primary alkyl chloride with alkyl Grignard reagents, [29, 30] in the presence of 1-phenylpropyne as additive.

In a remarkable way, in the presence of an alkyne or a diene as an additive, very high TON (up to 10^6) and TOFs (around $10^4\ h^{-1}$) were attained, as illustrated below by the copper-catalyzed cross-coupling of *n*-butyl magnesium chloride with 2-(4-bromobutyl)-thiophene (Fig. 6.9) [31].

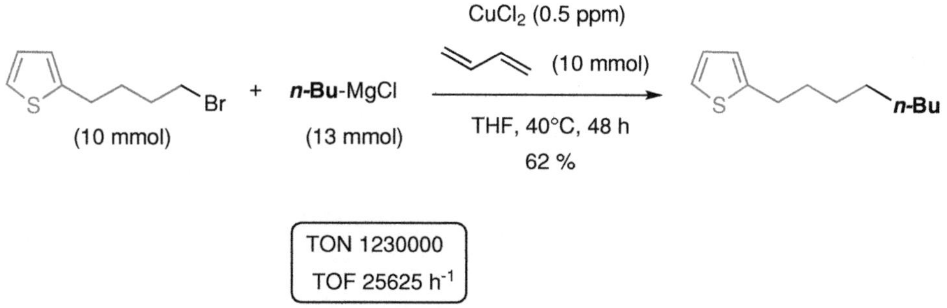

Fig. 6.9: Cu-catalyzed coupling of primary alkyl bromide with alkyl Grignard reagents with very high TON and TOFs (Kambe *et al.*).

Chung and co-workers also used an alkyne (1-phenylpropyne) as an additive for the coupling of primary alkyl bromides or chlorides with primary-, secondary- or tertiary-alkyl Grignard reagents in the presence of copper or copper oxide nanoparticles (Cu NPs) catalysts [32]. The Cu NPs employed in this study are commercially available and can be reused five times without any loss of activity (TON = 49).

Kambe and coworkers have proposed catalytic pathways to describe the copper-catalyzed alkyl–alkyl cross-coupling and also the side reactions potentially able to lower the selectivity in the expected products (Fig. 6.10) [31, 33]. The catalyst of the reaction, **A**, generated from a copper salt (CuCl or $CuCl_2$), a ligand **L** of butadienyl or alkyne type and one equivalent of **R'**MgX, would react with a second equivalent of the Grignard reagent to form a dialkyl cuprate intermediate **B**. The latter in the presence of the alkyl halide could allow the formation of a new intermediate **C** *via* a nucleophilic substitution. By a reductive elimination of *cis*-oriented alkyl partners, the expected product could be expelled and the catalyst **A** regenerated. The presence in the medium of the ligand **L** is a key factor to the selectivity for the expected coupling product. Coordinated to the copper, the ligand **L** strongly or fully disfavors the β-elimination pathway by blocking the β-hydride elimination from **A** and probably also by accelerating the competitive formation of the cuprate **B**. Thus, in the absence of **L** several side reactions are possible among which: (i) a β-hydrogen elimination from Cu**R'** leading to an alkene and CuH; (ii) a reaction of this latter with an alkyl halide to give an alkane and CuX and (iii) after formation of a Cu(0) from disproportionation of CuH and/or CuX, reaction of the former with an alkyl halide and β-hydrogen elimination leading to the formation of a new alkene (Fig. 6.10).

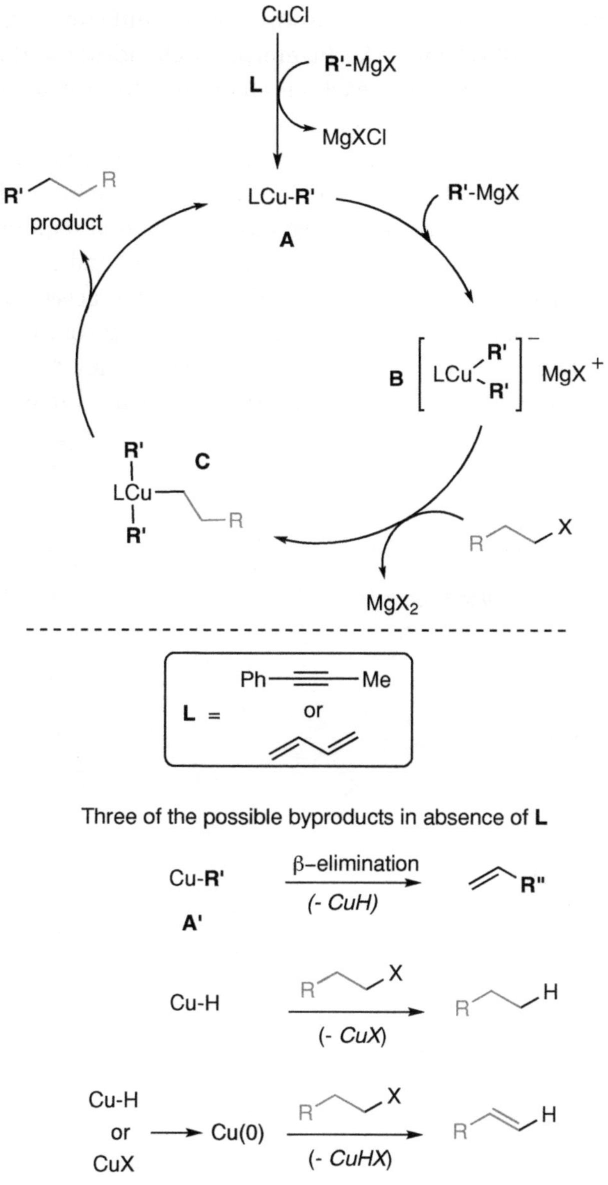

Fig. 6.10: A possible mechanism for the coupling of primary alkyl halide with alkyl Grignard reagents [31, 33]. Role of the ligand **L** and possible byproducts.

Whereas the copper-catalyzed coupling from primary alkyl halides is well described, the coupling involving secondary alkyl electrophiles, first described by Burns et Van Koten (*vide infra*) [20–22], has been scarcely reported. In 2012, Liu *et al.* presented a method also based on a copper catalytic system, allowing for this difficult cross-coupling of non-activated secondary alkyl halides (X = Br) and pseudohalides (X = OTs) with alkyl Grignard reagents (Fig. 6.11). Since this reaction can result in the formation of tertiary carbons and stereogenic centers, it is potentially more interesting than the related one involving primary alkyl electrophiles. Additionally, the authors addressed the even more challenging coupling involving secondary alkyl electrophiles with secondary or tertiary alkyl nucleophiles (Grignard reagents). The method requires the presence of two types of additives, tetramethylene diamine (TMEDA) and LiOMe used respectively in catalytic and stoichiometric amounts. The reaction proceeds by a S_N2 mechanism with inversion of configuration as illustrated by the stereocontroled C–C bond formation resulting from the coupling of chiral tosylates (obtained from the related chiral alcohols) such as for example, cyclohexyl magnesium bromide (ee = 98–99 %) [34].

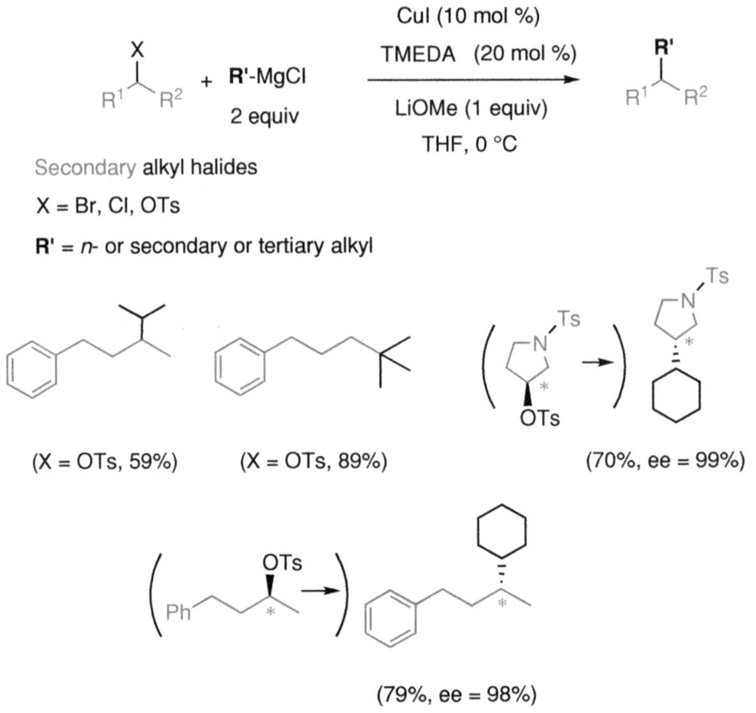

Fig. 6.11: Cu-catalyzed coupling of secondary alkyl halides with secondary or tertiary alkyl Grignard reagents [34]. Stereocontrolled C–C bond formation.

Hu *et al.* described, in 2012, a method for the cross-coupling of non-activated and functionalized primary alkyl halides and tosylates with secondary and tertiary Grignard reagents (Fig. 6.12). A wide scope was reported. The attempts to develop an enantioselective coupling in the presence of chiral ligands were not successful [35].

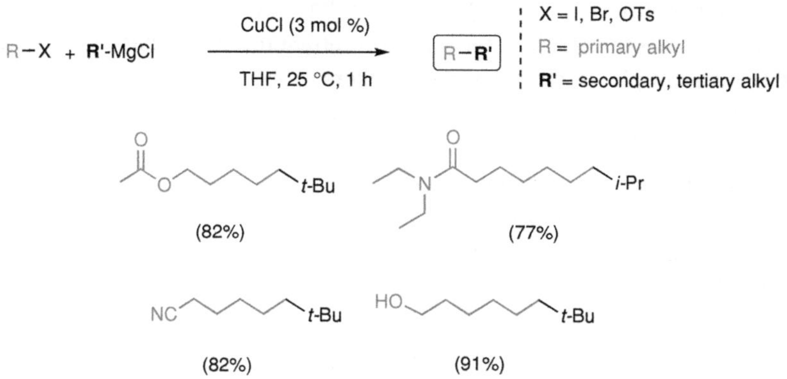

Fig. 6.12: Cu-catalyzed coupling of functionalized primary alkyl halides and tosylates with secondary or tertiary alkyl Grignard reagents [35].

Recently, Matsuoka *et al.*, based on the Kambe method [29], have reported the preparation of selectively isotope cluster-labeled long-chain fatty acids, *via* two consecutive copper-catalyzed cross-couplings of alkyl tosylates intermediates with alkyl magnesium bromides (Fig. 6.13) [36]. These NMR- and IR-active molecules of stearic acid type are interesting to determine the conformation of an alkyl chain of lipidic biomolecules upon interaction with proteins.

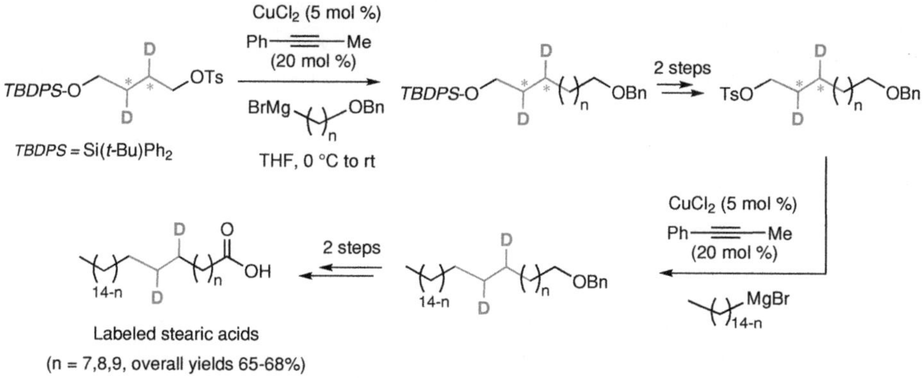

Fig. 6.13: Consecutive Cu-catalyzed cross-couplings of alkyl tosylates with alkyl magnesium bromides. Synthesis of stearic acids [36].

6.2.2 Coupling of alkyl Grignard reagents with vinyl halides or pseudo-halides derivatives (RX)

As early as 1943, Kharasch and coworkers reported the ability of copper chloride (5 mol %) to promote the coupling of PhMgBr and vinyl bromide. Using CuCl (5 mol %)

only resulted in poor product yield (11 %) and cobalt-based catalysts revealed much more efficient in this reaction [37]. Later on, more efficient copper-based systems were described. For example, Bäckwall and coworkers proposed a very general copper-catalyzed coupling reaction of dien-2-yl triflates with a wide array of alkyl- and also allyl- or aryl Grignard reagents (Fig. 6.14) [38]. The resulting 2-substituted 1,3-dienes were obtained in good to excellent yields in the presence of copper iodide (1–10 mol %) in THF at low temperature (–20 to 0 °C). The same group applied this method in the total synthesis of paenilactone A especially used in the treatment of abdominal pain [39].

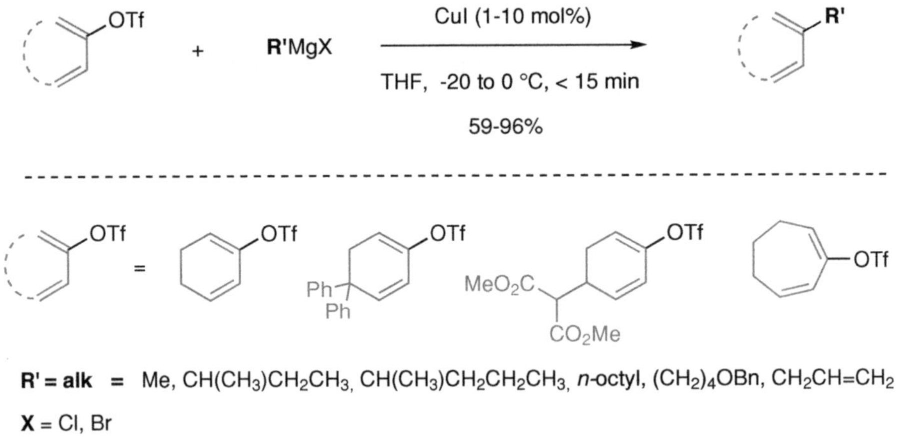

Fig. 6.14: Cu-catalyzed cross-couplings between alkyl Grignard reagents and dien-2-yl triflates [38].

6.2.3 Coupling of alkyl Grignard reagents with heteroaryl halides derivatives (RX)

It is worth noting that copper-catalyzed cross-couplings involving alkyl Grignard reagents and aryl or heteroaryl halides are very rare in the literature. For example, Hinterman and coworkers reported a copper-based catalytic system (CuI, 3–10 mol %) allowing the reaction between a wide range of heterocyclic electrophiles and bulky tertiary magnesium halides including dimethylalkyl, cycloaliphatic and cage-like Grignard reagents (Fig. 6.15) [40]. The cross-couplings selectively converted dichloroazacyles into monosubstituted monochloroazacyles and trichloroazacycles into either monosubstituted dichloroazacycles or disubstituted monochloroazacycles. All these extremely sterically hindered and highly functionalized targets, difficult to access otherwise, can be obtained in fair to excellent yields (51–95 % with one exception at 15 %) under mild reaction conditions.

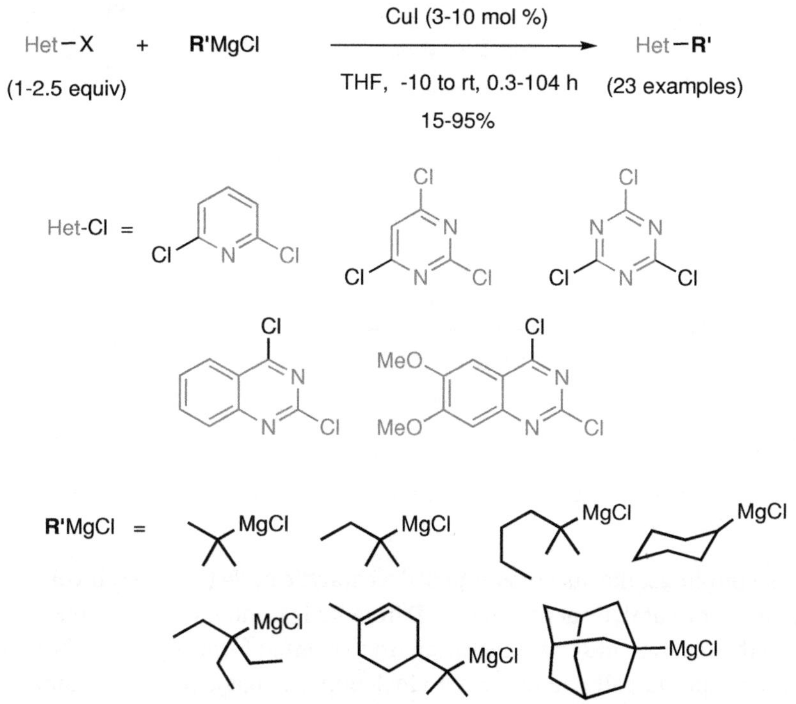

Fig. 6.15: Cu-catalyzed cross-couplings between heteroaryl chlorides and alkyl Grignard reagents [40].

6.2.4 Coupling of alkyl Grignard reagents with alkynyl halides or pseudo-halides derivatives (RX)

Until very recently, no method allowing the cross-couplings of alkynyl halides with aryl or alkyl Grignard reagents, in the presence of catalytic amounts of copper has been reported. In 2010, Cahiez and coworkers were the first authors to propose a copper-based catalytic system composed of copper(II) chloride (3 mol %) and NMP (4 mol %) able to promote the alkynylation of a wide array of alkyl Grignard reagents in THF at 0 °C (Fig. 6.16) [41]. The addition of NMP as a co-catalyst allowed to significantly increase the yield in the coupling products (Fig. 6.16, eq. 1), for example from 73 % to 91 % for the coupling of *n*-butyl magnesium bromide and heptynyl bromide. Noteworthy, this additive was not necessary when aromatic haloalkynes (e.g. phenylacetylene) and alkyl Grignard reagents were used as the coupling partners (Fig. 6.16, eq. 2).

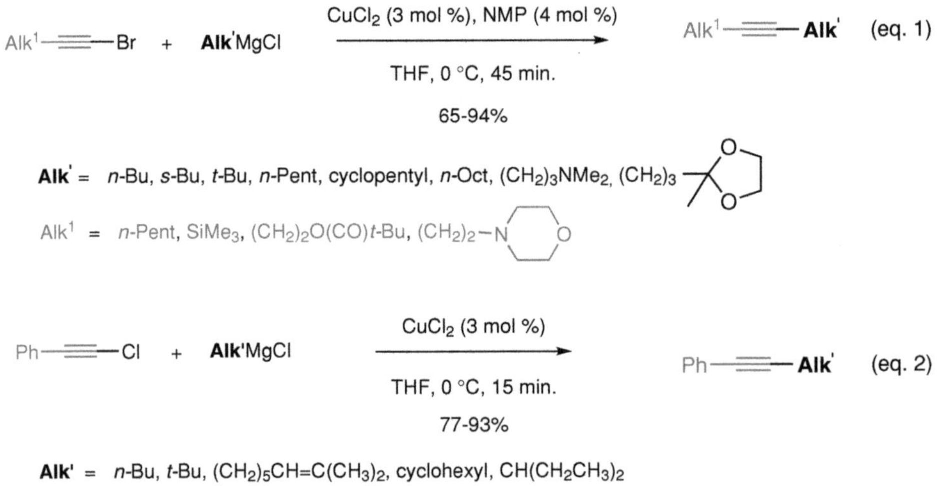

Fig. 6.16: Copper-catalyzed cross-couplings between alkyl Grignard reagents and alkylnyl-alkyl or alkynyl-aryl halides [41].

For these cross-couplings, the authors proposed a catalytic cycle (Fig. 6.17) involving: (i) the formation of the catalytic active species **D** from R'MgX and the copper precursor ($CuCl_2$ here); ii) the reaction of **D** with the Grignard reagent to form cuprate **E**; (iii) the reaction of **E** with the haloalkyne to form a vinylcopper reagent **G** *via* the complex **F/F'**. Formally, **G** results from a reductive elimination from the metallacyclopropene **F'** and (iv) the β-halogen elimination leading to the expected substituted product **H** and the regeneration of the catalytic active species **D.** This putative mechanism is consistent with the observations and with the catalytic cycles proposed for other copper-catalyzed cross-couplings of Grignard reagents and organic halides.

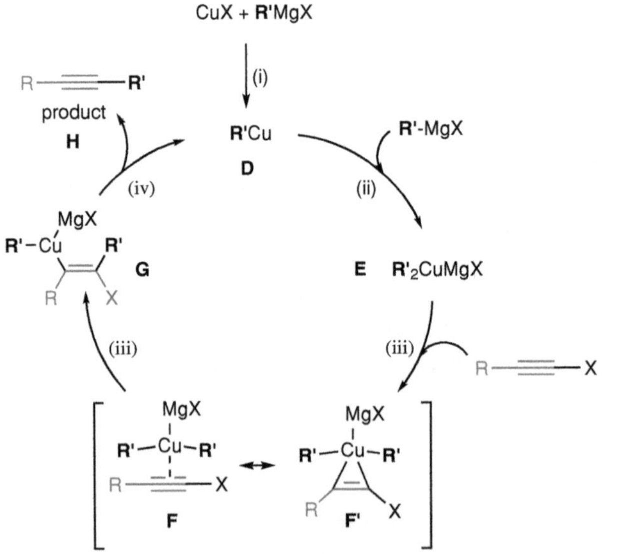

Fig. 6.17: Mechanism proposed by Cahiez *et al.* for the couplings between alkyl Grignard reagents and alkylnyl derivatives [41].

6.2.5 Coupling of alkyl Grignard reagents with allylic substrates

Alkyl Grignard reagents are involved in numerous couplings with allylic substrates (halides or esters including acetates, perfluorobenzoates, sulfonates, phosphates, etc.), under copper catalysis (Fig. 6.18) [42]. Such reactions involve the allylic rearrangements (S_N2' substitution) and are stereoselective [43, 44]. The reaction conditions can be tuned to obtain the branched isomer as the major compound and the association of chiral ligands to copper catalyst allows high enantioselectivity. From a mechanistic point of view, π-allylic copper complexes seem to be involved. Even if they display a great synthetic potential and represent a very active research field [43, 44], these transformations will not be discussed in deep details in this chapter. Indeed, they cannot be considered as cross-couplings, defined as σ-bond metathesis reactions between a nucleophilic and an electrophilic reagent [2].

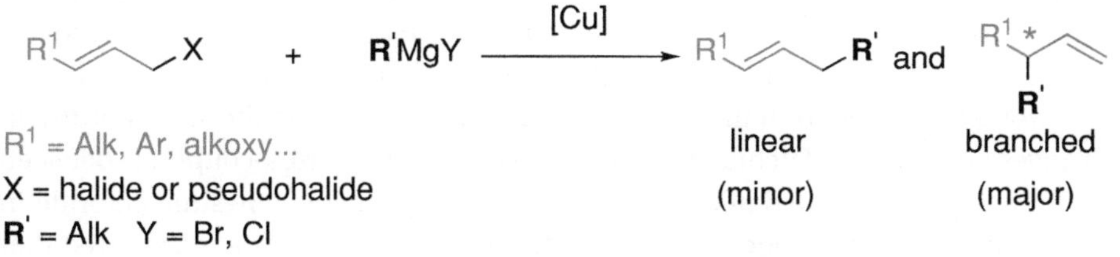

Fig. 6.18: Particular case of allylic alkylations.

6.3 Coupling reactions of aryl Grignard reagents [1, 2, 45]

6.3.1 Coupling of aryl Grignard reagents with alkyl derivatives

6.3.1.1 Coupling with alkyl halides or pseudo-halides (triflates, mesylates,...)

On the whole, aryl nucleophiles are generally less reactive than their alkyl counterparts in cross-couplings with alkyl halides and pseudohalides, the sp^2-hybridized carbon atom being less nucleophilic [1]. Therefore, there are few examples of cross-couplings involving aryl Grignard reagents compared to the alkyl ones. The effect of copper salts on the cross-couplings of aryl Grignard reagents with organic halides has been disclosed by Tamura and Kochi in 1971 [13]. Shortly afterwards, in 1974, Schlosser and coworkers reported the first cross-coupling involving the phenyl magnesium bromide and primary alkyl sulfonates in the presence of di-lithium tetrachlorocuprate [Li_2CuCl_4] (0.2–0.5 mol %) as the catalyst [17]. The reaction proceeded smoothly in THF at room temperature or below and this method could be successfully applied to the preparation of several natural products [46–48]. It is worth noting that the copper(I) counterpart of this precursor, CuCl.2LiCl, can also be used with at least the same efficiency [19]. The scope of this reaction was next broadened by Burns and coworkers who proposed, in 1997, a more complex catalytic system composed of $CuBr.SMe_2$ (6 mol %), LiSPh and LiBr to promote the cross-coupling of primary and secondary alkyl sulfonates (tosylates and mesylates) with various aryl Grignard reagents in a THF/HMPA mixture (94:6, v/v) at 67 °C [20]. This method was successfully applied for the synthesis of natural products [49] or for the construction of cyclophosphane architectures [21] with a broad field of applications as secondary and tertiary alkyl Grignard reagents could also be alkylated.

Couplings involving alkyl halides instead of sulfonates as the alkylating partners were also proposed. In 1986, Nishimura and coworkers demonstrated the ability of copper bromide (5 mol %) to promote couplings of phenyl or 2-naphthyl magnesium bromide and α,ω-dibromoalkanes with either one or two bromides being selectively replaced in a refluxing THF/HMPA mixture [50]. This method was used by Bruice and coworkers to prepare bulky key building blocks in the synthesis of porphyrin derivatives [51]. In 1996, Fuchikami and coworkers reported cross-couplings between β-perfluoroalkyl-substituted alkyl bromides with the phenyl magnesium chloride in the presence of copper iodide (3–6 mol %) [52]. Soon after, Cahiez *et al.* showed that dilithium tetrachlorocuprate [Li_2CuCl_4] (3 mol %) could efficiently catalyze the coupling of octyl bromide and phenyl magnesium bromide under mild conditions (room temperature, short reaction time, Fig. 6.19, eq. 1) [23]. Surprisingly, the addition of the polar co-solvent NMP inhibited the reaction although this additive is known to accelerate the parent cross-coupling of alkyl Grignard reagents mentioned above.

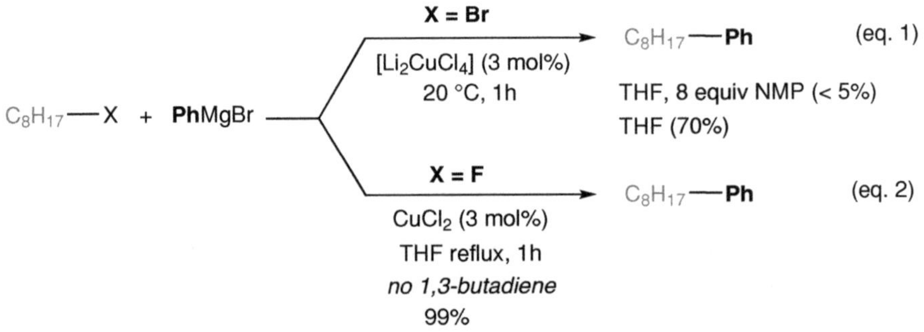

Fig. 6.19: Copper-catalyzed Kumada coupling by Cahiez et al. (eq. 1) [23] and Terao, Kambe *et al.* (eq. 2) [27].

In 2003, Terao, Kambe and co-workers described the copper-catalyzed cross-coupling of octyl fluoride with phenyl magnesium bromide (Fig. 6.19, eq. 2) [27] Contrary to the case of their parent alkyl–alkyl cross-coupling protocol (cf Fig. 6.7), the reactions were not accelerated by the addition of 1,3-butadiene. The reactivity of alkyl halides decreased in the order F > Br >> Cl, a trend that cannot be rationalized by considering the bond energies of the alkyl halides. No clear explanation was given for this effect by the authors.

Later, in 2007, Terao, Kambe and co-workers showed that alkyl chlorides could efficiently react with aryl Grignard reagents in the presence of $CuCl_2$ (2 mol %) and 1-phenylpropyne (10 mol %) as the additive in refluxing THF (Fig. 6.20) [29]. The copper-catalyzed reaction was shown to proceed *via* a S_N2-type mechanism by realizing a radical-clock study and 2D-labeled experiments. The method was also efficient starting from primary alkyl bromides and mesylates.

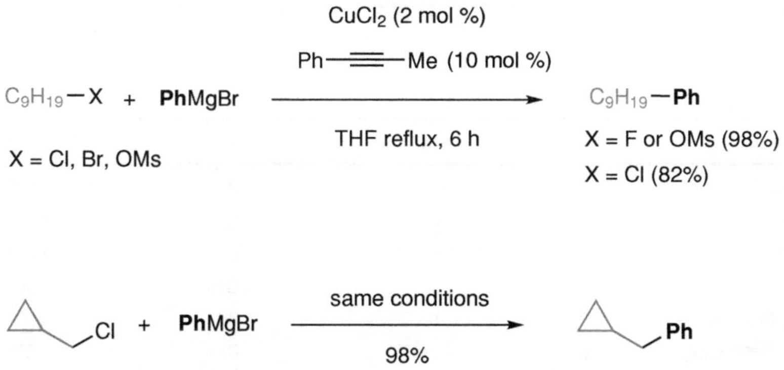

Fig. 6.20: Copper-catalyzed couplings of alkyl chlorides, bromides and mesylates with PhMgBr [29].

A general method was recently published by Kobayashi and coworkers (Table 6.1) [53]. Alkyl bromides, iodides and tosylates can react with a wide range of aryl Grignard reagents in the presence of copper iodide and lithium chloride as a co-catalyst under mild conditions and within short reaction times (<2 h). Lithium chloride was shown to accelerate the reaction which is in total accordance with the well-known efficiency of the dilithium tetrachlorocuprate [Li_2CuCl_4] [19, 17] and other catalytic systems including lithium-based additives (LiSPh/LiBr) in analogous cross-couplings [20]. Regarding the Grignard partner, the scope mainly includes aryl magnesium bromides (one example with PhMgCl, [Table 6.1, entry 2]) potentially substituted by electron-donating (Me, OMe) (Table 6.1, entries 5–10) or more electron-withdrawing groups (F) (Table 6.1, entry 11). Bulky substrates involving *ortho*-substituents (Me or OMe) could also be successfully used as arylating agents and the corresponding products were obtained in high yields (85–91 %) (Table 6.1, entries 8–10). Regarding the alkyl partner, bromides (Table 6.1, entries 1, 2, 5–11), iodides (Table 6.1, entries 3, 13, 14) and tosylates (Table 6.1, entry 4) act as effective leaving groups while alkyl chlorides were unreactive under the same conditions (Table 6.1, entry 12). Noteworthy, the method can allow the arylation of alkyl bromides possessing sensitive functional substituents such as cyanides or esters

(Table 6.1, entries 13, 14). The high amounts of copper precursor (30 mol %) and LiCl (200 mol %) required to get products in good yields constitute however the major drawback of this coupling. Interestingly, the authors demonstrated that the coupling of 3,5-dimethyl magnesium bromide and bromopentane could be successfully catalyzed by only 5 mol % of CuI (Table 6.1, entry 7, note d), the expected coupling product, a key intermediate in the preparation of cannabinoids, was obtained in 82 % yield (instead of 86 % by using 30 mol % of the same copper catalyst). Curiously, the extension of such improvement to other cross-couplings was not reported.

Table 6.1: Copper-catalyzed cross-couplings of aryl Grignard reagents with alkyl pseudo-halides.[a] [53]

$$\text{Alk–X} + \textbf{ArMgY} \xrightarrow[\text{THF, rt, 1-2 h}]{\text{CuI (30 mol \%), LiCl (200 mol \%)}} \text{Alk–}\textbf{Ar} \qquad X = \text{Br, Cl, I, OTs};\ Y = \text{Br, Cl}$$

Entry	Ar	Y	Alk	X	Yield (%)
1	Ph	Br	Ph-$(CH_2)_3$	Br	100[b]
2	Ph	Cl	Ph-$(CH_2)_3$	Br	100[b]
3	Ph	Br	Ph-$(CH_2)_3$	I	100[b]
4	Ph	Br	Ph-$(CH_2)_3$	OTs	100[b]
5	4-Me-C_6H_4	Br	Ph-$(CH_2)_3$	Br	91[c]
6	4-OMe-C_6H_4	Br	Ph-$(CH_2)_3$	Br	87[c]
7	3,5-OMe_2-C_6H_3	Br	C_5H_{11}	Br	86[c,d]
8	2-Me-C_6H_4	Br	Ph-$(CH_2)_3$	Br	91[c]
9	2-OMe-C_6H_4	Br	Ph-$(CH_2)_3$	Br	87[c]
10	2,6-Me_2-C_6H_3	Br	Ph-$(CH_2)_3$	Br	85[c]
11	4-F-C_6H_4	Br	Ph-$(CH_2)_3$	Br	89[c]
12	Ph	Br	Ph-$(CH_2)_3$	Cl	0
13	Ph	Br	NC-$(CH_2)_3$	I	83[c]
14	Ph	Br	EtO_2C-$(CH_2)_3$	I	76[c]

a) Conditions: ArMgX / AlkY (1.5:1), CuI (30 mol %), LiCl (2 equiv), THF, room temperature, 2 h.
b) Yields determined by ^1H NMR spectroscopy using anisole as an internal standard.
c) Isolated yield.
d) 55–60 °C. 82 % with 5 mol % CuI instead of 30 mol % at 55–60 °C.

The cross-couplings between benzylic electrophiles and aryl organometallics is a powerful method to prepare dissymmetric biarymethanes, moieties present in many biologically active compounds and drugs [54]. Noteworthy, Brandsma and coworkers reported that benzyl bromide and chloride reacted with phenyl magnesium bromide under copper catalysis (1 mol % CuBr). Interestingly, in these cross-couplings, palladium and nickel catalysts revealed much less efficient than the copper catalyst [55].

6.3.1.2 Coupling with alkyl phosphates

Besides benzylic halides, the reactivities and selectivities of other benzylic electrophiles have been investigated. Therefore, Knochel and coworkers reported in 2006 a general method for the arylation of highly reactive benzylic phosphates, readily available from the corresponding alcohols, by aryl Grignard reagents (Fig. 6.21) [56]. The catalytic system, composed of copper chloride (10 mol %), triethyl phosphite (20 mol %) and tetrabutylammonium iodide (10 mol %), was able to efficiently promote various cross-couplings in dimethylether at 60 °C. The triethylphosphite was necessary to stabilize the arylcopper reagents. Electron-poor and electron-rich aryl Grignard reagents as well as magnesiated indoles underwent alkylation with functionalized benzylic and heterocyclic phosphates in good to excellent yields. This method, allowing for the formation of highly functionalized carbocyclic and heterocyclic biarylmethanes in the presence of a catalytic amount of cheap copper chloride, offers an attractive alternative to the corresponding palladium-catalyzed cross-couplings. It was successfully applied, as the key step, to synthesize the antibiotic thimethoprim (overall yield: 52 %).

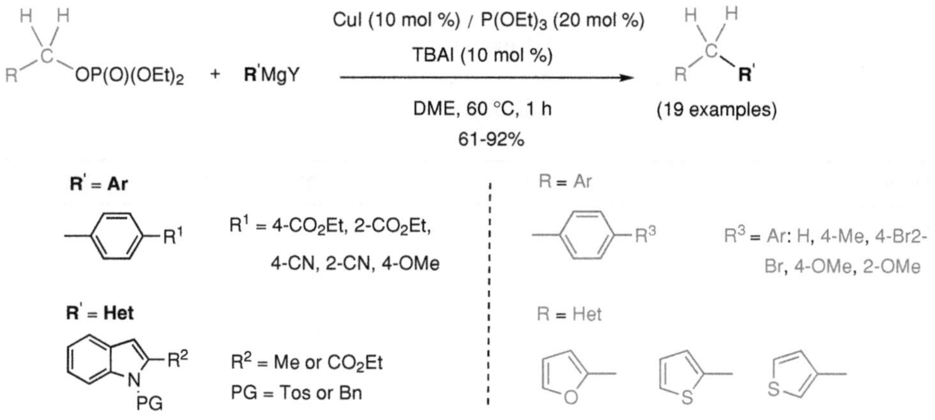

Fig. 6.21: Cross-couplings between aryl Grignard reagents and benzylic phosphates [56].

6.3.1.3 Coupling with alkyl sulfonalides and sulfonimides

Along the same lines, Tian and coworkers demonstrated in 2011 that Grignard reagents, and in particular aryls, were able to cleave the sp^3 carbon-nitrogen bonds of sulfonyl-activated benzylic, allylic, and propargylic amines in the presence of copper iodide (5 mol %) in refluxing THF (Table 6.2) [57]. Cross-couplings involving a large panel of double *p*-toluenesulfonyl-activated primary alkyl amines (R-CH_2-NTs_2, R' = Ts) and aryl Grignard reagents proceeded in very high yields with the tolerance of a wide range of functional groups such as halide, alkenes, alkynes, benzyl ethers and allyl ethers (Table 6.2, entries 1–18). Concerning the scope of Grignard reagent, mainly phenyl and electron-rich aryl groups (including bulky 2-tolyl) were shown to participate to the cross-couplings. Single *p*-toluenesulfonyl-activated secondary allylic and propargylic amines (Table 6.2, entries 19–20) and *p*-toluenesulfonyl-activated α-branched amines (Table 6.3) could also act as alkylating agents able to react with aryl magnesium bromides and chlorides under the same conditions. In the latter case, enantioenriched sulfonimide (Table 6.3, entry 4) and sulfonamide (Table 6.3, entry 5) were arylated with aryl Grignard reagents. Whatever the substrate, a configurational inversion of the chiral centers was observed which is in accordance with the occurrence of an S_N2-type mechanism in these cross-couplings. Moreover, epimerization of the sulfonamide occurred under the reaction conditions whereas the sulfonamides gave rise to a single isomer without loss of optical purity. Noteworthy, the authors also reported several examples of cross-couplings involving alkyl or allyl magnesium halides in the same publication. The method described by Tian and coworkers is therefore very broad in scope and the necessity to prepare the electrophilic alkyl partner can be seen as its main disadvantage.

Table 6.2: Cu-catalyzed cross-couplings of aryl Grignard reagents with sulfonimides and sulfonamides.[a] [57]

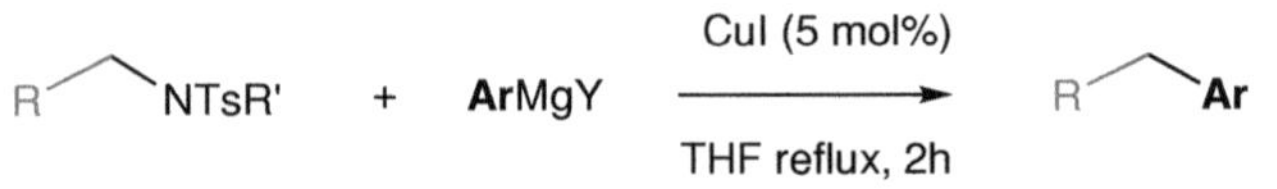

Entry	Ar	Y	R	R′	Time (h)	Yield (%)[b]
1	**Ph**	Br	4-OMe-C_6H_4	Ts	2	93
2	**Ph**	Br	2-OMe-C_6H_4	Ts	2	88
3	**Ph**	Br	1-heptynyl	Ts	2	88
4	**4-OMe-C_6H_4**	Br	Ph	Ts	2	75
5	**4-OMe-C_6H_4**	Br	4-Br-C_6H_4	Ts	2	72
6	**4-OMe-C_6H_4**	Br	4-Cl-C_6H_4	Ts	2	65
7	**4-OMe-C_6H_4**	Br	4-F-C_6H_4	Ts	2	74
8	**4-OMe-C_6H_4**	Br	1-naphtyl	Ts	2	74
9	**4-OMe-C_6H_4**	Br	2-furyl	Ts	2	80
10	**4-OMe-C_6H_4**	Br	2-thienyl	Ts	2	86
11	**4-OMe-C_6H_4**	Br	CH_2=CH	Ts	1	97
12	**4-OMe-C_6H_4**	Br	(*E*)-PhCH=CH	Ts	1	96
13	**4-OMe-C_6H_4**	Br	Me_2C=CH	Ts	1	99
14	**4-OMe-C_6H_4**	Br	CH_2=C(Me)	Ts	1	96
15	**2-Me-C_6H_4**	Cl	4-OMe-C_6H_4	Ts	2	92
16	**2-Me-C_6H_4**	Cl	(*E*)-BnOCH_2CH=CH[c]	Ts	1	83[c]
17	**2-Me-C_6H_4**	Cl	(*Z*)-BnOCH_2CH=CH[d]	Ts	1	85[d]
18	**2-Me-C_6H_4**	Cl	phenylethynyl	Ts	2	90

Entry	Ar	Y	R	R′	Time (h)	Yield (%)[b]
19	**4-OMe-C_6H_4**	Br	(*E*)-PhCH=CH	CH_2Ph	2	63
20	**2-Me-C_6H_4**	Cl	Ph-CC	Me	2	77

a) Conditions: **ArMgY** / AlkX (1.2:1), CuI (5 mol %), THF, 70 °C, 1–2 h.

b) Isolated yield.

c) *E*/*Z* = 98:2.

d) *E*/*Z* = 5:95.

Table 6.3: Copper-catalyzed cross-couplings of aryl Grignard reagents with *p*-toluene-sulfonyl-activated α-branched amines.[a] [57]

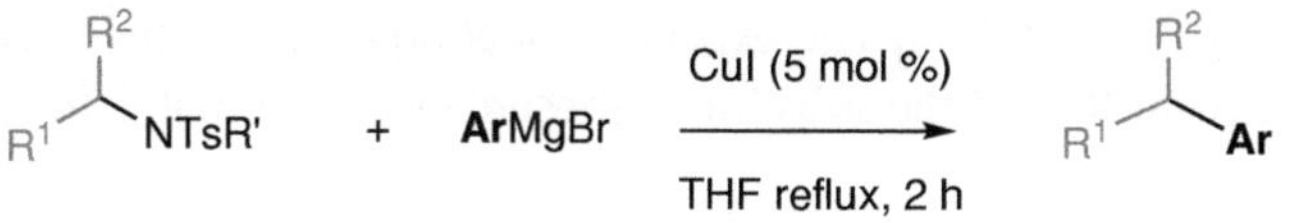

Entry	Ar	Starting Amine (ee)	Product	Yield (%)[b]
1	**4-OMe-C_6H_4**	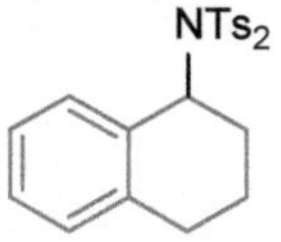	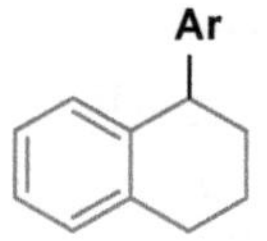	75
2	**4-OMe-C_6H_4**	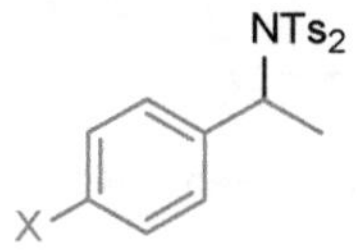	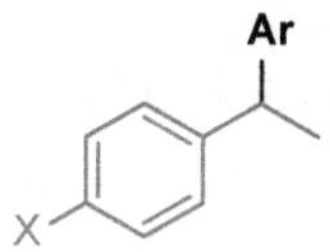	(X = Cl, 58) (X = F, 59)
3	**4-OMe-C_6H_4**	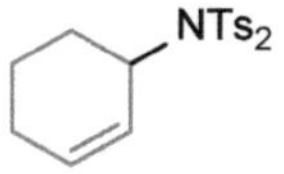	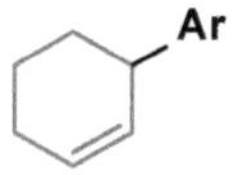	93
4	**4-OMe-C_6H_4**	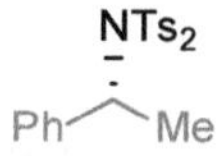 (ee = 99 %)	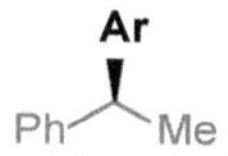 (ee = 44 %)	93
5	**Ph**	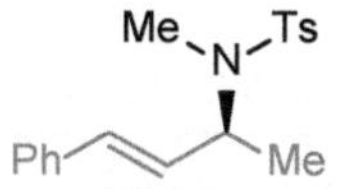 (ee = 95 %)	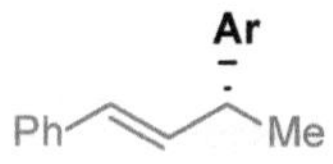 (ee = 95 %)	88

a) Conditions: ArMgY / AlkX (1.2:1), CuI (5 mol %), THF, 70 °C, 1–2 h.

b) Isolated yield.

More unusual Grignard reagents were also used as the coupling partners in C–C bond forming reactions. For example, Yorimitsu, Oshima and coworkers demonstrated that copper(II) triflate efficiently catalyzed the cross-coupling of cyclopentadienyl Grignard reagents with various electrophiles in mild reaction conditions (diethylether, 25 °C, 3 h) [58].

6.3.2 Coupling of aryl Grignard reagents with vinyl derivatives

Bäckwall method, described above and allowing the coupling of vinyl triflates with alkyl Grignard reagents (Fig. 6.14), was also used in the coupling of various aryl magnesium bromides with vinyl triflates (Fig. 6.22) [38]. The expected functionalized dienes were obtained in very good yields regardless of the substituent present on the aromatic ring.

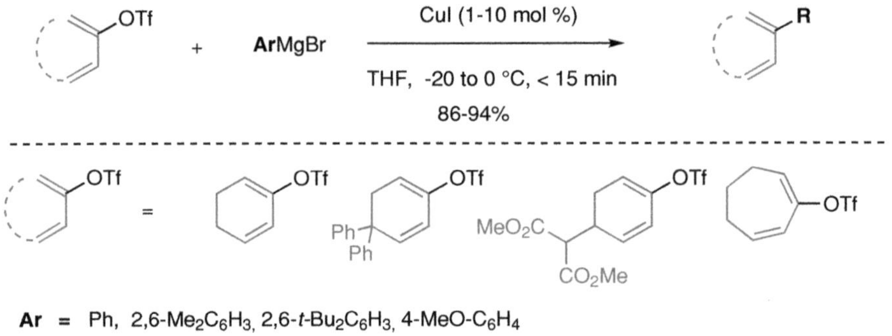

Fig. 6.22: Cross-couplings between aryl Grignard reagents and dien-2-yl triflates [38].

6.3.3 Coupling of aryl Grignard reagents with alkynyl derivatives

Cahiez *et al.* system described above and allowing the coupling of a wide range of alkynyl halides with alkyl Grignard reagents (Fig. 6.16) is also efficient to promote the coupling of aryl magnesium bromides with alkynyl halides (Fig. 6.23). The expected compounds were obtained in good to very good yields under very mild conditions [41].

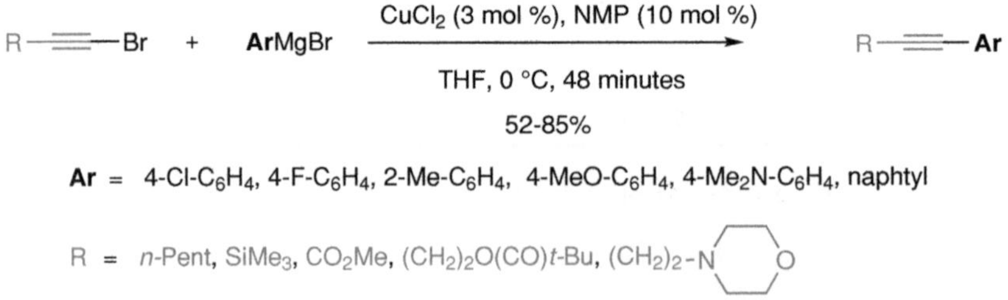

Fig. 6.23: Copper-catalyzed cross-couplings between aryl Grignard reagents and alkylnyl halides [41].

6.4 Coupling reactions of other Grignard reagents

6.4.1 Coupling reactions of vinyl Grignard reagents

Coupling of vinyl Grignard reagents and various alkyl halides or pseudohalides were also reported in the literature. In 1981, Nunomoto and coworkers demonstrated that copper iodide (20 mol %) was able to promote the coupling of 1,3-butadien-2-yl-magnesium chloride with alkyl halides [59]. Shortly afterwards, the same authors succeeded in decreasing the copper amounts by replacing CuI by Li_2CuCl_4 (1–5 mol %), which efficiently and selectively catalyzed the cross-coupling of the same vinyl Grignard reagent with primary alkyl iodides and bromides in THF (20–65 °C). The expected 2-substituted 1,3-dienes were obtained in fair to very good yields (Fig. 6.24) [18]. Fleming and coworkers also showed that CuBr (10 mol %) was able to catalyze the vinylation of ω-bromonitriles by the same Grignard reagent (1,3-butadien-2-yl-magnesium chloride) under analogous reaction conditions [60]. Vinyl magnesium bromide could also be used, however, with much lower efficiency. Noteworthy, the systems reported by Fuchikami [52] and Burns [20] allowing the coupling of aryl Grignard reagents with alkyl halides and sulfonates, respectively were also able to afford the vinylation of these alkyl substrates by vinyl magnesium bromide.

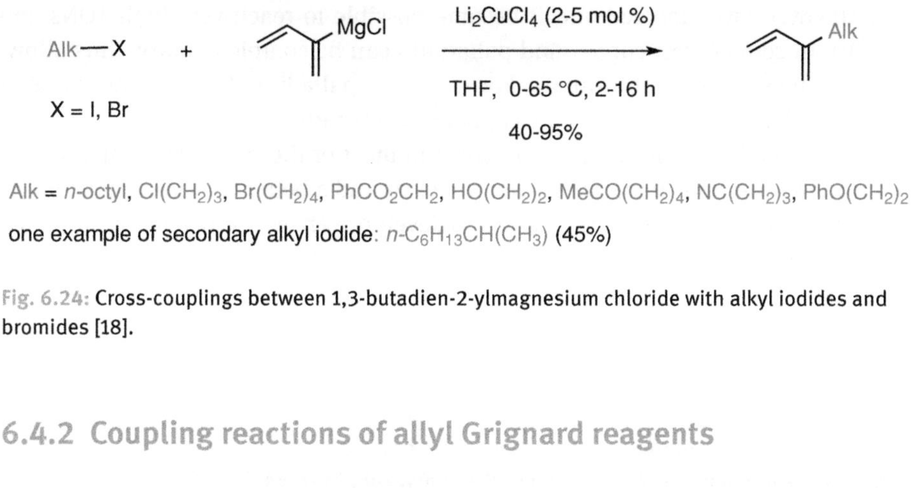

Fig. 6.24: Cross-couplings between 1,3-butadien-2-ylmagnesium chloride with alkyl iodides and bromides [18].

6.4.2 Coupling reactions of allyl Grignard reagents

Allyl Grignard reagents can also participate to a cross-coupling with various alkyl halides, including secondary and tertiary ones, in the presence of copper catalysts [19, 28, 61]. Noteworthy, the precursor [Li_2CuCl_4] (20 mol %) was successfully applied by Wang and co-workers at Novartis for the cross-coupling of allyl magnesium bromide and 1-bromo-4-chlorobutane on kilogram scale (Fig. 6.25) [62]. This reaction, leading to the expected product in high yield (83 %), constituted the first step in the multi-step synthesis of an HCV NS3 protease inhibitor.

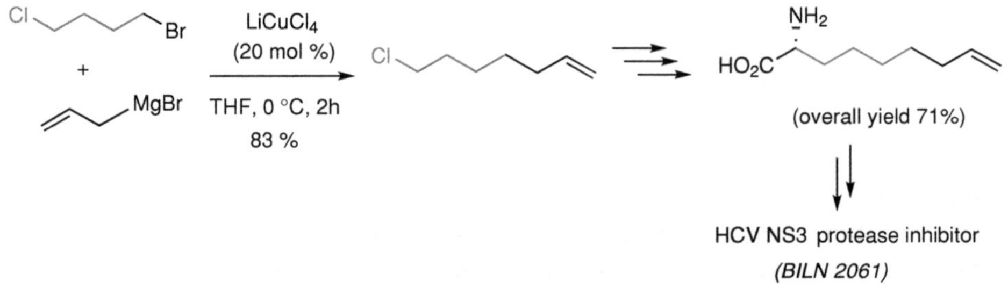

Fig. 6.25: Cross coupling of allyl magnesium bromide and 1-bromo-4-chlorobutane on kilogram scale [62].

6.5 Conclusion

Even if palladium, nickel and iron are often recognized as metals of choice to promote a lot of existing C–C cross-couplings, one have to remind that organocuprates were the first to perform cross-coupling reactions. Although, for a long time they often required stoichiometric amounts of metal [2, 63]. Today, very efficient and general copper-catalyzed protocols allowing the couplings of Grignard reagents (mainly alkyl, aryl, allyl, vinyl) and organic halides or pseudohalides (mainly alkyl, allyl, vinyl, alkynyl) have emerged and it became possible to reach very high TONs and TOFs [31]. In some cases, copper and palladium can be complementary, thus allowing for chemoselective subtitutions. For example, palladium-based catalysts favor the substitution of aryl halides while copper systems rather prefer the substitution of alkyl halides [59]. The mechanisms involved in most of these copper-catalyzed C–C bond forming reactions remain however not clear and a better comprehension of the processes will probably constitute the key to develop even more efficient Cu-promoted transformations.

References

[1] Frisch AC, Beller M, Angew Chem Int Ed, 2005, 44, 674–88.
[2] Beletskaya IP, Cheprakov AV, Coord Chem Rev, 2004, 248, 2337–64.
[3] Shinokubo H, Oshima K, Eur J Org Chem 2004, 2081–91.
[4] Kambe N, Iwasaki T, Terao J, Chem Soc Rev, 2011, 40, 4937–47.
[5] Grignard V, C R Académie des Sciences 1900, 1322–24.
[6] Kharasch MS, Fields EK, J Am Chem Soc, 1941, 63, 2316–20.
[7] Kharasch MS, Nudenberg W, Archer S, J Am Chem Soc, 1943, 65, 495–98.
[8] Kharasch MS, Reinmuth O, Grignard Reagents of Nonmetallic Substances, Prentice-Hall, Inc, New York, 1954.
[9] Tamao K, Sumitani K, Kumada M, J Am Chem Soc, 1972, 94, 4374–76.

[10] Corriu RJP, Masse JP, J Chem Soc, Chem Commun, 1972, 144.
[11] Kharasch MS, Nudenberg W, Archer S, J Am Chem Soc, 1943, 65, 495–98.
[12] Parker VD, Noller CR, J Am Chem Soc, 1964, 86, 1112–16.
[13] Tamura M, Kochi JK, Synthesis, 1971, 303–05.
[14] Tamura M, Kochi JK, J Am Chem Soc, 1971, 93, 1485–87.
[15] Tamura M, Kochi JK, J Organomet Chem, 1972, 42, 205–28.
[16] Fouquet G, Schlosser M, Angew Chem Int Ed, 1974, 13, 82–83.
[17] Schlosser M, Fouquet G, Chem Ber, 1974, 107, 1162–70.
[18] Nunomoto S, Kawakami Y, Yamashita Y, J Org Chem, 1983, 48, 1912–14.
[19] Johnson DK, Ciavarri JP, Ishmael FT, Schillinger KJ, Vangeel TAP, Stratton SM, Tetrahedron Lett, 1995, 36, 8565–68.
[20] Burns DH, Miller JD, Chan H-K, Delaney MO, J Am Chem Soc, 1997, 119, 2125–33.
[21] Burns DH, Chan H-K, Miller JD, Jayne CL, Eichhorn DM, J Org Chem, 2000, 65, 5185–96.
[22] Donkervoort JG, Vicario JL, Jastrzebski JTBH, Gossage RA, Cahiez G, van Koten G, J Organomet Chem, 1998, 558, 61–69.
[23] Cahiez G, Chaboche C, Jézéquel M, Tetrahedron, 2000, 56, 2733–37.
[24] Terao J, Watanabe H, Ikumi A, Kuniyasu H, Kambe N, J Am Chem Soc, 2002, 124, 4222–23.
[25] Terao J, Naitoh Y, Kuniyasu H, Kambe N, Chem Lett, 2003, 32, 890–91
[26] Kambe N, Terao J, Iwasaki T, J Synth Org Chem Jpn, 2011, 69, 1271–81.
[27] Terao J, Ikumi A, Kuniyasu H, Kambe N, J Am Chem Soc, 2003, 125, 5646–47.
[28] Shen R, Iwasaki T, Terao J, Kambe N, Chem Commun, 2012, 48, 9313–15.
[29] Terao J, Todo H, Begum SA, Kuniyasu H, Kambe N, Angew Chem Int Ed, 2007, 46, 2086–89.
[30] Terao J, Kambe N, Acc Chem Res, 2008, 41, 1545–54.
[31] Iwasaki T, Imanishi R, Shimizu R, Kuniyasu H, Terao J, Kambe N, J Org Chem, 2014, 79, 8522–32.
[32] Kim JH, Chung YK, Chem Commun, 2013, 11101–03.
[33] For a general review on mechanisms involving organocopper compounds in organic chemistry see: Nakamura E, Yoshikai N, Chem Rev, 2012, 112, 2339–72.
[34] Yang C-T, Zhang Z-Q, Liang J, Liu J-H, Lu X-Y, Chen H-H, Liu L, J Am Chem Soc, 2012, 134, 11124–27.
[35] Ren P, Stern L-A, Hu X, Angew Chem Int Ed, 2012, 51, 9110–13.
[36] Lethu S, Matsuoka S, Murata M, Org Lett, 2014, 16, 844–47.
[37] Kharasch MS, Fuchs CF, J Am Chem Soc 1943, 65, 504–07.
[38] Karlström ASE, Rönn M, Thorarensen A, Bäckwall JE, J Org Chem 1998, 63, 2517–22.
[39] Jonasson C, Rönn M, Bäckwall JE, J Org Chem 2000, 65, 2122–26.
[40] Hintermann L, Xiao L, Labonne A, Angew Chem Int Ed 2008, 47, 8246–50.
[41] Cahiez G, Gager O, Buendia J, Angew Chem Int Ed. 2010, 49, 1278–81.
[42] Harutyunyan SR, den Hartog T, Geurts K, Minnaard AJ, Feringa BL, Chem Rev 2008, 108, 2824–52.
[43] Giannerini M, Fananas-Mastral M, Feringa BL, J Am Chem Soc 2012, 134, 4108–11.
[44] Hornillos V, Pérez M, Fananas-Mastral M, Feringa BL, J Am Chem Soc 2013, 135, 2140–43.
[45] Nakamura M, Ito S. Metal-catalyzed arylations of nonactivated alkyl (pseudo)halides *via* cross-coupling reactions. In: Modern arylation methods. ed. L. Ackermann. Wiley-VCH Verlag GmbH & Co. KGaA, Weinheim, 2009,155–81.
[46] Seki M, Mori K, Eur J Org Chem 2001, 3797–09.
[47] Organ MG, Wang J, J Org Chem 2003, 5568–74.
[48] Vyvyan JR, Holst CL, Johnson AJ, Schwenk CM, J Org Chem 2002, 2263–65.
[49] This methodology was applied in the preparation of natural products, see: Tamagawa H, Takikawa H, Mori K, Eur J Org Chem 1999, 973–78.
[50] Nishimura J, Yamada N, Horiuchi Y, Ueda E, Bull Chem Soc Jpn 1986, 59, 2035–37.

[51] Zhang HY, Blasko A, Yu JQ, Bruice TC, J Am Chem Soc 1992, 114, 6621–30.
[52] Shimizu R, Yoneda E, Fuchikami T, Tetrahedron Lett 1996, 37, 5557–60.
[53] Nakata K, Feng C, Tojo T, Kobayashi Y, Tetrahedron Lett 2014, 55, 5774–77.
[54] McPhail KL, Rivett DEA, Lack DE, Davies-Coleman MT, Tetrahedron 2000, 56, 9391–96.
[55] de Lang RJ, van Hooijdonk MJCM, Bransma L, Tetrahedron 1998, 54, 2953–66.
[56] Kofink CC, Knochel P, Org Lett 2006, 8, 4121–24.
[57] Li MB, Tang XL, Tian SK, Adv Synth Catal 2011, 353, 1980–84.
[58] Sai M, Someya H, Yorimitsu H, Oshima K, Org Lett 2008, 10, 2545–47.
[59] Nunomoto S, Kawakami Y, Yamashita Y, Bull Chem Soc Jpn 1981, 54, 2831–32.
[60] Fleming FF, Jiang T, J Org Chem 1997, 62, 7890–91.
[61] Sai M, Yorimitsu H, Oshima K, Bull Chem Soc Jpn 2009, 82, 1194–96.
[62] Wang XJ, Zhang L, Smith-Keenan LL, Houpis IN, Farina V, Org Proc Res Dev 2007, 11, 60–63.
[63] Lipschutz BH, Acc Chem Res 1997, 30, 277–82.

Janine Cossy

7 Grignard Reagents and Silver

7.1 Introduction

Transition metal-catalyzed cross-couplings are among the most powerful methods in organic synthesis to form carbon–carbon bonds and these reactions have received considerable attention from chemists over the past eighty years. Among these coupling reactions, Grignard reagents were used extensively and were coupled with Grignard reagents, halides or pseudo-halides. To induce these couplings, different transition metal catalysts can be used such as palladium, nickel, iron, cobalt, manganese, copper, as well as silver and more scarcely, titanium, chromium, zirconium, rhodium or ruthenium. Among these metal catalysts, the most commonly used metals are probably palladium and nickel, however in the context of "sustainable chemistry", chemists have to develop methods that are eco-friendly and other metals, than palladium and nickel catalysts have to be utilized. As iron and cobalt catalysts are more eco-friendly than palladium and nickel catalysts, they have been extensively used. Even if silver catalysts have not been extensively studied, these catalysts should not be neglected as some reactions can be efficiently be performed with silver salts and not with other metal catalysts. Silver-catalyzed couplings have been reported to form carbon–carbon bonds in between two sp^2 carbons, in between one sp^2 and one sp^3 carbons and mainly in between two sp^3 carbons.

7.2 Coupling reactions of aryl Grignard reagents

7.2.1 With aryl Grignard reagents

7.2.1.1 Homocoupling

Since the early work of Gardner and Borgstrom, in 1929 [1], it has been shown that silver(I) salts and, particularly silver bromide (AgBr), are working as stoichiometric oxidants to realize the synthesis of symmetrical biaryls from aryl magnesium halides. The coupling was realized by the addition of an aryl magnesium halide to a well-stirred suspension of AgBr in ether, followed by the addition of benzene. After heating, biphenyl (66 %), 4,4'-dimethoxy biphenyl (48 %) and bi-*p*-tolyl (72 %) were obtained in satisfactory purity and yields when the coupling reaction was performed on small scale (0.05 mol). However, when the reaction was scaled up, the yield in the bi-aryl derivatives was reduced. For example, the yield in bi-*p*-tolyl was reduced from 72 %

to 25 %, when the reaction was performed on 0.35 mol of Grignard reagent instead of 0.05 mol [1] (Fig. 7.1).

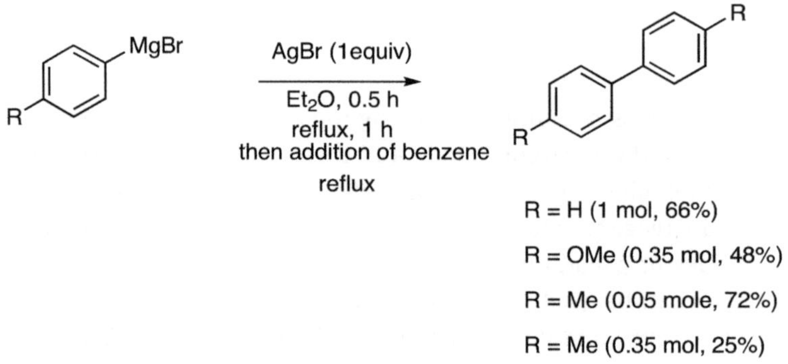

Fig. 7.1: Cross-coupling of aryl Grignard reagents.

Gardner and Borgstrom [1] reported that radical intermediates are responsible for the formation of the coupling products and that these radicals are formed by the interaction of the Grignard reagents with silver bromide, assuming that the reaction takes place according to eqs. 1 and 2 represented in Fig. 7.2.

$$RMgX + AgBr \longrightarrow R^{\cdot} + Ag + MgXBr \qquad \text{(eq 1)}$$

$$R^{\cdot} + R^{\cdot} \longrightarrow R-R \qquad \text{(eq 2)}$$

Fig. 7.2: Radicals involved in the coupling reactions.

7.2.1.2 Heterocoupling

In 1937, Gardner, Joseph and Gollub realized a coupling in between two different Grignard reagents by treating a mixture of phenylmagnesium bromide and *p*-anisole magnesium bromide with AgBr [2]. To separate and identify the resulting formed compounds, the mixture was boiled with hydroiodic acid and acetic acid and, after a base/acidic treatment, the products were isolated by fractional crystallization to afford the coupling products. 4-Hydroxybiphenyl was isolated in 8.2 % yield, accompanied by the homocoupling products, biphenyl and 4,4'-dihydroxybiphenyl which were isolated in 46 % yield and 19.4 %, respectively. Even if the yield in the heterocoupling product is low, it was demonstrated for the first time that a heterocoupling was possible when two different aryl Grignard reagents were treated with a silver catalyst (Fig. 7.3).

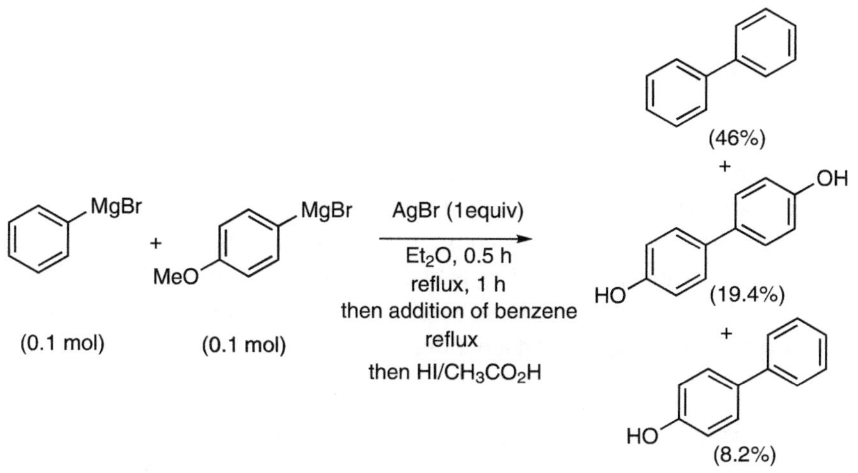

Fig. 7.3: Heterocoupling between two different Grignard reagents.

7.2.2 With alkyl halides

The alkylation of an aryl Grignard reagent by an alkyl halide can be realized in the presence of silver catalysts. Oshima *et al.* showed that the cross-coupling of an aryl magnesium bromide with an alkyl bromide was possible in the presence of AgBr (10 mol %) and triphenylphosphite [$P(OPh)_3$ (10 mol %)] as the additive (Fig. 7.4). It is worth noting that acyclic primary, secondary and tertiary alkyl bromides as well as cyclic bromides can be involved in a coupling reaction and that the coupling products were isolated in good yields (> 60 %). However for tertiary bromides, the reaction took 10 h instead of 5 h for primary and secondary halides (Fig. 7.4). It seems that the coupling reaction is not sensitive to steric hindrance as the reaction of *o*-tolyl magnesium bromide with bromocyclohexane furnished the coupling product in 90 % yield [3] (Fig. 7.4).

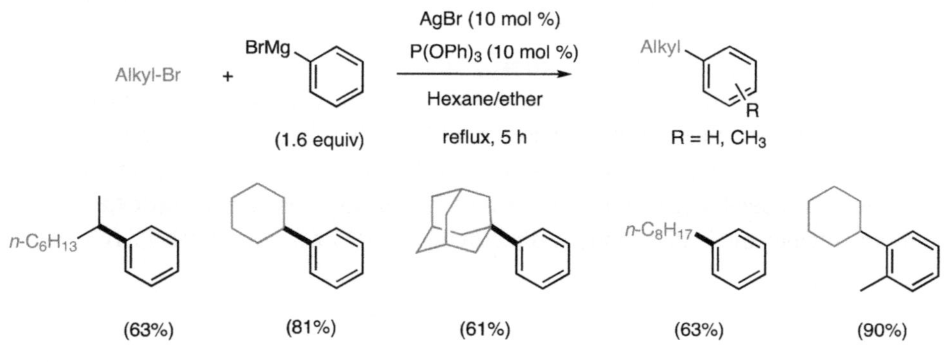

Fig. 7.4: Coupling between an aryl Grignard and an alkyl halide.

7.3 Coupling of vinyl Grignard reagents with alkyl halides

Only one example was reported in the literature for the coupling of a vinyl Grignard reagent with a halide, catalyzed by silver salts. When *cis*-propenyl magnesium bromide was treated with methyl bromide, in the presence of a silver(I) salt, *cis*-but-2-ene was produced with retention of configuration of the vinyl Grignard showing that under these conditions, the coupling is highly stereoselective [4] (Fig. 7.5). To our knowledge, no other coupling reactions, catalyzed by silver salts, were reported in between a vinyl Grignard reagent and a halide.

Fig. 7.5: Coupling between a vinyl Grignard and an alkyl halide.

7.4 Coupling of alkyl Grignard reagents

The coupling of an alkyl Grignard reagent with another alkyl Grignard reagent or with an alkyl halide, catalyzed by silver salts, is the most commonly reported coupling in the literature.

7.4.1 With alkyl Grignard reagents

7.4.1.1 Intermolecular homocoupling

In their early work, in 1929, Gardner and Borgstrom have reported that a homocoupling can take place when an alkyl Grignard is treated with a catalytic amount of AgBr. *n*-Butyl, benzyl-, cyclohexyl- Grignard reagents respectively led to *n*-octane, bi-benzyl, bi-cyclohexyl in moderate to good yields (40–72 %) [1]. With *iso*-butyl-, and *sec*-butyl Grignard reagents, these authors reported, at first, that they do not observe any coupling products [1]. However, 10 years later, in 1939, Gardner and Joseph reported that *iso*-butyl, and *sec*-butyl Grignard were reacting in the presence of AgBr to produce 2,5-dimethylhexane (37.5 %) and 3,4-dimethylhexane (13 %), respectively [5] (Fig. 7.6).

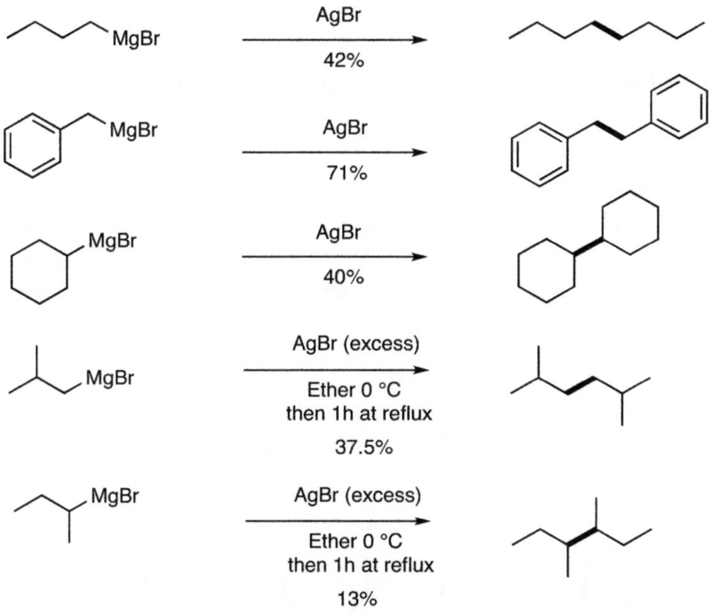

Fig. 7.6: Homocoupling of alkyl Grignard reagents catalyzed by AgBr.

When the coupling was achieved with AgBr in excess, the yields in the homocoupling products were moderate to low. On the contrary, when the reaction was realized with a catalytic amount of AgOTs (1 mol %) in the presence of a reoxidant such as dibromoethane (1.2 equiv), the yields in the homocoupling products were excellent even with Grignard reagents prepared from secondary halides [6] (Fig. 7.7).

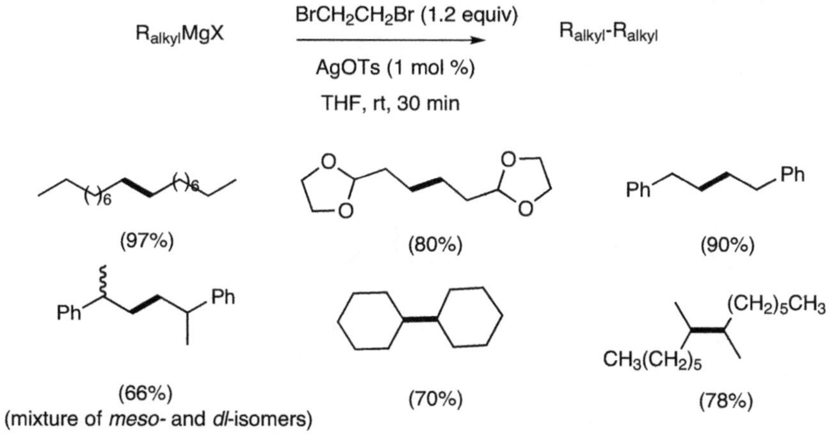

Fig. 7.7: Homocoupling of akyl Grignard catalyzed by AgOTs/dibromoethane.

7.4.1.2 Intramolecular homocoupling

The coupling of two alkyl Grignard reagents was realized in an intramolecular version. When α,ω-di-Grignard derivatives were prepared in THF, by addition of magnesium onto α,ω-di-halides and, after treatment of the resulting α,ω-di-Grignard with a catalytic amount of tetrakis iodo(tri-*n*-butylphosphine)silver(I) $[AgIPBu_3]_4]$, cyclobutane, cyclopentane and cyclohexane were formed in good yields (83–93 %). On the contrary, cyclooctane was obtained in only 23 % yield. It is worth nothing that the homocoupling is not effecient to form cyclooctane and cyclodecane as these latter were only detected (yields inferior to 1 %) [7] (Fig. 7.8).

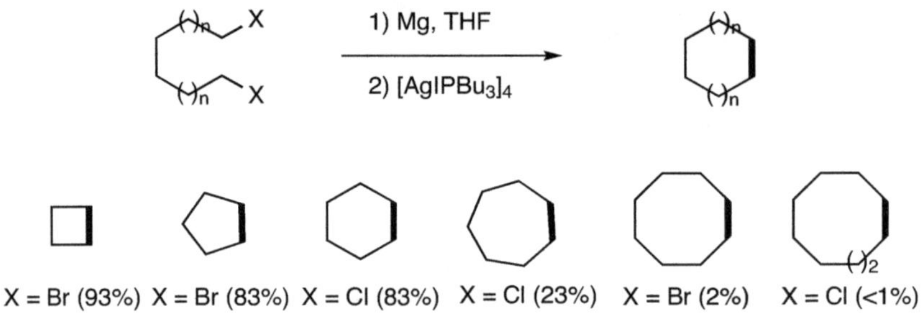

Fig. 7.8: Formation of cyclic compounds by Intramolecular homocoupling of α,ω-di-Grignard reagents.

Other silver salts such as silver triflate (AgOTf), silver bromide (AgBr) as well as silver nitrate ($AgNO_3$) can be used for an intramolecular coupling. Bicyclic compounds were formed in good yields when α,ω-dichlorides were treated with magnesium and then with Ag(OTf) [7] (Fig. 7.9).

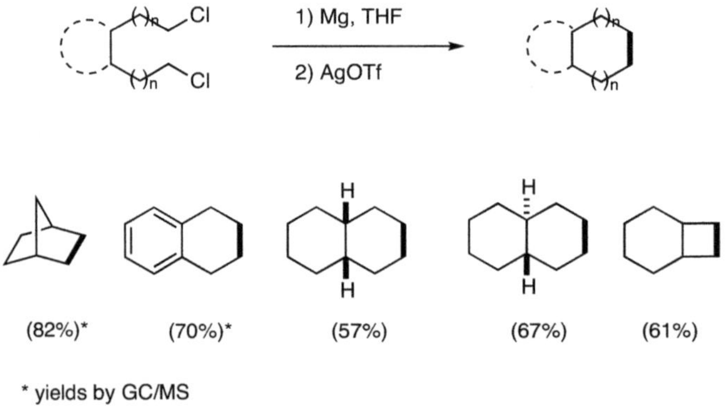

Fig. 7.9: Formation of bicyclic compounds by intramolecular coupling of α,ω-di-Grignard reagents.

7.4.2 With alkyl halides

Silver salts are also effective catalysts for the coupling of Grignard reagents with alkyl halides to produce the coupling products in good yields if the alkyl group of the Grignard is identical to the alkyl group of the halide (Fig. 7.10, eq 1). On the contrary, when the alkyl group of the Grignard is different from the alkyl group of the halide, a mixture of the coupling products is formed (Fig. 7.10). For example, when a Grignard reagent is formed from a primary halide and is reacting with a primary halide different from the one that was used to prepare the Grignard, a mixture of three coupled products was obtained in an equimolar ratio [8] (Fig. 7.10, eq 2).

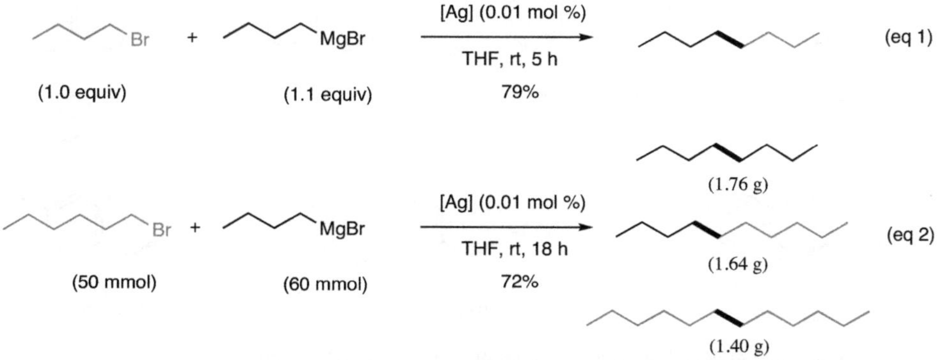

Fig. 7.10: Coupling of an alkyl Grignard with an alkyl halide.

In 1971, it was demonstrated by Tamura and Kochi that the rate of the catalytic coupling of EtMgBr with an alkyl halide is much faster for tertiary halides (*tert*-BuBr) than for secondary halides (*i*PrBr) than for primary halides (*n*-PrBr) in a rate of 20:3:1. It is worth noting that structural changes in the Grignard reagents do not induce an apparent systematic variation in the reaction rate [9]. This observation was used several years later by Oshima *et al.* to realize a coupling between benzylmagnesium bromide and secondary as well as tertiary bromides in the presence of $AgNO_3$ in ether, to produce the corresponding cross-coupling products in high yields [10] (Fig. 7.11).

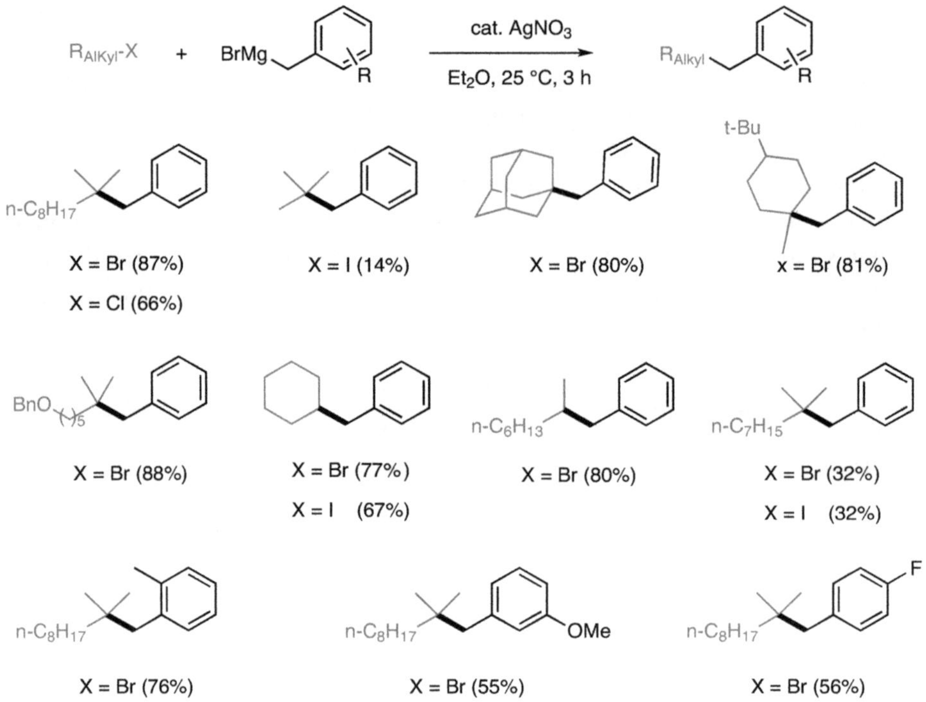

Fig. 7.11: Coupling between benzyl magnesium bromide and alkyl halides.

As secondary halides react faster than primary halides, the coupling of halides with Grignard reagents is very chemioselective [3]. The reaction can be performed by using AgCl, AgBr, AgBr/KF or AgBr/P(OPh)$_3$, in a mixture of ether/CH_2Cl_2 [3]. For example, when **1** was treated with Grignard **2** and AgBr/P(OPh)$_3$ as the catalystic system, compound **3** was formed in 55 % yield and **4** was not detected (Fig. 7.12). When the Grignard reagent is phenylmagnesium bromide, under the silver-catalyzed arylation conditions (AgBr/KF), the monophenylated product **6** and the diphenylated one, **7**, were isolated in 43 % and 16 % yields, respectively. With phenylmagnesium bromide, the reaction is less chemioselective than with alkyl Grignard reagents. (Fig. 7.12).

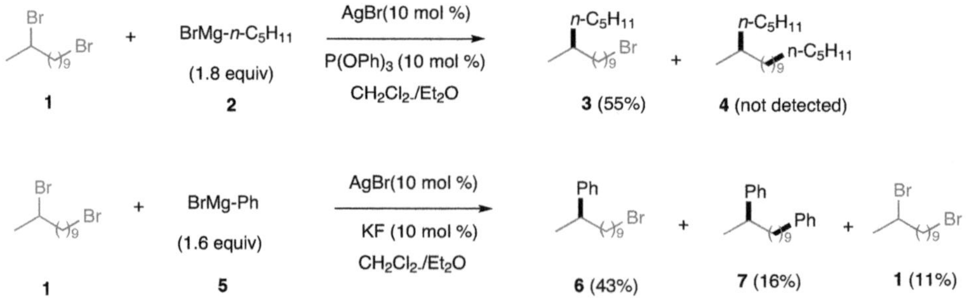

Fig. 7.12: Chemoselective coupling of phenylmagnesium bromide with dibromo alkyl derivatives.

The coupling reaction between an alkyl Grignard and an alkyl halide, catalyzed by Ag(0), was used to prepare glycerol triethers in good yields. It is worth noting that the Grignard reagent was prepared from an iodide and that the alkyl halide was an alkyl chloride [11] (Fig. 7.13).

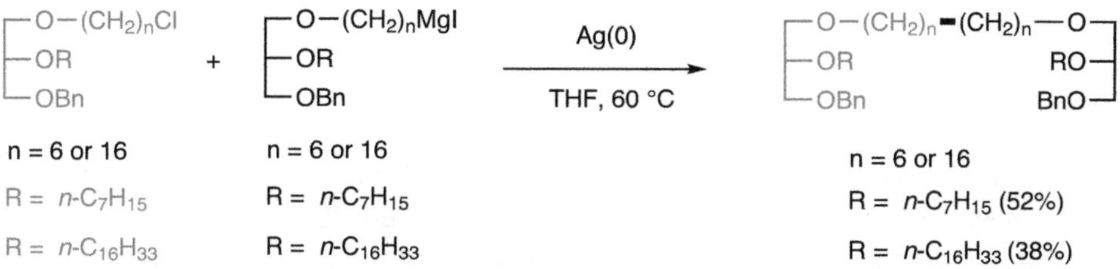

Fig. 7.13: Synthesis of glycerol triethers by coupling of alkyl Grignard reagents with alkyl halides.

The silver-catalyzed coupling can proceed also with allylic Grignard reagents (Fig. 7.14). Allylations and methallylations of tertiary and secondary allyl bromides were realized under smooth conditions. Unfortunately, silver-catalyzed prenylation and crotylation of alkyl halides led to a poor regioselectivity [10] (Fig. 7.14).

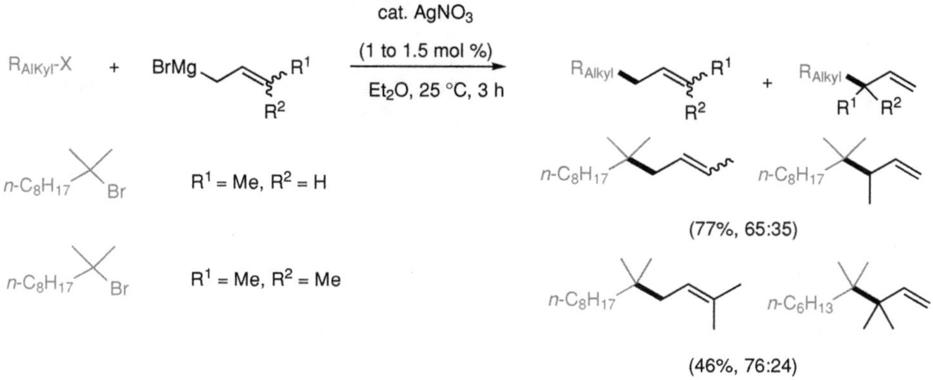

Fig. 7.14: Coupling of allyl and methallyl Grignard reagents with alkyl halides.

7.4.5 With vinyl halide

If *Z*-vinyl Grignard reagents were coupled with alkyl halides with retention of configuration, on the contrary the reverse combination, for example the coupling of an alkyl Grignard reagent with a *cis*-vinyl halide, did not lead to the substituted olefin with the retention of configuration of the vinyl halide. For example, the coupling of *cis*-propenyl bromide with CH_3MgBr, in the presence of Ag(I), led to a mixture of *cis*- and *trans*-but-2-ene [11] (Fig. 7.15).

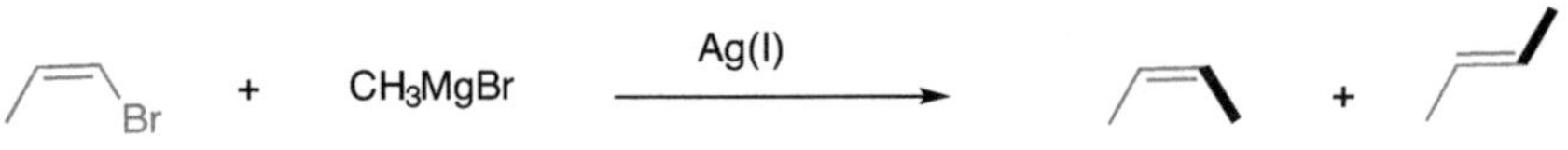

Fig. 7.15: Coupling of alkyl Grignard reagents with vinyl halides.

7.5 Mechanism

According to the early work of Gardner and Borgstrom [1] and then Kochi and Tamura [9], it has been postulated that radical intermediates were formed when a Grignard reagent and a halide were treated with silver salts to form the coupling products. In 2008, radical intermediates have been highlighted by Oshima *et al.* [10] by using a radical clock. Thus, when **8** was treated with $PhCH_2MgBr$ in the presence of $AgNO_3$ in ether at 25 °C (3 h), the functionalized cyclized product **9** was isolated and then transformed to **10**, proving that radical intermediates were formed (Fig. 7.16).

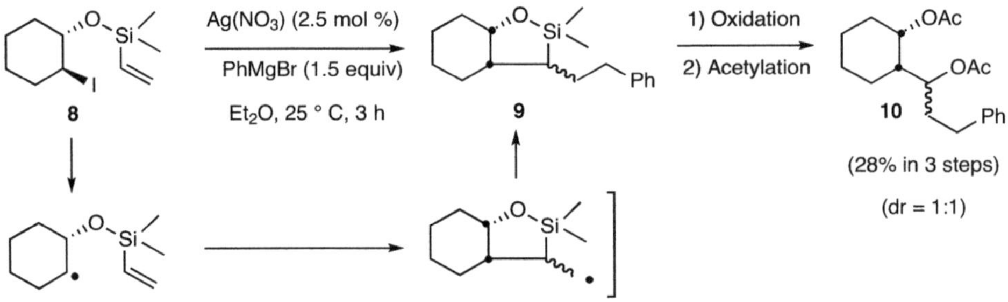

Fig. 7.16: Trapping of the radical intermediates.

Based on the studies on silver-catalyzed couplings [9, 10, 12], a mechanism has been proposed by Oshima *et al.* [10]. A Grignard reagent such as, for example, benzyl magnesium bromide would first react with the silver(I) salt to produce silver(0) (which can be multi-atomic or clustered and might not be single atom), and diphenylethane (steps **A** and **B**). The zero valent silver would affect a Single Electron Transfer (SET) to the alkyl halide to form the corresponding alkyl halide radical and silver halide (step **C**). The alkyl radical would couple with another zero valent silver to yield the alkyl silver intermediate (step **D**). The silver halide formed by the SET would then react with benzylmagnesium bromide to generate a benzyl silver intermediate (step **E**). Reductive coupling of the benzyl silver with the alkyl silver intermediate will afford the coupling product and will regenerate the zero valent silver (step **F**). We have to point out that the reasons for the predominant cross-coupling over a possible homocoupling are not clear at this stage. In addition, the formation of a carbocation cannot be excluded (Fig. 7.17).

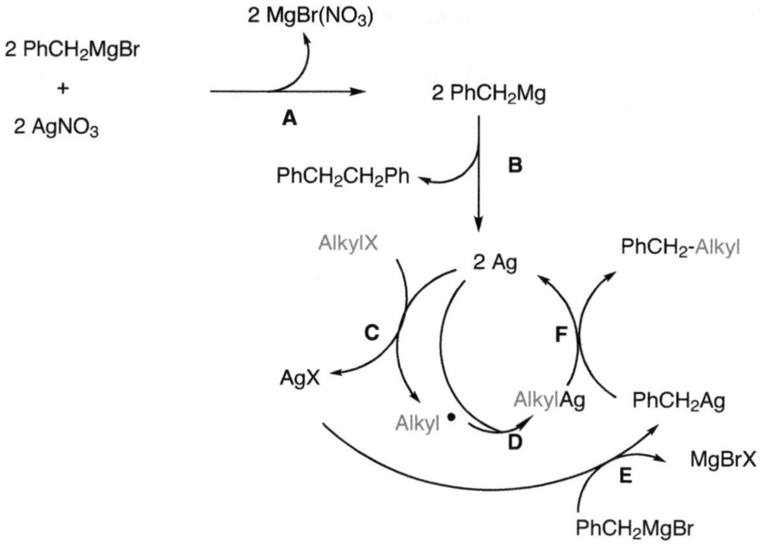

Fig. 7.17: Proposed mechanism.

7.6 Conclusion

Since the preliminary results of Gardner and then Kochi, only a few examples of coupling reactions catalyzed by silver salts were reported in the literature. Silver salts can catalyze the coupling of alkyl, allyl, vinyl and aryl Grignard with alkyl, vinyl and aryl Grignard reagents or halides. These couplings are very stereoselective when a vinyl Grignard is used, as well as chemoselective as tertiary halides are more reactive than secondary halides and more reactive than primary halides. As all the coupling possibilities using silver-catalyzed cross-couplings, in between a Grignard reagent and another Grignard reagent or a halide, have not been examined extensively, useful synthetic methods could be developed in the future. In addition, the development of enantioselective coupling reactions is highly desirable.

References

[1] Gardner JH, Borgstrom PA. J Am Chem Soc 1929;51:3375–3377.
[2] Gardner JH, Joseph L, Gollub F. J Am Chem Soc 1937,59:2583–2584.
[3] Someya H, Yorimitsu H, Oshima K. Tetrahedron Lett 2009, 50:3270–3272.
[4] Kochi JK. J Organomet Chem 2002,653:11–19.
[5] Gardner JH, Joseph, L. J. Am. Chem. Soc. 1939, 61:2551–2552.
[6] Nagano T, Hayashi T. Chem. Letters 2005, 34:1152–1153.

[7] Whitesides GM, Gutowski FD. J Org Chem 1976, 41:2882–2885.
[8] Tamura T, Kochi J. Synthesis 1971, 303–305.
[9] Tamura M, Kochi, J. J Amer Chem 1971, 93:1483–1485.
[10] Someya H, Ohmiya O, Yorimitsu H, Oshima K. Org Lett 2008, 10:969–971.
[11] Berkowitz WF, Pan D, Bittman R.Tetrahedron Lett 1993, 34:4297–4300.
[12] Kochi JK. J Organomet Chem 2002, 653:11–19.

Index

Bei Fragen zur Produktsicherheit wenden Sie sich bitte an:
If you have any questions regarding product safety,
please contact:

Walter de Gruyter GmbH
Genthiner Straße 13
10785 Berlin
productsafety@degruyterbrill.com